FLUORINE-CONTAINING MOLECULES

MOLECULAR STRUCTURE AND ENERGETICS

Series Editors

Joel F. Liebman
University of Maryland Baltimore County

Arthur Greenberg
New Jersey Institute of Technology

Advisory Board

Other Volumes in the Series

Chemical Bonding Models
Physical Measurements
Studies of Organic Molecules
Biophysical Aspects
Advances in Boron and the Boranes
Modern Models of Bonding and Delocalization
Structure and Reactivity
Mechanistic Principles of Enzyme Activity
Environmental Influences and Recognition in Enzyme Chemistry

FLUORINE-CONTAINING MOLECULES

STRUCTURE, REACTIVITY, SYNTHESIS, AND APPLICATIONS

Edited by

Joel F. Liebman
Arthur Greenberg
William R. Dolbier, Jr.

VCH

Joel F. Liebman
Department of
  Chemistry
University of Maryland
  Baltimore County
Baltimore, Maryland
  21228

Arthur Greenberg
Chemistry Division
New Jersey Institute of
  Technology
Newark, New Jersey
  07102

William R. Dolbier, Jr.
Department of
  Chemistry
University of Florida
Gainesville, Florida
  32611

Library of Congress Cataloging-in-Publication Data

Fluorine-containing molecules : structure, reactivity, synthesis, and
  applications / edited by Joel F. Liebman, Arthur Greenberg, William
  R. Dolbier, Jr.
      p.   cm. — (Molecular structure and energetics ; v. 8)
  Includes index.
  ISBN 0-89573-705-1
  1. Fluorine compounds.   I. Liebman, Joel F.   II. Greenberg,
Arthur.   III. Dolbier, William R.   IV. Series.
QD461.M629 1986 vol. 8
[QD181.F1]
540 s—dc19
[546'.7312]
                                                                88-19227
                                                                CIP

© 1988 VCH Publishers, Inc.

This work is subject to copyright.

Printed in the United States of America.

ISBN-0-89573-705-1 VCH Publishers
ISBN-3-527-26906-1 VCH Verlagsgesellschaft

Distributed in North America by:

VCH Publishers, Inc.
220 East 23rd Street
Suite 909
New York, New York 10010

Distributed Worldwide by:

VCH Verlagsgesellschaft mbH
P.O. Box 1260/1280
D-6940 Weinheim
Federal Republic of Germany

Contributors

Leland C. Allen, Department of Chemistry, Princeton University, Princeton, New Jersey 08540

Donald E. Bergstrom, Department of Chemistry, University of North Dakota, Grand Forks, North Dakota 58202

Carl L. Bumgardner, Department of Chemistry, North Carolina State University, Raleigh, North Carolina 27695

Donald J. Burton, Department of Chemistry, University of Iowa, Iowa City, Iowa 52242

William R. Dolbier, Jr., Department of Chemistry, University of Florida, Gainesville, Florida 32611

Seetha Eswarakrishnan, Department of Chemistry, State University of New York, Albany, New York 12222

Robert Filler, Department of Chemistry, Illinois Institute of Technology, Chicago, Illinois 60616

Albert W. Jache, Department of Chemistry, Marquette University, Milwaukee, Wisconsin 53233

Heinz F. Koch, Department of Chemistry, Ithaca College, Ithaca, New York, 14850

Judith G. Koch, Department of Chemistry, Ithaca College, Ithaca, New York, 14850

Henryk Koroniak, Adam Mickiewicz University, Grunwaldzka 6, Poznan, Poland

Joel F. Liebman, Department of Chemistry, University of Maryland Baltimore County, Baltimore, Maryland 21228

Michael L. McKee, Department of Chemistry, Auburn University, Auburn, Alabama 36849

Nancy J. S. Peters, Natural Science Division of Southampton College, Long Island University, Southampton, New York 11968

M. Rahman, Department of Chemistry, Auburn University, Auburn, Alabama 36849

Philip B. Shevlin, Department of Chemistry, Auburn University, Auburn, Alabama 36849

Anne Skancke, Department of Mathematical and Physical Sciences, University of Tromsø, N-9001 Tromsø Norway

Daniel J. Swartling, Department of Chemistry, University of North Dakota, Grand Forks, North Dakota 58202

John T. Welch, Department of Chemistry, State University of New York, Albany, New York 12222

Myung-Hwan Whangbo, Department of Chemistry, North Carolina State University, Raleigh, North Carolina 27695

Series Foreword

Molecular structure and energetics are two of the most ubiquitous, fundamental and, therefore, important concepts in chemistry. The concept of molecular structure arises as soon as even two atoms are said to be bound together since one naturally thinks of the binding in terms of bond length and interatomic separation. The addition of a third atom introduces the concept of bond angles. These concepts of bond length and bond angle remain useful in describing molecular phenomena in more complex species, whether it be the degree of pyramidality of a nitrogen in a hydrazine, the twisting of an olefin, the planarity of a benzene ring, or the orientation of a bioactive substance when binding to an enzyme. The concept of energetics arises as soon as one considers nuclei and electrons and their assemblages, atoms and molecules. Indeed, knowledge of some of the simplest processes, e.g., the loss of an electron or the gain of a proton, has proven useful for the understanding of atomic and molecular hydrogen, of amino acids in solution, and of the activation of aromatic hydrocarbons on airborne particulates.

Molecular structure and energetics have been studied by a variety of methods ranging from rigorous theory to precise experiment, from intuitive models to casual observation. Some theorists and experimentalists will talk about bond distances measured to an accuracy of 0.001 Å, bond angles to 0.1°, and energies to 0.1 kcal/mol and will emphasize the necessity of such precision for their understanding. Yet other theorists and experimentalists will make equally active and valid use of such seemingly ill-defined sources of information as relative yields of products, vapor pressures, and toxicity. The various chapters in this book series use as their theme "Molecular Structure and Energetics," and it has been the individual authors' choice as to the mix of theory and of experiment, of rigor and of intuition that they have wished to combine.

As editors, we have asked the authors to explain not only "what" they know but "how" they know it and explicitly encouraged a thorough blending of data and of concepts in each chapter. Many of the authors have told us that writing their chapters have provided them with a useful and enjoyable (re)education. The chapters have had much the same effect on us and we trust readers will share our enthusiasm. Each chapter stands autonomously as a combined review and tutorial of a major research area. Yet clearly there are interrelations between them and to emphasize this coherence we have tried to have a single theme in each volume. Indeed the first four volumes of this series were written in parallel, and so for these there is an even higher degree of unity. It is this underlying unity of molecular structure and energetics with all of chemistry that marks the series and our efforts.

Another underlying unity we wish to emphasize is that of the emotions and of the intellect. We thus enthusiastically thank Alan Marchand for the opportunity to write a volume for his book series, which grew first to multiple volumes, and then became the current, autonomous series for which this essay is the foreword. We also wish to emphasize the support, the counsel, the tolerance and the encouragement we have long received from our respective parents, Murray and Lucille, Murray and Bella; spouses, Deborah and Susan; parents-in-law, Jo and Van, Wilbert and Rena; and children, David and Rachel. Indeed, it is this latter unity, that of the intellect and of emotions, that provides the motivation for the dedication for this series:

"To Life, to Love, and to Learning."

Joel F. Liebman
Baltimore, Maryland

Arthur Greenberg
Newark, New Jersey

Introduction

The twelve chapters of this volume seek to aid the reader's understanding of the diverse, often unprecedented, chemistry of fluorinated species. They provide experimental results, conceptual models, and theoretical calculations. Both organic and inorganic compounds are explored and explained. The inherent chemical complexity ranges from diatomic molecules and their associated ions in the gas phase to organometallics and carbanions in solution and nucleic acids and their monomeric components in the organism.

Chapter 1, by Rahman, McKee, and Shevlin, discusses the chemical phenomena that result from reactions of the nearly inert perfluorocarbons with the ravenous atomic carbon. The unique chemical behavior of the resulting fragments, for example, CF, singlet and triplet CF_2, CF_3, and CF_2O_2, arises from their intermediate degrees of reactivity and selectivity. Insights and techniques taken from high temperature chemistry, ab-initio quantum chemical calculations, and classical product analysis, are combined cogently and coherently.

Chapter 2, by Filler, discusses both inter- and intramolecular arene-polyfluoroarene interactions. The use of gas phase photoelectron spectroscopy, absorption maximum shifts upon complexation in solution, and X-ray crystallography of the organic solid state, as well as qualitative theory results in a unified picture of these chemical/physical interactions. The chapter concludes with the author's novel interpretations of diverse literature reactions of fluorinated organic compounds.

Chapter 3, by Skancke, presents the effects of fluorination on the energetics of aromatic species. The surprising similarities and differences of the effects on benzenoid and nonbenzenoid compounds are highlighted. The effects of the net molecular charge are also addressed. Numerous challenges and suggestions to the experimentalist are found in this largely theoretical study.

Chapter 4, by Dolbier and Koroniak, contrasts the electrocyclic ring opening reactions of the parent cyclobutene, its perfluorinated analog, and diversely numerous substituted derivatives, all of which form the corresponding butadiene. Electronic and steric effects, rates and equilibria, rigorous theory and qualitative models, are all incorporated in the understanding of this archetypical and superficially simple class of organic reactions.

Chapter 5, by Bumgardner and Whangbo, contrasts hydrogen and fluorine-containing groups in electrocyclic reactions. Expressing their analysis in terms of secondary orbital interactions, these authors also discuss the effect of fluorination on relative isomer stabilities, especially that of Z vs. E olefins and of ketones vs. their corresponding enols. Other molecular properties such as the effects of fluorination on the acidity of 1,3-diketones are also discussed.

Chapter 6, by H. F. Koch and J. G. Koch, describes the formation of, halide elimination from, and protonation of fluorinated carbanions in alcoholic media. The interrelationships of exchange rates, equilibria, and isotope effects are discussed. A comparison of the roles of solvation, ion pairing, and delocalization on carbanion stability and reactivity is also made. Both formal kinetics and qualitative structural models provide quantitative as well as qualitative understanding of these phenomena.

Chapter 7, by Welch and Eswarakrishnan, extends the discussion of fluorinated carbanions to that of enolates and such related anionic and neutral equivalents as silyl ethers. Effects of fluorine substitution on stability, stereo-specificity and -selectivity, and reactivity, are discussed, as are considerations of fundamental stereochemistry and of synthetic utility.

Chapter 8, by Burton, has a focus similarity to that of the preceding chapter, but deals explicitly with polyfluorinated alkenyl organometallics. The roles of the metal (cf., Li, Zn, Cd, Hg) in affecting thermal stability, product yields, and spectroscopic identification are emphasized. Detailed descriptions of the use of these organometallics in the synthesis of otherwise inaccessible classes of fluorinated organic compounds is also presented.

Chapter 9, by Jache, moves the reader into explicitly inorganic fluorine chemistry by presenting an extensive study of the *smallest* compound of fluorine, hydrogen fluoride. All forms of the pure species—from the diatomic molecule to associated oligomers, in both the gaseous and condensed phase—are discussed. The role of HF as a solvent is also presented at length—the reader will recall that liquid HF plays the role of water for much of fluorine chemistry.

Chapter 10, by Peters and Allen, describes the numerous exotic species that contain the elements nitrogen, oxygen, and fluorine. Though the formulas, and even atomic attachment, are often analogous to the normal hydrogen-containing species, the relative stabilities, energetics, geometries, and orbital patterns of the fluorinated species are unrecognizable from these analogs. This does not mean that systematics are absent—rather, the authors demonstrate new structure/energy patterns for these superficially simply recognized compounds.

Chapter 11, by Bergstrom and Swartling, considers fluorine-containing analogs of nucleic acids and their monomeric components. These derivatives, say by replacing an H or OH in a nucleoside by F, often vary less from their unsubstituted parent than those in the previous chapter. Nonetheless the substituted species often exhibit profoundly new biochemical activities. Synthesis and spectroscopy augment the discussion.

Chapter 12, by Liebman, closes this book with an almost heretical study of which nonfluorine containing substituents *mimic* those that contain fluorine. While discussion is limited to the basic physical/chemical properties of heats of vaporization, of sublimation, of formation, and of protonation (proton affinities), application of the simple, but reliable, mimicry rules in this chapter is made to numerous classes of organic compounds.

Contents

CHAPTER 1

The Reactions of Atomic Carbon with Fluorocarbons

**M. Rahman, Michael L. McKee, and
Philip B. Shevlin**

Department of Chemistry, Auburn University, Alabama

CONTENTS

1. INTRODUCTION

One of the most interesting reactive intermediates in organic chemistry is atomic carbon. The high energy of zero valent carbon atoms makes possible a wide variety of interesting chemical reactions.[1] For example, atomic carbon has been observed to insert into C—H bonds (Equation 1–1), to add to double bonds to generate cumulenes (Equation 1–2), and to abstract oxygen from carbonyl compounds in a facile synthesis of carbenes (Equation 1–3).

1

$$C + CH_4 \rightarrow H{-}\overset{\cdot\cdot}{C}{-}CH_3 \rightarrow CH_2{=}CH_2 + CH{\equiv}CH \qquad (1\text{--}1)$$

$$C + \overset{\diagup}{\underset{\diagdown}{C}}{=}\overset{\diagup}{\underset{\diagdown}{C}} \rightarrow \overset{\overset{\overset{\cdot\cdot}{C}}{\diagup\diagdown}}{C{=}C} \rightarrow \overset{\diagup}{\underset{\diagdown}{C}}{=}C{=}\overset{\diagup}{\underset{\diagdown}{C}} \qquad (1\text{--}2)$$

$$C + \overset{\diagup}{\underset{\diagdown}{C}}{=}O \rightarrow CO + \overset{\diagup}{\underset{\diagdown}{C}}{:} \qquad (1\text{--}3)$$

In view of this extreme reactivity of atomic carbon it is of interest to consider its reaction with a fluorocarbon.[2] It is well known that the strength of C—F bonds renders them inert to many reactions that are facile with C—H bonds. For example, methylene, itself an extremely indiscriminate reagent, fails to undergo intermolecular insertion into C—F bonds.[3] Thus, the reaction of C atoms with C—F bonds can be considered a classic case of an irresistible force meeting an immovable object. In this chapter, we report an investigation of the reactions of atomic carbon with simple fluorocarbons in which we elucidate the structure and reactivity of the various intermediates involved.

A consideration of the possible modes of reaction between C atoms and CF_4 leads to the conclusion that the processes in Equations 1–4 through 1–6 could occur. The insertion in Equation 1–4 leads to tetrafluoroethanylidene, **1,** which can either react intermolecularly or rearrange to tetrafluoroethylene. Equations 1–5 and 1–6 depict two possible abstraction modes in which either one fluorine is removed to generate fluoromethylidyne, CF, or two fluorines are removed to give difluoromethylene, CF_2. A consideration of the heats of formation of CF,[4] CF_2,[5] and CF_3[6] indicates that, when carbon is in the $C(^1D)$ state,[7] the abstraction in Equation 1–5 will be exothermic by 30 kcal/mol while the formation of two molecules of $^1CF_2$ in Equation 1–6 is exothermic by 75 kcal/mol. The intermediacy of CF in the reactions of ^{11}C atoms, generated by nuclear recoil, was postulated by Blaxell, MacKay, and Wolfgang who reacted carbon–11 atoms with fluorocarbons and observed organic products which were thought to contain a single fluorine atom.[8] Finn, Ache, and Wolf reported high yields of $^{11}CO_2$ when ^{11}C was reacted with fluorocarbons and O_2 and proposed that an intermediate, perhaps CF, was oxidized to CO_2.[9]

$$C + CF_4 \rightarrow F{-}\overset{\cdot\cdot}{C}{-}CF_3 \rightarrow CF_2{=}CF_2 \qquad (1\text{--}4)$$
$$\mathbf{1}$$

$$C + CF_4 \rightarrow F{-}\overset{\cdot\cdot}{C}{\cdot} + CF_3 \qquad (1\text{--}5)$$

$$C + CF_4 \rightarrow 2CF_2 \qquad (1\text{--}6)$$

In order to evaluate the viability of the reactions in Equations 1–4 through 1–6, we have carried out ab initio molecular orbital studies of the reaction between carbon and CF_4. We also examined the reaction of carbon with fluorocarbons experimentally in the presence of various trapping agents. The results of this combined experimental and theoretical approach will be presented in subsequent sections.

2. MOLECULAR ORBITAL CALCULATIONS

In these calculations, which utilized the GAUSSIAN 82 program,[10] geometries were optimized using the 3–21G basis set at the single configurational level with the restricted Hartree–Fock formalism for closed-shell systems and the unrestricted formalism for open-shell systems. Single point calculations were made at the 6–31G* level and at the MP2/6–31G level in order to estimate relative energies at the MP2/6–31G* level by using the additivity approximation in which the effects of electron correlation and polarization functions on a 6–31G calculation are simply added.[11]

The reactions considered in these calculations were the insertion of carbon into the C—F bond to generate tetrafluoroethanylidene (Equation 1–4), fluorine abstraction to give CF (Equation 1–5), and concerted formation of two molecules of CF_2 (Equation 1–6). The calculated energies of the various products and transition states along these reaction coordinates, relative to $C(^3P) + CF_4$, are shown in Table 1–1.

In agreement with experimentally measured heats of formation, the calculations indicate that the processes in Equations 1–4 through 1–6 are generally thermodynamically favorable. Thus, $C(^3P)$ inserts into the C—F bond to give triplet **1** with a ΔH of −49.6 kcal/mol while $C(^1D)$ forms singlet **1** with $\Delta H = -87.5$ kcal. We calculate that **1** has a singlet ground state with a singlet–triplet splitting of 7.9 kcal/mol. Dixon, using a two configuration wave function and a double zeta plus polarization basis set, has calculated a singlet–triplet splitting of 9.1 kcal/mol for the same carbene.[12] Abstraction by $C(^3P)$ to give CF is slightly endothermic by 5.0 kcal/mol while abstraction by $C(^1D)$ is exothermic by 25.0 kcal/mol in good agreement with the experimental value of 30 kcal/mol.[4,5,6] The abstraction of two fluorines in Equation 1–6 to give two molecules of $^1CF_2$ is exothermic by 57.6 kcal/mol. However, this reaction is calculated to have a higher barrier on both the triplet and singlet energy surfaces than the processes in Equations 1–4 or 1–5. These calculations indicate that $C(^3P)$ will react with carbon via the transition structure **2,** which lies 5 kcal/mol above $C(^3P) + CF_4$ at the [UMP2/6–31G*] level. Removal of two fluorines by $C(^3P)$ to give either $2CF_2$ or C_2F_4 is calculated to have a barrier of 38.1 kcal/mol (entry 12, Table 1–1). A treatment of the reaction between CF_4 and $C(^1D)$ is more complex in that this state of carbon has both open- and closed-shell components and is not well treated by a single configurational method. Thus, we have evaluated the reactivity of $C(^1D)$ by separately considering both open- and closed-shell reaction paths.[13]

TABLE 1-1. Relative Energies (kcal/mol) of Various Species on the C_2F_4 Potential Energy Surface

Molecule	Electronic state	6–31G	MP2/6–31G	6–31G*	[MP2/6–31G*][a]
1 $C + CF_4$	$^3P, {}^1A_1$	0	0	0	0
2 $C + CF_4$	$^1D, {}^1A_1$				30.0[b]
3 $CF + CF_3$	$^2\pi, {}^2A_1$	14.9	3.7	16.2	5.0
4 $2(CF_2)$	$^1A_1$	11.9	20.5	1.8	−27.6
5 $FCCF_3$	$^1A'$	−23.0	−52.7	−27.8	−57.5
6 $FCCF_3$	$^3A''$	−37.9	−52.7	−34.8	−49.6
7 $F_2C{=}CF_2$	^1A_g	−58.4	−97.0	−55.6	−94.2
8 $F_2C{=}CF_2$	$^3A_1$	−22.6	−41.4	−14.0	−32.8
9 $C{-}CF_3$ (F bridge)	$^1A'$	23.7	80.5	131.4	88.2
10 $C(F,F){-}CF_2$	$^1A_1$	89.4	35.5	97.2	43.0
11 $C{-}CF_3$ (F bridge)	$^3A''$	18.3	4.8	18.5	5.0
12 $C(F,F){-}CF_2$	$^3B_2$	65.3	31.4	72.0	38.1
13 $C(^3P) + CF_3{}^c$	$^3P, {}^2A_1$	0	0	0	0
14 $CF + {}^3CF_2$	$^2\Pi, {}^3B_1$	19.5	10.9	21.0	12.4
15 $C{-}CF_3$	$^2A''$	−37.6	−58.2	−45.3	−65.9
16 $C(F,F){-}CF_2$	$^2A''$	5.0	−33.8	−6.8	−45.6[d]
17 $FC{=}CF_2$	$^2A'$	−66.4	−91.8	−67.2	−92.6

[a] Additivity approximation Reference 11.
[b] Taken from experimental $C(^3P)-C(^1D)$ splitting. Moore, C. E. *Natl. Bur. Stand. (U.S.) Circ.* **1949,** *1,* No. 467.
[c] Energies in entries 14–17 are relative to the energy in entry 13.
[d] The force constant matrix has 2 negative eigenvalues, indicating the existence of a lower symmetry transition state at the 3–21G level. A transition state of C_1 symmetry could yield the C_2F_3 radical (**17**) as the reaction product.

When this is done, the closed-shell singlet is calculated to form tetra-fluoroethylene directly with a barrier of 13.0 kcal/mol. In order to evaluate the open-shell singlet surface, we have assumed that the energy of open shell **2,** with its widely separated unpaired spins, is only slightly higher than that of triplet **2.** Since $C(^1D)$ is 30 kcal higher in energy than $C(^3P)$, this assumption leads to the conclusion that $C(^1D)$ will react with CF_4 to form **2** with little or no barrier. The fact that the spins are indeed separated in triplet **2** is indicated by the fact that the C—C distance is 1.1 Å longer than in triplet **1.**

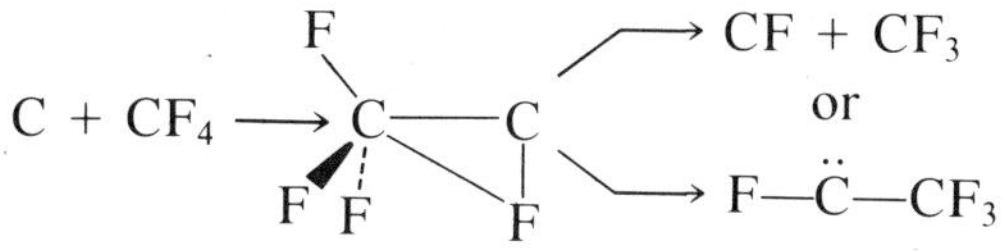

$$C + CF_4 \longrightarrow \underset{2}{\overset{F}{\underset{F}{\overset{|}{C}}}} \!\!-\!\! \overset{F}{\underset{F}{\overset{|}{C}}} \!\! \begin{array}{l} \longrightarrow CF + CF_3 \\ \qquad \text{or} \\ \longrightarrow F\!-\!\ddot{C}\!-\!CF_3 \\ \qquad \qquad \mathbf{1} \end{array}$$

In addition, the spin densities on carbons in **2** (1.17, 1.15; 6–31G*) are essentially uncoupled. This is in contrast to **1** where both electrons are on the carbene carbon (0.06, 2.18; 6–31G*).

Thus, the calculations predict that the lowest energy process for both singlet and triplet carbon will be the formation of **2**. However, an examination of the geometry of **2** does not allow us to determine if it will proceed directly to $CF + CF_3$ or collapse to tetrafluoroethanylidene, **1**. We will now turn to an experimental evaluation of this point.

3. THE REACTION OF ARC GENERATED CARBON ATOMS WITH CF₄

A. Reaction in the Absence of Added Trapping Agents

In order to examine this reaction experimentally, we have generated atomic carbon in a carbon arc and cocondensed it with CF_4 at 77K.[14] Reaction of carbon with CF_4 alone gives no volatile products. Extraction of the reactor bottom with acetone yields a polymer with an ir absorbance at 1215 cm^{-1}. It is interesting that this ir band is the same position as that reported for $(CF)_n$.[15] An extensive examination of reaction products did not reveal the presence of tetrafluoroethylene or the dimer of carbene **1**.

B. The Reaction of C with CF₄ in the Presence of Added Alkenes

The lack of volatile products in the $C + CF_4$ reaction induced us to follow the lead of Wolfgang and co-workers[8] and add alkenes to trap any intermediates that may be present. Since the products of the reaction of atomic carbon with alkenes themselves are generally well understood,[1] we have focused our attention on products containing fluorine. Two common products always observed are fluorocyclopropanes and 1,1–difluoroalkanes (Equation 1–7) whose yields are shown in Table 1–2. In no case could tetrafluoroethylene or adducts of carbene **1** be detected among the products.

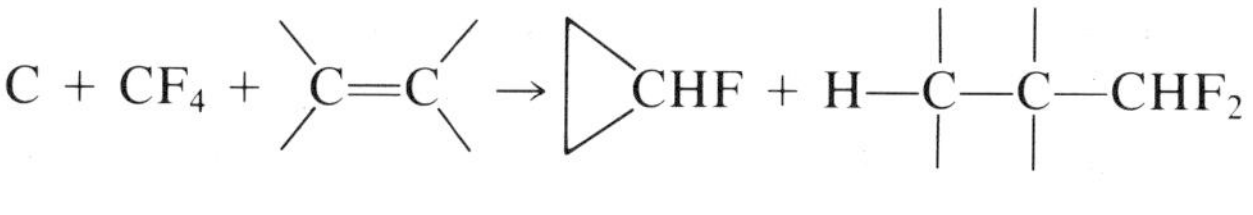

$$C + CF_4 + \underset{/}{\overset{\backslash}{C}}\!\!=\!\!\underset{\backslash}{\overset{/}{C}} \rightarrow \underset{}{\triangleright}\!CHF + H\!-\!\overset{|}{\underset{|}{C}}\!-\!\overset{|}{\underset{|}{C}}\!-\!CHF_2 \qquad (1\text{--}7)$$

TABLE 1–2. Yields of Fluorocyclopropanes and 1,1–Difluoroalkanes in the Reaction of C + CF$_4$ + Alkenes

Reactants[a] Alkene carbon[c] (mmol)		Product yields (mmol × 10³)[b]	
		Fluorocyclopropane	1,1-Difluoroalkane
==	8.36	6.84	1.28
	13.10	Cis 8.79 Trans 6.2	1.23
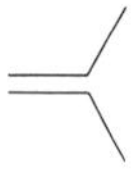	8.40	2.95	2.78
	2.88	Cis 0.87 Trans 1.14	0.27
	5.79	Cis 2.28 Trans 2.30	1.97
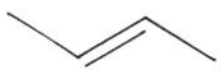	1.76	2.34	0.44
	3.27	Cis 0.95 Trans 1.60	0.30
Ph—==	1.55	Cis 0.77 Trans 0.94	[d]
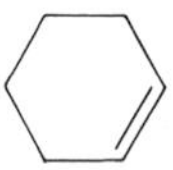	1.62	Syn 1.96 Anti 3.86	1.10

[a] In all cases 15 mmol each of alkene and CF$_4$ were cocondensed with carbon.

[b] Although CHF$_3$ is detected among the products it is difficult to quantitate due to its volatility under the reaction conditions.

[c] This is the amount of carbon lost from the graphite electrodes. Since some carbon is physically removed, the actual amount of carbon vaporized is less.

[d] No 3–phenyl–1,1–difluoropropane could be detected. Perhaps the HCF$_2$ polymerizes the styrene.

The structures of the fluorine-containing products may be deduced from their ^{19}F NMR spectral data. The stereochemistry of the fluorocyclopropanes can be determined from their F—H couplings and ^{19}F chemical shifts.[16] The vicinal protons trans to the F in fluorocyclopropanes couple with $J_{H—F}$ = 8–12 Hz while cis protons have $J_{H—F}$ = 18–23 Hz. In addition, fluorines cis to alkyl groups in fluorocyclopropanes always resonate at higher fields than trans fluorines. The 1,1-difluoroalkanes generated in these reactions are characterized by the presence of a doublet with $^2J_{H—F}$ = 55–57 Hz which is further split to show appropriate $^3J_{H—F}$ at δ = 116–124 ppm relative to CFCl$_3$. In order to further characterize the fluorine-containing products, they were separated by gas chromatography and their structures confirmed by mass spectrometry. While the fluorocyclopropanes show a parent ion in their mass spectrum, the 1,1–difluoroalkanes generally exhibit a peak corresponding to the loss of HF from the parent. In all cases, the spectra of our products were identical to those of authentic samples.

C. The Mechanism of Formation of the Fluorocyclopropanes

Since the fluorocyclopropane products represent the addition of both a carbon and a fluorine to the olefinic trapping agent, it is reasonable to assume the involvement of CF in the mechanism of their formation. CF can be considered as either a carbene or a free radical and it is of interest to determine if its reactions are those of free radicals or carbenes. If CF exhibits carbenic-type reactivity, addition to the double bond to give a fluorocyclopropyl radical is expected. Subsequent hydrogen abstraction generates the fluorocyclopropane as shown in Equation 1–8. Alternatively, the CF could react like a radical and abstract hydrogen to generate HCF which then adds to the alkene to give the fluorocyclopropane (Equation 1–9). However, an examination of the enthalpy of the hydrogen abstraction in Equation 1–9 reveals that the process is endothermic for both vinyl (ΔH = +37 kcal/mol) and allylic C—H bonds (ΔH = +10 kcal/mol). Hence, this mechanism is unlikely. In contrast, we calculate (3–21G) that the hydrogen abstraction by the fluorocyclopropyl radical in Equation 1–8 is favorable for vinyl and allylic C—H bonds by 2.5 and 26.4 kcal/mol, respectively. When the C + CF_4 + ethylene reaction is run in the presence of ethane, a source of more readily abstractable hydrogen, the fluorocyclopropane yield doubles.

$$CF + C{=}C \rightarrow \overset{F}{\underset{\triangle}{\cdot C}} \xrightarrow{R-H} \overset{H\quad F}{\underset{\triangle}{C}} \qquad (1\text{–}8)$$

$$CF + R{-}H \rightarrow H\ddot{C}F \xrightarrow{C{=}C} \overset{H\quad F}{\underset{\triangle}{C}} \qquad (1\text{–}9)$$

When substituted alkenes are used as substrates in the C + CF_4 reaction, both cis and trans substituted fluorocyclopropanes are generated with the trans generally predominating. In contrast, when CHF is reacted with alkenes in the gas phase[17] or in solution,[18] the cis : trans ratios in the product cyclopropanes are invariably close to unity. Table 1–3 compares cis/trans ratios from reaction of fluoromethylene with alkenes to those observed in the present study. These data indicate that the reactivity of fluoromethylene is different from that observed here and provide evidence that the intermediate responsible for fluorocyclopropane formation is not HCF but CF.

Table 1–3 demonstrates that, when propene is added to the C + CF_4 + cyclohexene system, the syn : anti ratio in 7–fluorobicyclo[4.1.0]heptane

TABLE 1–3. Cis–trans Ratios of Fluorocyclopropanes Formed in the Reactions of C + CF$_4$ and CHF with Alkenes

Alkene substrate(s)	Cis–trans ratios in product fluorocyclopropanes		
	C + CF$_4$	CTF (gas)[a]	CHF (soln)[b]
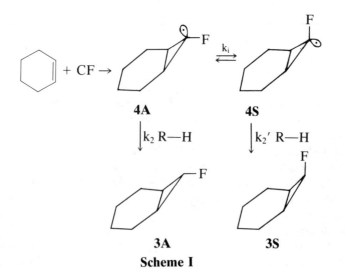	0.76	1.09	—
	0.99	1.0	—
	0.59	1.0	1.1
	0.51	—	1.0
+ (1:1)	0.23		
+ (1:1.5)	0		
+ (1:2)	0		

[a] Generated by the reaction of energetic tritium atoms with difluoromethane.[15]
[b] From the photolysis of fluorodiiodomethane.[16]

4A $\rightleftharpoons^{k_i}$ **4S**

k_2 R—H k_2' R—H

3A   **3S**

Scheme I

(3-S and 3-A) decreases from 0.51 to 0. This result is consistent with the mechanism in Scheme I involving initial addition of CF to cyclohexene to give the anti radical 4-A. Radical 4-A can either invert to the syn radical 4-S or abstract a hydrogen. Since propene is expected to be more mobile on the low temperature matrix than cyclohexene and hence a better hydrogen donor, the ratio $k_2[R—H] : k_i$ is expected to increase as propene is added to the matrix resulting in an increase in 3-A. Since we calculate ([UMP2/6–31G*]) a barrier of 13.7 kcal/mol for inversion of the fluorocyclopropyl radical,[19] the interconversion of syn and anti radicals is possible if these species are formed with a moderate amount of excess energy on the matrix. A variation of the 3-S : 3-A ratio as a function of hydrogen donor concentration would not be anticipated if CF were to first abstract hydrogen to give CHF.

D. The Stereochemistry of CF Additions

The use of $E–$ and $Z–2–$butene as substrates allows an evaluation of the stereochemistry about the alkene C—C bond during addition of CF. If CF were to add as a radical, one might expect the open-shell intermediate 5 to have sufficient lifetime to permit rotation about the C—C bond and consequently nonstereospecific formation of the fluorocyclopropyl radical. Alternatively, addition in the same manner as a singlet carbene is expected to generate the fluorocyclopropyl radical stereospecifically. When this reaction is carried out, the stereospecific formation of trans–trans and cis–cis 2,3–dimethylfluorocyclopropane from $Z–2–$butene and only *cis–trans–2,3–di*methylfluorocyclopropane from $E–2–$butene rules out an open-shell intermediate such as 5 in which there is free rotation about the C—C bond.

$$CF + \overset{:CF}{\diagup\!=\!\diagdown} \rightarrow \underset{\mathbf{5}}{\diagdown\!\diagdown\!\cdot} \rightarrow \overset{F}{\triangle} + \overset{F}{\triangle}$$

It is interesting that the fluorocyclopropyl radicals generated in these reactions do not undergo electrocyclic ring opening to allyl radicals. Although this reaction is exothermic, it has been observed that the barrier to ring opening in cyclopropyl radicals is high compared to that for hydrogen abstraction.[20]

E. The Reaction of CF with Benzene

Since the preceding experiments demonstrate that CF undergoes initial reaction as a carbene and carbenes are known to react with benzene to give norcaradienes which subsequently ring open to cycloheptatrienes, we expect that the initial product of the reaction of CF with benzene will be the 7–

fluoronorcaradien–7–yl radical, **6.** It is then possible that **6** will ring open to the fluorotropyl radical, **7.** As just stated, it has been well documented in the literature that electrocyclic ring openings of cyclopropyl radicals to allyl radicals are difficult in the absence of a large thermodynamic driving force. The delocalization energy of the tropyl radical may well provide this driving force in this case. However, several studies of benzonorcaradienyl radicals show that they require temperatures of 180°C for ring opening[21] or they do not open at all.[22] While radical **6** will undoubtedly abstract hydrogen from donors such as isobutane and propene, a consideration of the heat of forma-tion of the tropyl radical indicates that such hydrogen abstractions by **7** will be endothermic by 15 kcal/mol and should not occur.[23] Thus if **6** is trappable, reaction of CF with benzene in the presence of isobutane should lead initially to 7–fluoronorcaradiene, **8,** which will ring open to 7–fluorocyclohepta-triene, **9.** Compound **9** has not been reported in the literature and it is expected to be in equilibrium with tropylium fluoride, which itself should be extremely labile. However, it should be possible to trap **9** as the very stable tropylium fluoroborate, **10,** by adding BF_3 to the low temperature matrix upon which it is formed as shown in Equation 1–10.

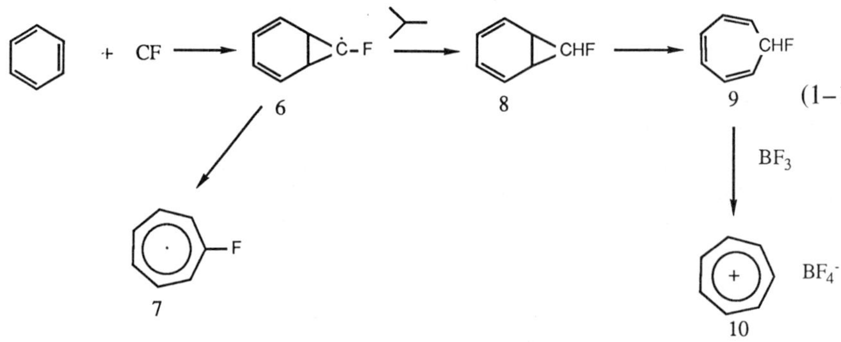

Cocondensation of $C + CF_4 +$ benzene + isobutane followed by addition of BF_3 after the condensation results in the formation of tropylium fluoro-borate. These experiments demonstrate that CF adds to benzene to generate **6** which will abstract hydrogen before it can ring open to the fluorotropyl radical, **7.** It is unlikely that **7** is the precursor to either **9** or **10** as hydrogen abstraction by **7** from isobutane will be endothermic.[23]

F. Attempts to React CF with Saturated Hydrocarbons

Since carbenes are observed to insert into C—H bonds, we have investi-gated the potential for CF to undergo this reaction. However, cocondensa-tion of carbon, CF_4, and either ethane or isobutane failed to produce any products corresponding to C—H insertion. A significant difference between a C—H insertion by a carbene and by a methylidyne is the fact that the latter reaction does not generate a low-energy stable product but an energetic free radical. Perhaps it is this factor that precludes C—H insertion by CF.

G. The Mechanism of Formation of 1,1–Difluoroalkanes in the Reaction of C + CF$_4$ + Alkenes

Although the presence of 1,1–difluoroalkanes in the C + CF$_4$ reaction appears to be indicative of the intermediacy of CF$_2$, no products in which CF$_2$ has added to the alkene to give a difluorocyclopropane are ever observed. Instead, the results are best accommodated by invoking a triplet CF$_2$ which abstracts a hydrogen to generate CHF$_2$ radical which then adds to the alkene in a radical manner as shown in Equation 1–11. The 1,1–difluoroalkanes have the regiochemistry that is expected for addition of CHF$_2$ radical to a substituted alkene to give the most stable free radical. Since hydrogen abstractions by carbenes are characteristic of the triplet state, we feel that the reactive form of CF$_2$ in this system is the excited triplet state which has been calculated to lie 49 kcal/mol above the singlet ground state of CF$_2$.[24] However, the reaction shown in Equation 1–6 in which either C(^{3}P) or C(^{1}D) reacts with CF$_4$ to give exclusively ^{3}CF$_2$ is endothermic. As an alternative to the reaction in Equation 1–6, we have postulated that the ^{3}CF$_2$ arises by reaction of carbon with the CF$_3$ radical (Equation 1–12).

$$^3CF_2 + R—H \rightarrow HCF_2 \xrightarrow{=—R}$$

$$HCF_2\text{—}\overset{\bullet}{\diagdown}\,R \xrightarrow{R—H} HCF_2\text{—}\diagdown\,R \quad (1–11)$$

$$C + CF_3 \rightarrow CF + {}^3CF_2 \quad (1–12)$$

An evaluation of the thermodynamics of the process in Equation 1–12 is complicated by the fact that a direct experimental determination of the heat of formation of ^{3}CF$_2$ has not been carried out. This value must be estimated by adding the calculated singlet–triplet energy difference of 56.6 kcal/mol[24] to the observed heat of formation of the singlet.[6] With this fact in mind, we conclude that if C(^{1}D) is the reactant, the process in Equation 1–12 is exothermic by 22.4 kcal/mol. The results of our theoretical analysis of the reaction in Equation 1–12, which are shown in Table 1–1, indicate that reaction of C(^{1}D) with CF$_3$ to generate ^{3}CF$_2$ is exothermic by 17.6 kcal/mol while the formation of ^{3}CF$_2$ by C(^{3}P) is endothermic by 20.0 kcal/mol. However, the calculations did not locate a transition state for the process in Equation 1–12 and the kinetics of this process are unknown. The energies in Table 1–1 demonstrate that an alternate reaction between C and CF$_3$ is simply bond formation to give C—CF$_3$ which can rearrange to the trifluorovinyl radical as shown in Equation 1–13.

$$C + CF_3 \rightarrow \cdot\ddot{C}-CF_3 \rightarrow F\dot{C}=CF_2 \qquad (1-13)$$

In order to determine if the reaction in Equation 1–12 leads to the formation of the $^3CF_2$ in this system, we have reacted atomic carbon with perfluoroisobutane in the presence of ethylene and propene. This reaction resulted in the formation of fluorocyclopropanes but no 1,1–difluoroalkanes. Since perfluoroisobutane cannot give CF_3 upon fluorine abstraction by atomic carbon, this result indicates that CF_3 is the precursor to $^3CF_2$ and ultimately the 1,1–difluoroalkanes. Thus, the reaction of atomic carbon with CF_3 appears to represent an interesting method of generating $^3CF_2$. Previous routes to $^3CF_2$ have generally involved passing fluorocarbons through electric discharges or high temperature furnaces.[25,26]

The formation of triplet rather than singlet CF_2 by the abstraction in Equation 1–12 may be a consequence of the fact that the reaction is S_H2 in character and will thus yield a CF_2 with two electrons in different orbitals. The most stable open shell CF_2 is the triplet.

4. THE REACTION OF C WITH CF_4 IN THE PRESENCE OF O_2

A. The Products of the Reaction of CF and CF_2 with Oxygen

The reaction of nucleogenic carbon–11 atoms with CF_4 has been reported to generate $^{11}CO_2$.[9] This is in contrast to the reaction of carbon atoms with oxygen alone which yields CO.[1] When arc generated carbon atoms are co-condensed with CF_4 containing 5% oxygen, carbon dioxide and carbonyl fluoride are formed in a 1.1 : 1 ratio as shown in Equation 1–14. The carbon dioxide, which has also been reported in the ^{11}C system,[9] may be rationalized by addition of CF to O_2 followed by ring opening to a fluorocarboxy radical which loses a fluorine atom (Equation 1–15).

$$C + CF_4 + O_2 \rightarrow CO_2 + CF_2=O \qquad (1-14)$$

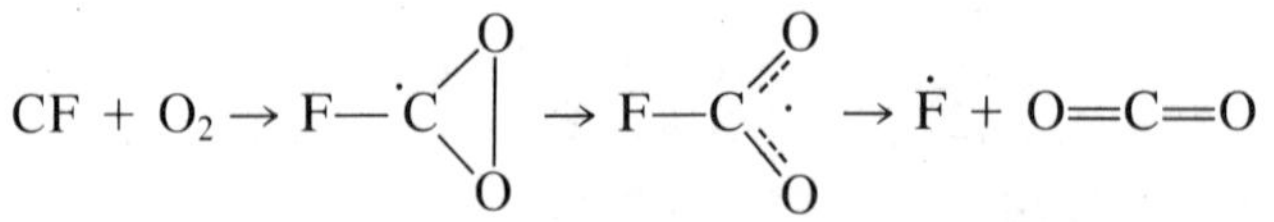

$$CF + O_2 \rightarrow F-\underset{O}{\overset{O}{C}} \rightarrow F-\overset{O}{C}\cdot \rightarrow \dot{F} + O=C=O \qquad (1-15)$$

The fact that carbonyl fluoride is generated in this reaction is additional evidence for the intermediacy of $^3CF_2$ in this system. The formation of carbonyl fluoride in the reaction between $^3CF_2$ and O_2 has been reported and is thought to be a reaction characteristic of $^3CF_2$.[27] The mechanism of carbonyl fluoride formation in the reaction of triplet CF_2 with oxygen has not been

TABLE 1–5.   Relative Energies on the CF_2O_2 Potential Energy Surface

	Electronic state	6–31G	MP2/6–31G	6–31G*	[MP2/6–31G*](ZPC)[a]
	$^1A'$ [b]	9.1	42.7	31.4	65.0(64.0)
O	$^1A'$ [c]	35.7	75.2	64.2	103.7(101.0)
	$^3A'$ [d]	38.9	85.6	70.1	116.8(113.8)
O					
	$^1A_1$	0.0	0.0	0.0	0.0(0.0)
	$^3B_1$	3.3	13.4	16.5	26.6(25.3)
	$^3A_2$	47.2	29.8	63.6	46.1(42.2)
	$^1A_1, \, ^3\Sigma_g^-$	26.1	43.8	41.0	58.8(54.2)
	$^3B_1, \, ^3\Sigma_g^-$	47.9	82.8	73.2	108.0(103.3)
O	$^1A_1, \, ^3P$	−63.7	−8.4	−40.8	14.5(12.5)

correction made using vibrational frequencies from the 3–21G level.

ell solution.

solution with two singly occupied π orbitals.

21G level the $^3A''$ electronic state is dissociative, decomposing to $F_2C{=}O + O(^3P)$ ighly exothermic.

culations lead us to conclude that the species responsible for epox- tion in the reaction of C + CF_4 + O_2 + alkenes is the difluorodiox- As a tentative mechanism for the oxygen transfer reaction, we hat **12** reacts with alkenes to generate an initial zwitterionic inter- **13**, in a stereospecific manner. Loss of carbonyl fluoride from **13** en generate the corresponding epoxide stereospecifically. Interme- ould also be involved in the formation of the carbonyl compounds in these reactions if a hydrogen migration were to occur during loss bonyl fluoride (Equation 1–23).

elucidated and could involve direct reaction to generate carbonyl fluoride and an oxygen atom as shown in Equation 1–16. An alternate mechanism is the second order reaction between two CF_2O_2 molecules to generate carbonyl fluoride and an oxygen molecule shown in Equation 1–17.

$$^3CF_2 + O_2 \rightarrow CF_2{=}O + O \tag{1–16}$$

$$^3CF_2 + O_2 \rightarrow CF_2O_2$$

$$2CF_2O_2 \rightarrow 2CF_2{=}O + O_2 \tag{1–17}$$

B. Reaction of Carbon with CF_4 and O_2 in the Presence of Olefinic Trapping Agents

The formation of carbonyl fluoride in the reaction of oxygen with $^3CF_2$ by the mechanisms in either Equation 1–16 or Equation 1–17 raises the possibility that either oxygen atoms or CF_2O_2 could be trapped by the addition of appropriate trapping agents. In order to attempt these trapping experiments we have carried out the reaction of carbon with CF_4 and oxygen in the presence of olefinic trapping agents. In these experiments, we have focused our attention on products in which an oxygen has been added to the alkene and we have ignored those products resulting from addition of CF or $^3CF_2$ to the alkene. These experiments gave a mixture of oxygenated products consisting of the olefin epoxide along with aldehyde and/or ketone as shown in Equations 1–18 through 1–21. Thus the alkenes are being oxygenated under the reaction conditions and it is of interest to attempt to determine the nature of the oxygenating species.

It is well known that triplet carbon atoms react with oxygen to generate carbon monoxide and (presumably) an oxygen atom as shown in Equation 1–22.[1] Hence, it is possible that the oxygenated products of Equations 1–18 through 1–21 result from reaction of O atoms, generated in the C + O_2

reaction, with O_2. In order to test this possibility, we have reacted carbon atoms with oxygen in the presence of the alkene trapping agents but in the absence of any CF_4. This reaction again resulted in the formation of epoxides and carbonyl compounds but with different product ratios and stereochemistries than the reaction in which CF_4 is included as a reactant. The relative yields of products in this reaction and those in the presence of CF_4 are shown in Table 1–4. An inspection of the data in Table 1–4 reveals that the ratio of carbonyl compounds to epoxide is invariably higher when CF_4 is included among the reactants and that the epoxides are formed stereospecifically when CF_4 is included. The results of the reactions in the absence of CF_4 may be rationalized by invoking the presence of free oxygen atoms which are generated by the reaction in Equation 1–22. Previous investigations of the reaction of $O(^3P)$ with $E-$ and $Z-2$–butene at 77K have demonstrated nonstereospecific epoxide formation and have revealed epoxide to carbonyl product ratios very similar to those observed here.[28] While the formation of carbon monoxide in the reaction of carbon atoms with oxygen is well documented, these studies mark the first example of trapping oxygen atoms generated in this reaction.

$$C(^3P) + O_2 \rightarrow CO + O(^3P) \qquad (1-22)$$

TABLE 1–4. Yields of Epoxides and Carbonyl Compounds in the Reaction of C + O_2 + Alkenes and in the Reaction C + O_2 + CF_4 + Alkenes

Reactants (mmol)				Product yields (mmol × 10^2)			
Alkene[a]	CF_4	O_2	C	Epoxide	Aldehyde	Ketone	Carbonyl/Epoxide
=	—	1.25	5.7	4.0	5.1		1.3
=	25	2.5	5.1	1.2	6.7		5.6
⋀	—	1.25	4.0	5.0	2.8	1.9	0.9
⋀	25	2.5	5.9	0.2	1.9	1.3	14.6
⬠	—	1.25	5.8	2.4[b]	1.2[c]	4.6	1.3
⬠	25	2.5	5.5	5.5[b]	—	15.8	2.9
⋀	—	1.25	6.2	0.5[b]	2.5[c]	5.1	1.7
⋀	25	2.5	5.4	—	4.0[c]	16.0	4.0

[a] In all cases, 15 mmol of alkene were reacted.
[b] This is the yield of cis-2–butene oxide.
[c] The yield of trans-2–butene oxide.

The preceding results demonstrate that fr be involved in the formation of oxygenated O_2 are condensed with alkenes. Thus, we p the reaction of $^3CF_2$ with O_2, acts as an oxy Possible structures for CF_2O_2 include the ca dioxirane, **12**. Carbonyl oxides have long b ozonolysis[29] and in the reactions of carbe oxygen.[30,31] Calculations indicate that the rately described as 1,3 singlet biradicals as i now been isolated and characterized by sp

$$CF_2 \overset{+}{=} \overset{-}{O}-O \text{ or } \dot{C}F_2-O-$$

11a   **11b**

It is of interest that both carbonyl oxides observed to react with alkenes to generate of tetrafluoroethylene, which may be expec CF_2O_2, leads to high yields of tetrafluoroet dation of alkenes with carbonyl oxides ge rather than stereospecific epoxide formatio stereochemistry of epoxide formation has determining if the reagent transferring the or a dioxirane.[33a]

We have carried out ab initio molecular o energies of **11** and **12**. The results of these indicate that, at the [MP2/6–31G*] level, sir singlet or triplet **11** by 64.0 and 113.8 kcal are consistent with previous theoretical wo the parent dioxirane is more stable than the mol.[32a] In the present calculations, the carbo to be too unstable with respect to **12** as bo forms are known to be important[32] and thes present computational technique. Howeve ground state is the dioxirane, **12**. The increa to triplet **11** (S—T = −49.8 kcal/mol) as co in the parent carbonyl oxide of −27.7 kca considers that fluorine stabilizes single (S—T(CH$_2$) = 9 kcal/mol; S—T(CF$_2$) = calculations on the parent carbonyl oxide, stitution stabilizes the zwitterionic form o ble 1–5).

The stereospecific epoxide formation in

The oxygen–oxygen bond length in **12** is calculated to be 1.561 Å. In contrast, we calculate an O—O bond length of 1.522 Å in the parent dioxirane. These data indicate a weaker O—O bond in **12** and presumably indicate that **12** is a better oxygen donor than dioxirane itself.

5. CONCLUSIONS

These investigations demonstrate that fluorocarbons, which are inert to most reactive intermediates, react readily with atomic carbon by a fluorine abstraction to give CF. This reaction appears to be a convenient way of generating this monovalent carbon intermediate. CF is observed to react in a manner similar to that of a singlet carbene by adding to double bonds in a stereospecific manner to give 1–fluorocyclopropyl radicals. This mode of reactivity is similar to that which has been observed for carbethoxymethylidyne.[35] Although there have been spectral,[36] theoretical,[37] and kinetic[38] studies of CF, the investigations discussed here provide the first examples which elucidate the reactions and products of this interesting species.

In a secondary reaction of this system, the CF_3 radicals, generated in the initial abstraction, react further with carbon to give CF and $^3CF_2$. The $^3CF_2$ reacts in a radical manner abstracting hydrogen to give the HCF_2 radical which adds to alkenes. The reaction of oxygen with $^3CF_2$ generates an intermediate, perhaps the dioxirane, which acts as an oxygen transfer reagent to convert olefins to epoxides stereospecifically.

ACKNOWLEDGMENTS

The authors of this chapter gratefully acknowledge support of this research by the National Science Foundation (Grant CHE–8401198). We thank the Auburn University Computation Center and Mass Spectrometry Facility for generous allotments of time.

REFERENCES

1. For reviews of the chemistry of atomic carbon see: (a) Skell, P. S.; Havel, J.; McGlinchey, M. J. *Acc. Chem. Res.* **1973,** *6*, 97. (b) MacKay, C. In "Carbenes"; Moss, R. A.; and Jones, M., Jr., Eds.; Wiley–Interscience: New York, 1975; Vol. II, pp. 1–42. (c) Shevlin, P. B. In "Reactive Intermediates", Abramovitch, R. A., Ed.; Plenum Press: New York, 1980; Vol. I, pp. 1–36.
2. Rahman, M.; McKee, M. L.; Shevlin, P. B. *J. Am. Chem. Soc.* **1986,** *108*, 6296.
3. Kirmse, W. "Carbene Chemistry". Academic Press: New York, 1971.
4. Hildenbrand, D. L. *Chem. Phys. Lett.* **1975,** *32*, 523.
5. Golden, D. M. *Ann. Rev. Phys. Chem.* **1982,** *33*, 493.
6. Lias, S. G.; Karpas, Z.; Liebman, J. F. *J. Am. Chem. Soc.* **1985,** *107*, 6089 and references cited therein.

7. The lowest energy singlet state of atomic carbon, C(^{1}D), which is the state thought to be involved in many of the reactions of carbon with organic molecules, has a heat of formation of 200 kcal/mol and lies 30 kcal/mol above the ^{3}P ground state.

8. Blaxell, D.; MacKay, C.; Wolfgang, R. *J. Am. Chem. Soc.* **1969,** *92,* 50.

9. Finn, R. D.; Ache, H. J.; Wolf, A. P. *J. Phys. Chem.* **1970,** *74,* 3194.

10. Binkley, J. S.; Frisch, M.; Raghavachari, K.; Fluder, E.; Seeger, R.; Pople, J. A. "GAUSSIAN 82", Carnegie-Mellon University.

11. McKee, M. L.; Lipscomb, W. N. *J. Am. Chem. Soc.* **1981,** *103,* 4673. Nobes, R. H.; Bouma, W. J.; Radom, L. *Chem. Phys. Lett.* **1982,** *89,* 497. McKee, M. L.; Lipscomb, W. N. *Inorg. Chem.* **1985,** *24,* 762.

12. Dixon, D. A. *J. Phys. Chem.* **1986,** *90,* 54.

13. Ahmed, S. N.; McKee, M. L.; Shevlin, P. B. *J. Am. Chem. Soc.* **1983,** *105,* 3942. Shevlin, P. B.; McKee, M. L. *J. Am. Chem. Soc.* **1985,** *107,* 5191.

14. The carbon arc reactor is based on one described by Skell, P. S.; Wescott, L. D., Jr.; Goldstein, J. P.; Engel, R. R. *J. Am. Chem. Soc.* **1965,** *87,* 2829.

15. Lagow, R. J.; Shimp, L. A.; Lam, D. K.; Baddour, R. F. *Inorg. Chem.* **1972,** *11,* 2568.

16. Ando, T.; Yamanaka, H.; Namigata, F.; Funaska, W. *J. Org. Chem.* **1970,** *35,* 33.

17. Tang, Y-N.; Rowland, F. S. *J. Am. Chem. Soc.* **1967,** *89,* 6420.

18. Hahnfeld, J. L.; Burton, D. J. *Tetrahedron Lett.* **1975,** 1819.

19. Lien, M. H.; Hopkinson, A. C. *J. Comput. Chem.* **1985,** *6,* 274 have recently reported a calculated barrier of 16.7 kcal/mol for this inversion.

20. Beckwith, A. L. J.; Ingold, K. U. In "Rearrangements in Ground and Excited States"; de Mayo, P., Ed.; Academic Press: New York, 1980; pp. 161–310 and references cited therein.

21. Pomerantz, M.; Dassanayake, N. L. *J. Am. Chem. Soc.* **1980,** *102,* 678.

22. Barmetler, A.; Ruchardt, C.; Sustmann, R.; Sustmann, S.; Verhulsdonk, R. *Tetrahedron Lett.* **1974,** 4389.

23. DeFrees, D. J.; McIver, R. T., Jr.; Hehre, W. J. *J. Am. Chem. Soc.* **1980,** *102,* 3334.

24. (a) Koda, S. *Chem. Phys. Lett.* **1978,** *55,* 353. (b) Koda, S. *Chem. Phys.* **1982,** *66,* 383.

25. Margrave, J. L.; Wieland, K. *J. Chem. Phys.* **1953,** *21,* 1552.

26. Heicklen, J.; Cohen, N.; Saunders, D. *J. Phys. Chem.* **1965,** *69,* 1174.

27. Johnson, T.; Heicklen, J. *J. Chem. Phys.* **1967,** *47,* 475.

28. Scheer, M. D.; Klein, R. *J. Phys. Chem.* **1969,** *73,* 597.

29. Bailey, P. S. "Ozonation in Organic Chemistry", Vol. 1. Academic Press: New York, 1978.

30. Murray, R. W.; Suzui, A. *J. Am. Chem. Soc.* **1973,** *95,* 3343.

31. Higley, D. P.; Murray, R. W. *J. Am. Chem. Soc.* **1974,** *96,* 3330.

32. (a) Harding, L. B.; Goddard, W. A., III. *J. Am. Chem. Soc.* **1978,** *100,* 7180. (b) Kahn, S. D.; Hehre, W. J.; Pople, J. A. *J. Am. Chem. Soc.* **1987,** *109,* 1871. (c) Cremer, D. *J. Am. Chem. Soc.* **1979,** *101,* 7199.

33. (a) Murray, R. W.; Jeyaraman, J. *J. Org. Chem.* **1985,** *50,* 2847. (b) Suenram, R. D.; Lovas, F. J. *J. Am. Chem. Soc.* **1978,** *100,* 5117.

34. Agopovich, J. W.; Gillies, C. W. *J. Am. Chem. Soc.* **1980,** *102,* 7572.

35. Strausz, O. P.; Kennepohl, G. J. A.; Garneau, F. X.; DoMinh, T.; Kim, B.; Valenty, S.; Skell, P. S. *J. Am. Chem. Soc.* **1974,** *96,* 5723.

36. (a) Dyke, J. M.; Lewis, A. E.; Morris, A. *J. Chem. Phys.* **1984,** *80,* 1382. (b) Griemann, F. J.; Droege, A. T.; Engleking, P. C. *J. Chem. Phys.* **1983,** *78,* 2248. (c) Kawaguchi, K.; Yamada, C.; Hamada, Y.; Hirota, E. *J. Mol. Spectrosc.* **1981,** *86,* 136. (d) Van Der Heuvel, F. C.; Meerts, W. L.; Dymanus, A. *Chem. Phys. Lett.* **1982,** *88,* 59. (e) Saito, S.; Endo, Y.; Takami, M.; Hirota; E. *J. Chem. Phys.* **1983,** *78,* 116. (f) Jacox, M. E. *Chem. Phys.* **1981,** *59,* 199. See also references cited in all of the above.

37. Gutsev, G. L.; Zyubina, T. S. *Chem. Phys.* **1984,** *83,* 89.

38. Ruzsicska, B. P.; Jodhan, A.; Choi, H. K. J.; Strausz, O. P.; Bell, T. N. *J. Am. Chem. Soc.* **1983,** *105,* 2489.

CHAPTER 2

Nonbonded Interactions Between Arenes and Polyfluoroarenes: Structure and Dynamics

Robert Filler

Department of Chemistry, Illinois Institute of Technology, Chicago, Illinois

CONTENTS

1. INTRODUCTION

The unexpected observation, from freezing-point diagrams, that hexafluoro-benzene forms 1:1 solid complexes with benzene, mesitylene, and 2-methyl-naphthalene[1] was later confirmed by X-ray diffraction measurements.[2] The identification of such intermolecular complexes in the solid phase is not, however, proof of their existence in solution. Indeed, evidence for such complexes in solution is far more equivocal. These early reports led to a spate of investigations involving negative excess free energies or enthal-pies,[3–5] phase diagrams,[1,6–8] excess volumes,[9] and anomalously high solubili-ties of polyfluoroaromatics in hydrocarbons,[10] all of which strongly suggest favorable interactions between aromatic fluorocarbons and arenes. A de-crease in activity in the series hexafluorobenzene, pentafluorobenzene, and 1,2,4,5-tetrafluorobenzene is observed. This would be the expected order of stabilities of charge–transfer complexes involving the fluoroaromatic as π-electron acceptor. These experimental observations do not, however, nec-essarily require the existence of long-lived molecular complexes. In the absence of independent evidence, it is very difficult to ascribe the formation of complexes from thermodynamic data alone. The question then arises as to the nature of the special interactions involved in these mixtures. Despite the original anticipation[1] and subsequent unsubstantiated statements,[11] no firm evidence of π—π "charge–transfer" intermolecular interactions have been found to be associated with any mixture of aromatic fluorocarbons and hy-drocarbons. Thus, ^{19}F NMR chemical shifts[12] and ultraviolet absorption spectra[13] fail to reveal any such interactions between donor arenes, including the powerful hexamethylbenzene, and the acceptor, hexafluorobenzene. There is suggestive evidence for a π—π intermolecular complex, to be dis-cussed later.

By contrast, C_6F_6 and strong n-donors like N-substituted aryl amines, form solid 1:1 complexes[13–16] and new bands in the ultraviolet region, which have been attributed to charge-transfer transitions of n-π donor–acceptor complexes.[17–20]

2. THE STRUCTURES OF BENZENE AND HEXAFLUOROBENZENE DIMERS

If π—π complexes are not formed by C_6F_6 and arenes, what other attractive interactions seem reasonable? Before we consider the geometry of the dis-parate molecules relative to each other, it would be useful to know the structures of the benzene self-complex (benzene dimer) and hexafluorobenzene dimer. Both C_6H_6 and C_6F_6 are planar molecules,[21] each possessing sixfold axes of symmetry. Narten[22] demonstrated that the average molecular rearrangement in liquid benzene is similar to that of the solid and that in the dimer, the axes of each molecule of the pair are almost at right angles. The molecules form corrugated sheets perpendicular to the c-axis. Klemperer and co-workers[23,24] subsequently showed by electric deflection of molecular beams that $(C_6H_6)_2$ vapor is a polar species with perpendicular planes, as previously observed for nearest neighbors in crystalline and liquid benzene.[22] The other homomolecular dimer, $(C_6F_6)_2$, also possesses an orthogonal structure.[24]

It is likely that homomolecular pairing in $(C_6H_6)_2$ is stabilized by a favorable interaction involving a polarization effect in which the slightly positive hydrogen of the C—H dipole on one aromatic ring is attracted to and perhaps, penetrates the π–electron cloud above the second ring, forming dihedral angles that approach 90° (**1**). Conversely, the C_6F_6 molecule may be viewed as rings of negative charge ($C^{\delta+}$—$F^{\delta-}$ dipoles) surrounding regions of somewhat lower electron density. In the dimer, the $C^{\delta+}$—$F^{\delta-}$ dipole is similarly attracted to the relatively positive "hole" of the second C_6F_6 ring (**2**). These aromatic–aromatic interactions seem to be very significant, eg, in the binding of proteins.[25]

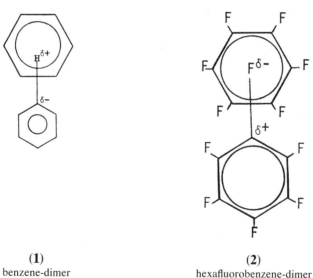

(1)
benzene-dimer

(2)
hexafluorobenzene-dimer

3. THE HETEROMOLECULAR DIMER, C_6H_6—C_6F_6

In contrast to the T-shaped structures of $(C_6H_6)_2$ and $(C_6F_6)_2$, Klemperer's studies[24] on the gas-phase heteromolecular dimer, C_6H_6—C_6F_6, led to the conclusion that this species, with a dipole moment of 0.44 ± 0.04 D, exhibits a structure featuring alternate stacking, with molecules arranged in a parallel plane orientation. Since this alignment closely resembles the structures of the crystalline "addition compounds" of C_6F_6 with p-xylene,[26] mesitylene,[27] durene,[28] and hexamethylbenzene,[29,30] it is reasonable to expect that C_6H_6—C_6F_6 also possesses the parallel plane geometry. This structure had been predicted from "wide-line" nuclear magnetic resonance studies[31] and from melting characteristics of a wide variety of arene–polyfluoroarene mixtures.[8]

As just mentioned, Dahl, in an elegant series of studies,[26–30] determined the crystal structures of the 1:1 addition compounds between C_6F_6 and aromatic hydrocarbons. Thus, at $-35°$, mesitylene (MES) and C_6F_6[27] are stacked alternately along the c-axis, with a distance of 3.56Å between nearly parallel planes. The rings are twisted 30° about the normal plane, relative to each other. An interaction between F atoms and CH_3 groups seems likely, since one C—C bond to a CH_3 points almost directly toward an F atom. The p-xylene-C_6F_6 complex, (3)[26] closely resembles that of MES-C_6F_6, with a mean separation of 3.55Å within the stack and the aryl rings twisted 30°.

Hexamethylbenzene (HMB) and C_6F_6 form two different 1:1 "addition compounds," depending on the temperature. In the trigonal form (5°), the partner molecules are alternately stacked in infinite columns, with some disorder in the stacking sequence.[29] The C_6F_6 ring is tilted more than HMB. Reorientation occurs at $-40°$ and HMB-C_6F_6 forms triclinic crystals, with an angle of 3.6° between molecular planes.[30] This complex differs significantly from that of MES-C_6F_6 at that temperature, with a mean separation of 3.43Å within the stack, 0.13Å shorter than in MES-C_6F_6, but with no shortening of distance between CH_3 and F. Durene (1,2,4,5-tetramethylbenzene) also

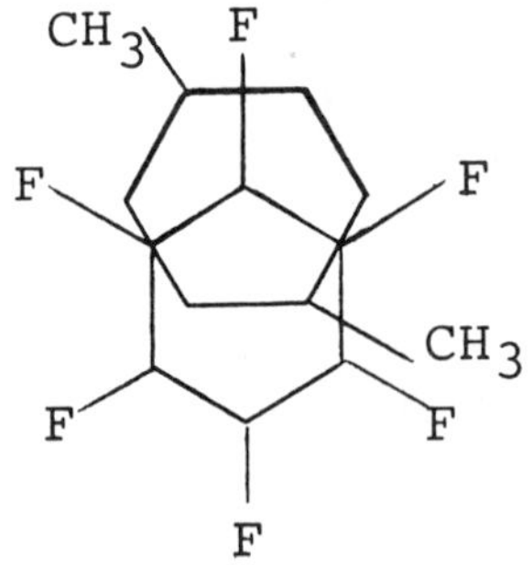

(3)

p-xylene-C_6F_6 complex

forms addition compounds which more closely resemble HMB-C_6F_6 than MES-C_6F_6.

These results indicate a close relationship between the number of methyl groups and the interplanar distance and relative orientation of the molecules. Four or more methyl groups increasingly favor shorter interplanar distances and a parallel orientation of the aryl rings, which overlap with a C—C bond of one ring approximately above the center of the proximate ring. These features are characteristic of charge–transfer complexes.[32] Increasing methyl substitution also increases the π-donor strength of aromatic hydrocarbons.

The variations in the crystal structures of these complexes cannot be readily explained by van der Waals forces alone. A reasonable interpretation is that charge–transfer forces in these crystals make significant contributions to stability when more than three methyl groups are present on the nonfluorinated ring.

In seeking an explanation for the observation that 2,3,4,5,6-pentafluorobiphenyl (4) exhibits an unexpectedly high melting point relative to those of biphenyl and decafluorobiphenyl, a detailed study of its crystal structure was carried out.[33,34] In the solid state, in which the aryl rings are non-coplanar, 4 is a self-complexing molecule which forms mixed stacks of phenyl and pentafluorophenyl rings (5). The mean distance between the C_6H_5 and C_6F_5 ring planes is 3.43Å, the same interplanar distance observed for HMB-C_6F_6.[30] Packing diagrams reveal stacks of molecules along the c-axis in which the disparate rings alternate. The rings are nearly parallel with each C_6H_5 ring interacting more strongly with one half of the partner ring than with the other half. Again, there is no evidence of π-donor/π-acceptor interactions, but the

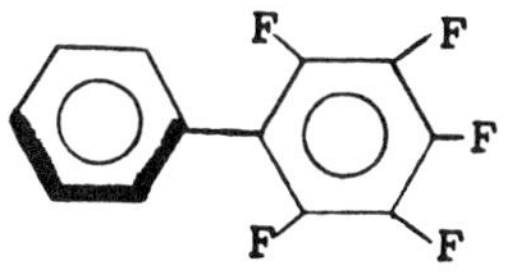

(4) 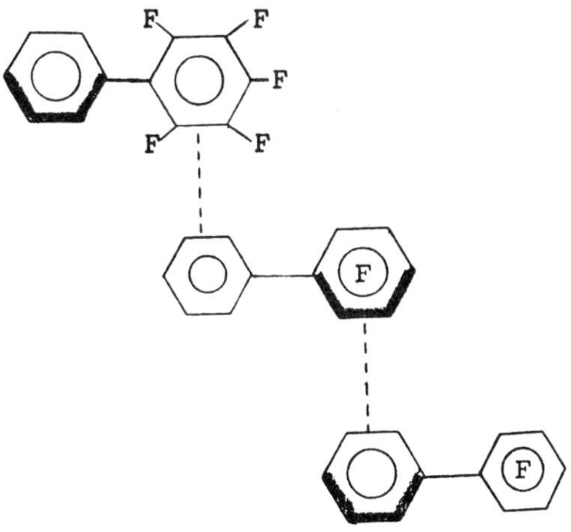   (5)

data strongly suggest the presence of important directional intermolecular forces, possibly including London dispersion forces and attractions between C—H and C—F dipoles.

In a similar fashion, biphenyl and decafluorobiphenyl form a 1:1 molecular complex, with the molecules packed alternately in infinite columns.[35]

4. THE NATURE OF THE INTERACTIONS BETWEEN ARENES AND POLYFLUOROARENES

While there seems to be no experimental evidence in support of π—π intermolecular charge–transfer complexes in the C_6H_6—C_6F_6 couple, or in Ar_HH—Ar_FF, generally, other explanations for nonbonded, donor–acceptor interactions have been postulated.

Since Dewar had argued previously[36] that charge–transfer need not provide the principal stabilization for donor–acceptor complexes, it has been proposed[17] that the interaction between the two highly polarizable molecules, C_6H_6 and C_6F_6, should be sufficiently stabilized by polarization of C—H and C—F bonds and dispersion forces of attraction between fluctuating induced dipoles in the closely disposed, nearly superimposable molecules. Using theoretical studies of C_6H_6—TCNE complexes[37–38] and data from the measurement of dielectric constants and refractive indices of liquid mixtures of C_6F_6 with p-xylene and mesitylene, Baur[39] concluded that the contribution to the energy of formation of C_6F_6—ArH complexes by electrostatic (quadrupole–quadrupole), induction, and dispersion interactions and weak hydrogen bonding, is comparable in magnitude to that of charge–transfer.

5. INTRAMOLECULAR INTERACTIONS INVOLVING PLANAR RINGS

In the discussion thus far, we have focused on intermolecular interactions between planar phenyl and polyfluorophenyl rings. It would also be interest-

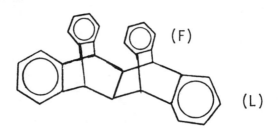

(F)

(L)

ing to explore the properties of an intramolecular system wherein these planar rings are in close proximity and are parallel or nearly so. These conditions can be met only in a sterically constrained molecule which dictates the required geometry. The bridged polycyclic molecule, janusene (6),[40] possesses the desired structural features. The two facial (F) aryl rings are coplanar and parallel and are separated by only 2.5Å, while the lateral (L) rings are far apart.

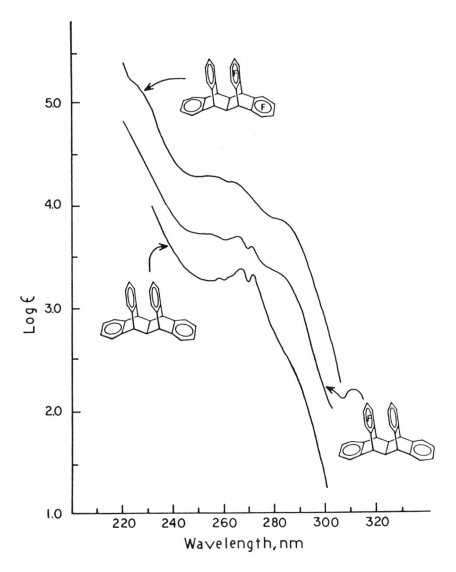

Figure 2–1. Ultraviolet absorbances of janusene, F-tetrafluorojanusene (Log $\varepsilon + \frac{1}{2}$), and Octafluorojanusene (Log $\varepsilon + 1$) in cyclohexane.

If one could construct a janusene analog in which one facial ring is tetra-fluorophenyl, we would essentially meet the specified structural requirements. Fortunately, two examples (**7,8**) of such a molecule have been described very recently.[41] Although X-ray crystallographic studies of **7** and **8** have not been reported, it is reasonable to expect a geometry very similar to that of **6**. Comparison of the ultraviolet spectra of **6,7,** and **8** in cyclohexane solution reveals absorption maxima (shoulders) at 283–284 nm in **7** and **8**, absent in **6**. (Figure 2–1). The presence of these bands strongly suggests attractive donor–acceptor interactions, perhaps of the charge–transfer type, between the closely disposed C_6H_4 and C_6F_4 rings. The progressive loss of fine structure in the ultraviolet spectra as the number of fluorines increase and the deshielding of the facial aryl protons in the ^{1}H NMR spectra of **7** and **8**, relative to **6**, also lend support for some type of internuclear interaction.

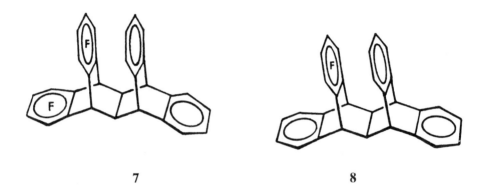

7   8

6. PHOTOELECTRON SPECTROSCOPY OF MOLECULAR COMPLEXES—THE π–FLUORO EFFECT

Photoelectron spectroscopy is a valuable experimental technique used to measure molecular ionization energies. In recent years, the method has been applied to an increasing degree in the study of intermolecular and intramolecular complexes. These complexes are frequently characterized by charge migration from donor to acceptor molecules. Consequently, the ionization energies of the complex often differ measurably from those of the uncomplexed molecule. These differences may provide information on the nature of the intermolecular or intramolecular interactions.[42]

Clark and Frost[43] conducted a low-resolution photoelectron study of benzene and four fluorine-substituted benzenes, including C_6F_6. The previously established value of 9.25 eV for the ionization potential (IP) of benzene[44,45] was confirmed and the π IP at 9.88 eV for C_6F_6 was determined.

In the first of a significant two-part study at high-resolution, Robin and co-workers[46] compared the successive ionization potentials in planar, nonaromatic hydrides with those of the corresponding perfluoro compounds and showed that "σ MOs are stabilized by 2.5–4.0 eV by perfluoro substitution, while stabilization can be an order of magnitude smaller for π MOs." This preferential stabilization of σ MOs was termed "the perfluoro effect." In the second study,[47] the σ- and π-ionization potentials of *planar* aromatic systems and their perfluoro counterparts were determined. The pairs examined included C_6H_6—C_6F_6, pyridine–pentafluoropyridine, s-triazine-cyanuric fluoride, p-benzoquinone-tetrafluoro-p-benzoquinone, borazine-B-trifluoroborazine, and naphthalene-octafluoronaphthalene. As in the nonaromatic compounds, the perfluoro effect is operative, with the σ IPs of the perfluoro compounds being larger than those of the perhydro analogs by several electron volts.

Comparison of π IPs should provide a good measure of the potential for charge–transfer between aromatic partners. Clearly, the larger the "gap" in π IPs of donor and acceptor, the greater the likelihood of a significant contribution of charge-transfer in increasing the stability of the complex. In sharp contrast to the σ IPs, the experimentally determined π IPs within each pair do not differ appreciably, with a shift in most cases of only 0.2–0.7 eV, eg, C_6H_6 (9.241 eV) vs C_6F_6 (9.93 eV).

The "perfluoro effect," which applies to partner molecules in which all hydrogens are replaced by fluorine, has subsequently been shown to be a much more general phenomenon. Thus, Liebman and co-workers[48] demonstrated that replacement of one or more hydrogens by fluorine in benzene and in a variety of substituted benzenes results in very small changes in π IP. Indeed, the range of $\Delta\pi$ IP is only 0.0–0.7 eV. Whereas the π IPs are relatively invariant to the degree of fluorination, the studies reveal that, in unsaturated aliphatic compounds, the σ IPs change significantly as fluorines replace hydrogen. Comparison of ethylene and tetrafluoroethylene illustrates the point: C_2H_4 (π IP 10.6 eV, σ IP 12.9 eV) vs C_2F_4 (π IP 10.7 eV, σ IP 16.0 eV). When fluorine substitution occurs on substituent groups attached to the aryl ring, σ—π mixing leads to a larger change in ionization potential. These investigators[48] have introduced a new term, "π-fluoro effect," which states that "to a high level of approximation, in a planar species, replacement of one or more hydrogens by fluorine causes minor effects on phenomena whose origin involves π-electrons, in contrast to the much larger effects observed with σ electrons." They then offer a speculative interpretation of the π-fluoro effect in terms of electrostatic potentials.

The π-fluoro effect provides a reasonable rationale for the experimental observations that there is an almost total lack of significant π—π charge–transfer interactions between planar aryl and polyfluoroaryl rings. At this stage, we must rely heavily on electrostatic, induction, and dispersion forces to explain the formation and relative stabilities of complexes between these disparate aromatic rings.

7. FLUOROAROMATIC CATION RADICALS

The removal or transfer of a π-electron from an aromatic ring, as in photo-electron spectroscopy, is an oxidative process and results in a cation radical, eg, $C_6H_6^+$ or $C_6F_6^+$: They are also familiar as transient species in mass spectrometry (molecular ion). For reactions involving fluoroaromatics, cation radicals have been invoked as reactive intermediates on several occasions, eg, in fluorinations (a) using CoF_3 and other high-valency metal fluorides[49,50] and (b) with xenon difluoride.[51]

A general discussion of cation radicals is beyond the scope of this review. It will suffice to point out that, although polyfluoroaryl radical cations have been studied to only a limited degree, they can be generated fairly readily, consistent with the π-fluoro effect, either by the reaction of the parent fluorocarbon with very strong Lewis acids or by electrochemical oxidation. Thus, octafluoronaphthalene forms the intensely green cation radical on reaction with oleum, SO_3, and SbF_5.[52] Treatment of C_6F_6, C_6F_5H, $C_6F_5CH_3$, and decafluorobiphenyl with FSO_3H—SbF_5 also produces the corresponding cation radicals.[53] Highly fluorinated aromatics, such as C_6F_6, C_6F_5H, $C_6F_5CH_3$, and $C_{10}F_8$, in FSO_3H containing 0.1 M acetic acid as base, undergo facile anodic one-electron oxidation at room temperature to form cation radicals,[54] which are stable on the timescale of cyclic voltammetry (half-lives: seconds to minutes), except for octafluoronaphthalene cation radical, which is stable indefinitely. If the oxidations are conducted at low temperatures or in FSO_3H—SbF_5 solution, several of the other cation radicals are stable indefinitely. The dioxygenyl cation (O_2^+ AsF_6^-) reacts with hexafluorobenzene in SO_2ClF solution to form the golden yellow salt $C_6F_6^+$ AsF_6^- and oxygen.[55]

8. TCNE COMPLEXES WITH POLYFLUOROAROMATICS

Benzene and especially the even more basic methyl- and methoxy-substituted benzenes behave as π-electron donors toward the powerful π-electron acceptor tetracyanoethylene (TCNE) to form species with distinct new charge–transfer bands.[56,57] The absorptions shift bathochromically and the complexes become stronger as the donor capacity of the aromatic increases.[56]

Just as the π-fluoro effect suggests that polyfluoroaromatics should form cation radicals almost as readily as do aromatic hydrocarbons, we can also predict that hexafluorobenzene and its congeners will form π—π charge–transfer complexes with TCNE. Indeed, C_6F_6 does form a weak complex.[57] In unpublished studies, Filler and Choe[58] measured the wavelengths of the charge–transfer bands of the TCNE complexes of a series of polyfluoroaromatic compounds (Table 2–1).

TABLE 2–1. Wavelengths of the Charge-Transfer Bands of the TCNE Complexes (CH_2Cl_2 Solution) of Polyfluoroaromatics and Selected Aromatics

π-Donor	λ_{max} (nm)
C_6H_6	384*
C_6F_6	end absorption
C_6F_5H	end absorption
 H—[F ring]—H	end absorption
C_6H_5Cl	379*
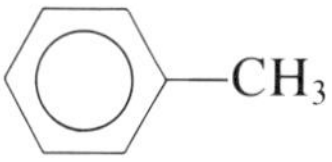 [ring]—CH_3	406*
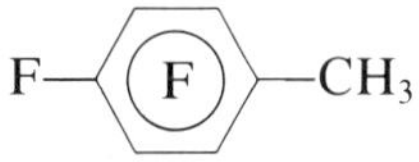 F—[F ring]—CH_3	325
 H—[F ring]—CH_3	340
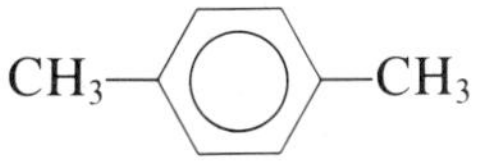 CH_3—[ring]—CH_3	460*
 CH_3—[F ring]—CH_3	360
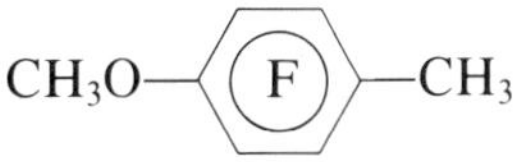 CH_3O—[F ring]—CH_3	370
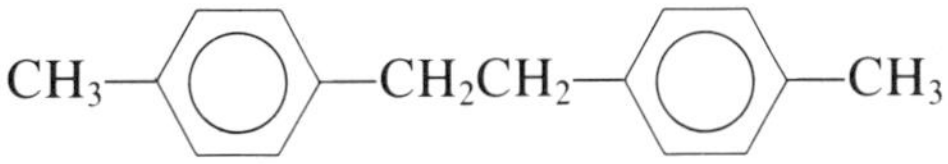 CH_3—[ring]—CH_2CH_2—[ring]—CH_3	420,460
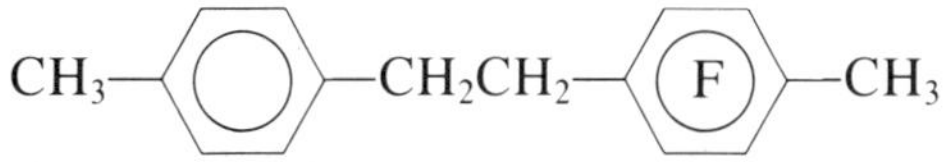 CH_3—[ring]—CH_2CH_2—[F ring]—CH_3	415,450
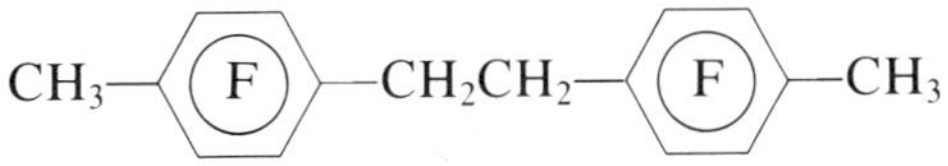 CH_3—[F ring]—CH_2CH_2—[F ring]—CH_3	358
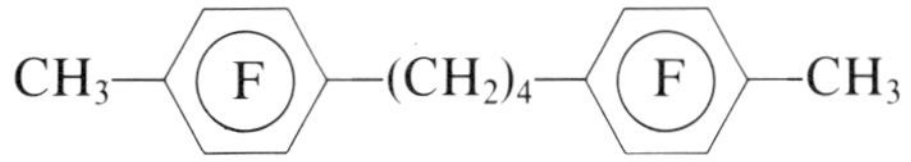 CH_3—[F ring]—$(CH_2)_4$—[F ring]—CH_3	370

* Ref. 56

9. NONPLANAR SYSTEMS

In the discussion to this point, we have focused on nonbonded intermolecular interactions between planar phenyl and polyfluorophenyl rings. The π-fluoro effect was shown to be applicable to planar systems.[48] Brief data on intramolecular interactions between closely disposed, vis-à-vis C_6H_4 and C_6F_4 rings in janusenes were also cited.[41]

It would be of interest to examine the properties of intramolecular systems in which the aryl rings are in close proximity, but are non-coplanar. An excellent model for such a molecule is [2.2] paracyclophane (9), which has been studied very extensively.

9

X-ray data of **9**[59–61] indicate that the aromatic rings of this highly strained molecule possess a boatlike shape which deviates from coplanarity by about 12.6°.[61] The benzene rings are much closer (2.78–3.09Å) than the sum of van der Waals radii (3.4–3.5Å). The rings are skewed to partially relieve coulombic repulsion and the strain energy of **9** is about 31–33 kcal/mol.[62,63] The subject has been reviewed succinctly, but effectively, by Liebman and Greenberg.[64] The ring distortions in **9** play a significant role in transannular, inter-ring chemical reactivity. Their influence on π-basicity is also reflected in its anomalous ultraviolet spectrum, as well as in the charge-transfer band in the visible region of its TCNE π-complex.[65–67]

Compound **10**, 4,5,7,8-tetrafluoro [2.2] paracyclophane, is the molecule corresponding to **9** which would permit the study of interactions between the transannular C_6H_4 and C_6F_4 rings in a sterically constrained system. This compound has been prepared by several routes,[68,69] and a detailed study of its spectral properties and chemical reactivity has recently been described.[70] Attempts to elaborate the structure of **10** by X-ray crystallographic analysis

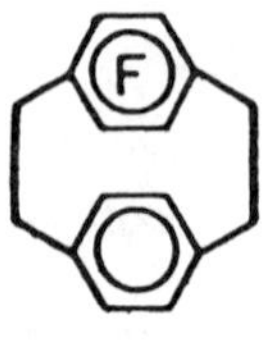

10

11

have been thwarted thus far by difficulty in obtaining a single crystal because of twinning.[71]

A. Spectral Properties

(1) Ultraviolet Spectra

A comparative study of the ultraviolet spectra of 9, 10, and 11 (the octafluoro analog) reveals the presence of a distinct absorption band at 297 nm in 10 and its absence in 11. Compound 9, however, exhibits a split spectrum, with a band at 286 nm and a shoulder of much lower intensity at 304 nm. Although it is tempting to attribute the 297 nm absorption in 10 to special attractive transannular interactions between the π-elecrons of the C_6H_4 and C_6F_4 rings, the distinctions are somewhat subtle and not clear-cut. The absence in 10 of a band near 243 nm, observed in both 9 and 11, may be ascribed to repulsions between like rings. Dilution studies showed that the interactions are intramolecular, since the extinction coefficients were independent of concentration.

(2) Fluorescence Spectra

Preliminary studies of the fluorescence spectra of 9, 10, and 11[72] indicate an interaction associated with the presence of fluorine substitutents in the paracyclophanes. Fluorine-substituted benzenes both absorb and fluoresce with energies similar to those of benzene itself. In 10 and 11, however, there is an appreciable shift in energies (20.8 kJ mol^{-1} for 10 and 32.6 kJ mol^{-1} for 11) over that for 9. Further studies seem desirable.

(3) Photoelectron Spectra

The He (1α) photoelectron (PE) spectrum of 9 has been examined in detail.[73-78] The lowest π-ionization energy is at 8.1 eV. One would expect, in general, that the PE spectrum of 11 would closely resemble that of 9 or of 1,1,2,2,9,9,10,10-octafluoro [2.2] paracyclophane (12), because all three molecules possess D_{2h} symmetry. The PE spectrum of 10 should be distinctly different if significant "through space" π-electron interactions between the

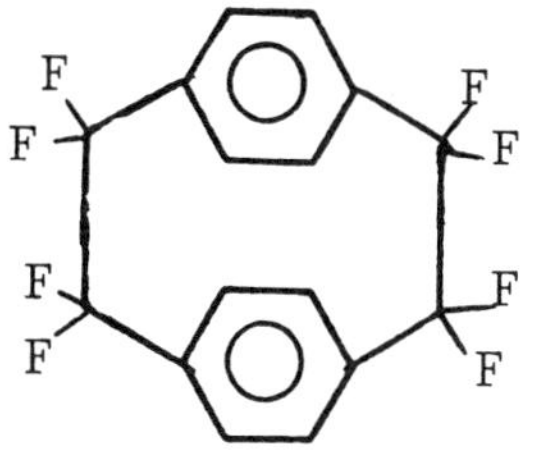

12

C_6H_4 and C_6F_4 rings are operating. In fact, preliminary studies[79] reveal the lowest ionization energies of **10** and **11** at 8.52 eV and 8.88 eV, respectively, rather small shifts from that of **9**. These data indicate only minor effects involving π-electrons and strongly suggest that, to a reasonable degree, the π-fluoro effect is also applicable to this non-coplanar [2.2] paracyclophane system.

(1) NMR

It was suggested in an earlier paper[68] that the presence of a quintet at δ 6.84 (aromatic, 4H), J = 0.8 Hz, in the ^{1}H NMR spectrum of **10** might be explained by a weak transannular "through-space" H—F coupling. More detailed examination of ^{1}H and ^{19}F NMR has now clearly established that there is no such transannular interaction.[70] A high-resolution 300 MHz spectrum resolved the previously reported δ 3.03 multiplet (methylene protons) into two complex multiplets, one centered at δ 3.07 (4H) and the other, at δ 2.98 (4H). The deshielded and less complex multiplet was assigned to the methylene protons adjacent to the nonfluorinated ring. Irradiation of this complex caused the 6.84 resonance to collapse to a singlet. Conversely, irradiation of the δ 6.84 resonance simplified the δ 3.07 complex. Therefore, the multiplicity of the aryl hydrogens was not due to transannular H—F coupling. Confirmation was found in the ^{19}F spectrum of **10**. A single resonance at 138.4 ppm was observed. Through-space coupling would have produced a quintet.

The nature of the interaction between the aryl and neighboring methylene protons in **10** is still unclear. X-ray crystallographic analysis may be helpful in resolving this issue.

Of considerable interest are the NMR spectra of the highly strained [2.2] metaparacyclophanes **13** and **14**, in which one ring is orthogonal to the other (cf. benzene and hexafluorobenzene dimers, vide supra). Compound **13**, prepared by acid-catalyzed rearrangement of **9**,[80] exhibits a doublet at δ 5.70 (H_b) and a broad singlet at δ 5.24, attributable to H_a (**13a**). These protons, especially H_a, are strongly shielded by ring currents of the transannular aryl nuclei, relative to H_c (δ 6.97) and H_d (δ 6.63). In **14a**,[69] the broad singlet of the

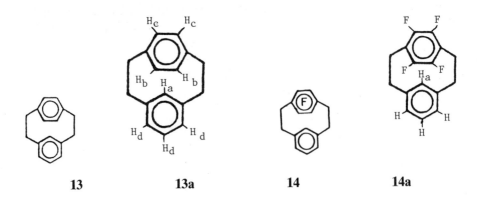

13   13a   14   14a

orthogonal proton H_a is much less shielded (δ 5.89) than in **13a**, because the π-electron ring currents are drawn, to some degree, toward the C—F bonds.

B. Chemical Reactivity

In the foregoing discussions, we dealt primarily with structural features and the energetics of electronic transitions in systems containing proximate phenyl and polyfluorophenyl rings. There is also considerable interest in evaluating how one of the disparate rings influences the chemical reactivity of the other, a special type of neighboring group effect.

When external reagents react with either aromatic nucleus of the Ar_H—Ar_F couple, new factors, such as induced polarization of π-electrons, the structures of transition states, and stability of intermediates, come into play and the role of the neighboring aryl group may be critical.

The [2.2] paracyclophanes provide good substrates for such chemical studies because the rigid geometry assures the close proximity of the aromatic nuclei. Although the aryl rings deviate somewhat from planarity, aromatic character is not lost, as shown by the retention of ring currents and the facile reaction of **9** with electrophiles.[66,81]

Filler and co-workers[70] have compared the electrophilic reactivity of tetrafluoro [2.2] paracyclophane **10** with that of **9**.[81] As mentioned earlier, **9** undergoes rapid acid-catalyzed (HCl, $AlCl_3$) rearrangement to the metaparacyclophane **13** and other products,[80] via a protonated σ complex. A measure of the influence of the C_6F_4 ring would be the rate of a comparable rearrangement of **10**. When the same procedure was used with **10**, an orange solution was observed, but there was no evidence of rearrangement, even under more rigorous conditions. Starting material was recovered almost quantitatively. Deuterium exchange confirmed that protonation of the phenyl ring occurs without difficulty, but the subsequent skeletal arrangement, which in **9** is facilitated by partial relief of strain energy associated with π—π repulsions and the lower energy of the conjugate acid of **13**, failed to occur. Also, this rearrangement requires initial protonation at an ipso carbon, ie, the para bridging carbons of the phenyl ring. In contrast to **9**, the neighboring C_6F_4 ring in **10** apparently causes substantial diminution of the basicity of the ipso carbons of the C_6H_4 ring relative to the other ring carbons (confirmed by ^{13}C NMR analysis[70a]), thus minimizing protonation at that site. The origin of this effect is yet unclear. Moreover, any energetically favorable attractive interactions between the aryl rings in **10** would be lost in tetrafluoro[2.2]metaparacyclophane, **14**, in which the rings are orthogonal. Compound **14** was prepared by an alternative route.[69]

While both electrophilic acetylation[82] and bromination[83] of **9** proceed readily, these reactions with **10** were totally unsuccessful. To demonstrate that this deactivation is due to some type of transannular interaction, an openchain analog (**15**), in which electronic effects must be transmitted through σ-bonds, was prepared and shown to undergo acetylation, although not as readily as **9**. A more rigorous test of reactivity would utilize another open-chain analog, **16**. This compound has now been prepared. Acetylation proceeds normally.[83a]

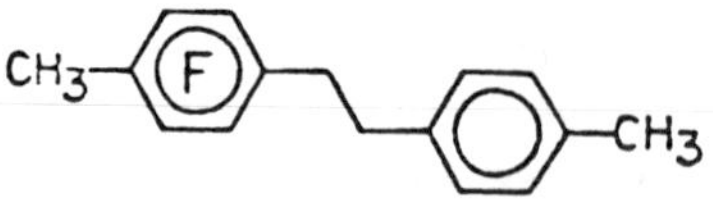

                    15 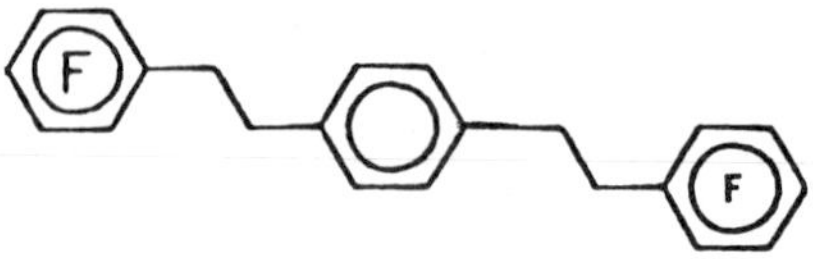   **16**

Despite the successful deuterium exchange at non-ipso aryl carbons, the otherwise lack of reactivity of the phenyl ring in **10** is not surprising. Previous studies[82–85] provide ample precedence when a phenyl ring possesses an electron-withdrawing group, as in [2.2] paracyclophanequinone, acetyl [2.2] paracyclophane, and even the acetyl [4.4] analog. The drain of π-electron density from the unsubstituted phenyl ring toward the substituted ring, as in **10,** renders the unsubstituted ring less susceptible to electrophilic attack.

C. TCNE Complexes

The long wavelength absorption bands in the visible region of the 1:1 TCNE-paracyclophane π-complexes and their open-chain analogs have been studied by Cram and Bauer.[86] The corresponding complexes of **10** and **11** and open-chain models have been reported recently.[70] The positions of the maxima and the stability of the complex correlate well with the electron-releasing ability of the base.[56]

Assuming that the longer of the two key bands in TCNE-**10** is associated with the nonfluorinated ring, which is reasonable,[58] comparison of these complexes and their open-chain models (Figure 2–2) lead to the following illuminating conclusions:

1. The difference of 10 nm (460 vs 450) in C and D is minimal. The inductive effect of the C_6F_4 ring through sigma bonds has a negligible effect on the π-basicity of the C_6H_4 ring.
2. The large difference of 61 nm (521 vs 460) between A and C demonstrates that the transannular effect of the non-complexed C_6H_4 ring substantially increases the π-basicity of the C_6H_4 ring which complexes with TCNE, as previously reported.[86]
3. The difference of 56 nm (521 vs 465) between A and B points to a significant diminution of the basicity of the C_6H_4 ring via a transannular drain of π-electron density toward the C_6F_4 ring (**17**).
4. λ_{max} 425 nm in TCNE-**10** (B) and 415 nm in its open-chain model (D) are probably due to complexing of TCNE with the less basic C_6F_4 rings. It is possible that 2:1 TCNE:aryl complexes are formed. Once the more

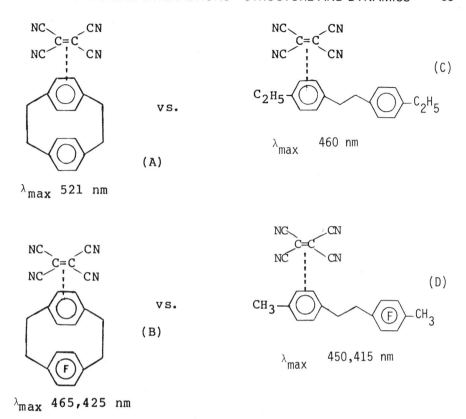

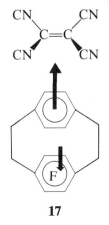

Figure 2–2. TCNE π-complexes of [2.2] paracyclophanes **9** and **10** and their open-chain models. (Visible spectra measured in CH₂Cl₂ solution)

basic C_6H_4 ring in **10** complexes with TCNE, it can no longer provide π-electrons to the C_6F_4 ring, which then behaves more like an independent entity, as in D.

17

10. CONSEQUENCES IN REACTION CHEMISTRY OF INTERMOLECULAR INTERACTIONS BETWEEN ARENES AND POLYFLUOROARENES

Attractive interactions, including complex formation, can play a major role in determining the course of organic reactions. The central feature is the entropy factor. When stable complexes are possible, the number of alignments between the reactants is very limited. In the absence of such interactions, there is a large number of possible alignments and reaction may occur in a different manner, or only with difficulty, if at all. Three examples are presented in this section.

A. Homolytic Arylation of Aromatic and Polyfluoroaromatic Compounds

In reactions such as Equation 2–1, *a* and *b*, attractive interactions between disparate rings are likely to exert a profound effect on relative rates of arylation and on isomer distribution.

$$\text{(a)} \quad C_6H_5\cdot + C_6F_6 \rightarrow C_6H_5\text{—}C_6F_5$$

$$\text{(b)} \quad C_6F_5\cdot + C_6H_6 \rightarrow C_6F_5\text{—}C_6H_5$$

$$(2\text{–}1)$$

Complex formation between phenyl radicals and polyfluorobenzenes were substantiated by a detailed study of phenylation of a series of ten polyfluorobenzenes, $C_6H_xF_{6-x}$ (x = 2–6).[87] The specific presence of C_6F_6 leads to increased selectivity in the orientation of products (isomer ratios) in phenylation of less fully fluorinated substrates, even in intramolecular competition.

Similar complexes can be formed between $C_6F_5\cdot$ and nonfluorinated arenes and comparable behavior was observed on pentafluorophenylation of benzene and a series of mono- and disubstituted benzenes.[88] The subject, including complex mechanisms, is discussed in a recent review.[89]

B. Reaction of Polyfluoroarylazides with Highly Basic Aryl Hydrocarbons

Perfluoroazidobenzene ($C_6F_5N_3$), perfluoro-4-azidopyridine and other members of the Ar_FN_3 family were thermolyzed in the presence of a large excess (10 : 1) of 1,3,5-trimethylbenzene (mesitylene) and 1,3,5-trimethoxybenzene to give, in some cases, remarkably high yields (80–92%) of diarylamines via insertion of the putative singlet nitrene at nuclear C—H bonds (Equation 2–2).[90] The reaction involves attack on the exceptionally electrophilic nitrenes by the strongly nucleophilic aromatic substrates.

$$\text{Ar}_F\text{N}_3 \xrightarrow[-\text{N}_2]{\text{heat}} \text{Ar}_F\ddot{\text{N}}: \xrightarrow{} $$

$$\text{Ar}_F\text{NHC}_6\text{H}_2\text{Me}_3\text{-}2,4,6 \qquad (2\text{-}2)$$

By contrast, thermolysis of Ar_FN_3 in the presence of benzene, a much weaker nucleophile, gives yields of only 2–13%.[90-93] The wide disparity in yields between the reactions involving benzene, toluene, and m-xylene and those with mesitylene and 1,3,5-trimethoxybenzene would not seem to be adequately explained by the differences in nucleophilicities of the aromatics. Although not previously recognized, this reaction would appear to be significantly enhanced by a favorable orientation of the reactants in the transition state, brought about by attractive interaction between the "faces" of the relatively electron-poor polyfluoroaryl nitrene and the strong π-bases, mesitylene, and 1,3,5-trimethoxybenzene. Support for this argument is found in the behavior of 4-azido-3-chlorotrifluoropyridine and 4-azido-3,5-dichlorodifluoropyridine, in which the yields with mesitylene are markedly reduced to 44% and 11%, respectively. A significant diminution in the electrophilicities of the resulting nitrenes would not be anticipated. However, steric interference, especially in the bulky dichloro compound, should preclude a favorable alignment of the rings for internuclear interactions.

C. Formation of the Facial Tetrafluorophenyl Isomer of Janusene

In an earlier section, we discussed the likelihood of internuclear attractions between the facial phenyl and tetrafluorophenyl rings in tetrafluorojanusene, **8,** as suggested from ultraviolet spectral studies. The synthesis of **8** may be approached by Diels-Alder reactions from two pairs of reactants (Figure 2–3). Two cycloaddition products are possible from these reactions, viz., (a) a janusene with two facial phenyl rings and a lateral tetrafluorophenyl ring or (b) a janusene with facial phenyl and tetrafluorophenyl rings (**8**).

In fact, only **8** is isolated, irrespective of mode of synthesis.[41] This extraordinary regiospecificity clearly indicates that attractive electronic interactions between —C_6H_4— and —C_6F_4— rings must play a critical role and may, indeed, be the driving force, in stabilizing the transition state and thereby, determining the structure of the cycloaddition product. Repulsive steric effects will be negligible because of the minimal space demands of the fluorines (cf. nitro and methyl, in which only regioselectivity, not specificity, is observed).[94]

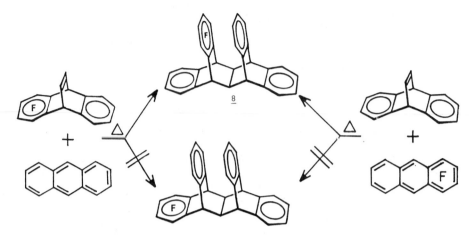

Figure 2–3. The preparation of the facial tetrafluorojanusene (**8**).

11. CONCLUDING REMARKS

Detailed examination of the literature reveals no credible experimental evidence to support the view that polyfluoroaromatics and arylhydrocarbons form π—π charge–transfer complexes. Instead, donor–acceptor complexes may be formed, with stabilization by electrostatic, induction, and dispersion forces. The experimental data support the hypothesis that, in planar species, replacement of aryl hydrogens by fluorines causes only minor effects on phenomena involving π-electrons, while exerting much larger effects on σ electrons (π-fluoro effect-Liebman). Even in the sterically constrained tetrafluoro [2.2] paracyclophane (**10**), in which the nonplanar rings are closely disposed, the data indicate relatively small, subtle differences from those of [2.2] paracyclophane, suggesting only minor contributions by transannular transfer of π-electrons.

In reaction chemistry, however, where factors such as induced polarization by external reagents and stabilization of intermediates may play important roles, the differences in reactivity between **9** and **10** are noteworthy. Stabilization of transition states by attractive interactions between phenyl and polyfluorophenyl rings can significantly alter molecular geometry and the course of a chemical reaction.

REFERENCES

1. Patrick, C. R.; Prosser, G. S. *Nature* **1960,** *187,* 1021.
2. Boeyens, J. C. A.; Herbstein, F. H. *J. Phys. Chem.* **1965,** *69,* 2153.
3. Fenby, D. V.; McLure, I. A.; Scott, R. L. *J. Phys. Chem.* **1966,** *70,* 602.

4. Gaw, W. J., Swinton, F. L. *Trans. Faraday Soc.* **1968**, *64*, 2023.
5. Dantzler, E. M.; Knobler, C. M. *J. Phys. Chem.* **1969**, *73*, 1602.
6. Duncan, W. A.; Swinton, F. L. *Trans. Faraday Soc.* **1966**, *62*, 1082.
7. Goates, J. R.; Ott, J. B.; Reeder, J. *J. Chem. Thermodynamics* **1973**, *5*, 135.
8. McLaughlin, E.; Messer, C. E. *J. Chem. Soc.* (*A*) **1966**, 1106.
9. Swinton, F. L. *Internat. Data Services, Selec. Data Mixtures*, Ser. A. **1973** (Publ. 1974), pp. 31–38; *Chem. Abs.* **1974**, *81*, 96847c.
10. Semenov, L. V.; Gaile, A. A.; Proskuryakov, V. A. *Zh. fiz. Khim.* **1972**, *46*, 2957; *Chem. Abs.* **1973**, *78*, 57562z.
11. Birchall, J. M.; Haszeldine, R. N.; Morley, J. O. *J. Chem. Soc.* (*C*) **1970**, 456.
12. Foster R.; Fyfe, C. A. *J. Chem. Soc. Chem. Commun.* **1965**, 642.
13. Brooke, G. M.; Burdon, J.; Stacey, M.; Tatlow, J. C. *J. Chem. Soc.* **1960**, 1768.
14. Ott, J. B.; Goates, J. R.; Reeder, J.; Shirts, R. B. *J. Chem. Soc., Faraday Trans. I* **1974**, *70*, 1325.
15. Armitage, D. A.; Brindley, J. M. T.; Hall, D. J.; Morcom, K. W. *J. Chem. Thermodynamics* **1975**, *7*, 97.
16. Goates, J. R.; Ott, J. B.; Reeder, J.; Shirts, R. B. *J. Chem. Thermodynamics* **1974**, *6*, 489.
17. Beaumont, T. G.; Davis, K. M. C. *J. Chem. Soc.* (B) **1967**, 1131.
18. Beaumont, T. G.; Davis, K. M. C. *Nature* **1968**, *218*, 865.
19. Hammond, P. R. *J. Chem. Soc.* (A) **1968**, 145.
20. Armitage, D. A.; Beaumont, T. G.; Davis, K. M. C.; Hall, D. J.; Morcom, K. W. *Trans. Faraday Soc.* **1971**, *67*, 2548.
21. Coulson C. A.; Stocker, D. *Molecular Physics* **1959**, *2*, 397.
22. Narten, A. H. *J. Chem. Phys.* **1968**, *48*, 1630.
23. Janda, K. C.; Hemminger, J. C.; Winn, J. S.; Novick, S. E.; Harris, S. J.; Klemperer, W. *J. Chem. Phys.* **1975**, *63*, 1419.
24. Steed, J. M.; Dixon, T. A.; Klemperer, W. *J. Chem. Phys.* **1979**, *70*, 4940.
25. Burley, S. K.; Petsko, G. A. *Science* **1985**, *229*, 23; Burley, S. K.; Petsko, G. A. *FEBS Letters* **1986**, *203*, 139; Burley, S. K.; Petsko, G. A. *J. Am. Chem. Soc.* **1986**, *108*, 7995.
26. Dahl, T. *Acta Chem. Scand.* Ser. A **1975**, *29*, 170.
27. Dahl, T. *Acta Chem. Scand.* **1971**, *25*, 1031.
28. Dahl, T. *Acta Chem. Scand.* Ser. A, **1975**, *29*, 699.
29. Dahl, T. *Acta Chem. Scand.* **1972**, *26*, 1569.
30. Dahl, T. *Acta Chem. Scand.* **1973**, *27*, 995.
31. Gilson, D. F. R.; McDowell, C. A. *Can. J. Chem.* **1966**, *44*, 945.
32. Herbstein, F. H. *Perspect. Struct. Chem.* **1971**, *4*, 166.
33. Brock, C. P.; Naae, D. G.; Goodhand, N.; Hamor, T. A. *Acta Cryst.* **1978**, *B34*, 3691.
34. Brock, C. P.; Naae, D. G. Personal communication.
35. Naae, D. G. *Acta Cryst.* **1979**, *B35*, 2765.
36. Dewar, M. J. S.; Thompson, C. C. *Tetrahedron* Supp. 7 **1966**, 97.
37. Hanna, M. W. *J. Am. Chem. Soc.* **1968**, *90*, 285.
38. Lippert, J. C.; Hanna, M. W.; Trotter, P. J. *J. Am. Chem. Soc.* **1969**, *91*, 4035.
39. Baur, M. E.; Knobler, C. M.; Horsma, D. A.; Perez, P. *J. Phys. Chem.* **1970**, *74*, 4594.
40. Cristol, S. J.; Lewis, D. C. *J. Am. Chem. Soc.* **1967**; *89*, 1476.
41. Filler, R.; Cantrell, G. L. *J. Fluorine Chem.* **1987**, *36*, 407.
42. Hillier, H. In "Molecular Interactions," Vol. 2; Ratajczak, H.; Orville-Thomas, W. J.; Eds.; Wiley: New York, 1981, pp. 493–507.
43. Clark, I. D.; Frost, D. C. *J. Am. Chem. Soc.* **1967**, *89*, 244.
44. El Sayed, M. F. A.; Kasha, M. *J. Chem. Phys.* **1961**, *34*, 334.
45. Al-Joboury, M. I.; Turner, D. W. *J. Chem. Soc.* **1964**, 4434.
46. Brundle, C. R.; Robin, M. B.; Kuebler, N. A.; Basch, H. *J. Am. Chem. Soc.* **1972**, *94*, 1451.
47. Brundle, C. R.; Robin, M. B.; Kuebler, N. A. *J. Am. Chem. Soc.* **1972**, *94*, 1466.
48. Liebman, J. F.; Politzer, P.; Rosen, D. C. In "Chemical Applications of Atomic and

Molecular Electrostatics Potentials"; Politzer, P.; Truhlar, D. G., Eds.; Plenum Press: New York, 1981, pp. 295–308.
49. Burdon, J.; Parsons, I. W.; Tatlow, J. C. *Tetrahedron* **1972,** *28,* 43.
50. Burdon, J.; Parsons, I. W. *Tetrahedron* **1975,** *31,* 2401.
51. Shaw, M. J.; Weil, J. A.; Hyman, H. H.; Filler, R. *J. Am. Chem. Soc.* **1970,** *92,* 5096.
52. Bazhin, N. M.; Akhmetov, N. E.; Orlova, L. V.; Shteingarts, V. D.; Shchegoleva, L. N.; Yakobson, G. G. *Tetrahedron Lett.* **1968,** 4449.
53. Bazhin, N. M.; Pozdnyakovich, Yu. V.; Shteingarts, V. D.; Yakobson, G. G. *Izv. Akad. Nauk SSSR* Seriya. Khim. **1969,** 2300; *Chem. Abs.* **1970,** *72,* 42447 t.
54. Coleman, J. P.; Fleischmann, M.; Pletcher, D. *Electrochim. Acta.* **1973,** *18,* 331.
55. Richardson, T. J.; Tanzella, S. L.; Bartlett, N. *J. Am. Chem. Soc.* **1986,** *108,* 4937.
56. Merrifield, R. E.; Phillips, W. D. *J. Am. Chem. Soc.* **1958,** *80,* 2778.
57. Voight, E. M. *J. Am. Chem. Soc.* **1964,** *86,* 3611.
58. Choe, E. W. Ph.D. dissertation, Illinois Institute of Technology, 1969.
59. Brown, C. J. *J. Chem. Soc.* **1953,** 3265.
60. Lonsdale, K.; Milledge, H. J.; Krishna Rao, K. W. *Proc. R. Soc. London* Ser. A. **1960,** *225,* 82.
61. Hope, H.; Bernstein, J.; Trueblood, K. N. *Acta Cryst.* Sect. B. **1972,** *28,* 1733.
62. Boyd, R. H. *Tetrahedron* **1966,** *22,* 119.
63. Rodgers, D. L.; Westrum, Jr., E. F.; Andrews, J. T. S. *J. Chem. Thermodyn.* **1973,** *5,* 733.
64. Liebman, J. F.; Greenberg, A. *Chem. Rev.* **1976,** *76,* 311.
65. Smith, B. H. "Bridged Aromatic Compounds". Academic Press: New York, 1964.
66. Cram, D. J.; Cram, J. M. *Accts. Chem. Res.* **1971,** *4,* 204.
67. Vogtle, F.; Neumann, P. *Top. Curr. Chem.* **1974,** *48,* 67.
68. Filler, R.; Choe, E. W. *J. Am. Chem. Soc.* **1969,** *91,* 1862.
69. Filler, R.; Cantrell, G. L.; Wolanin, D.; Naqvi, S. M. *J. Fluorine Chem.* **1986,** *30,* 399.
70. Filler, R.; Cantrell, G. L.; Choe, E. W. *J. Org. Chem.* **1987,** *52,* 511.
70a. Cantrell, G. L. Ph.D. dissertation, Illinois Institute of Technology, May 1985.
71. Hamor, T. A. Personal communication.
72. Maroncelli, M.; Longworth, J. W. Personal communication.
73. Pignataro, S.; Mancini, V.; Ridyard, J. N. A.; Lemka, H. J. *J. Chem. Soc. Chem. Commun.* **1971,** 142.
74. Boschi, R.; Schmidt, W. *Angew. Chemie, Int. Ed. Engl.* **1973,** *12,* 402.
75. Heilbronner, E.; Maier, J. P. *Helv. Chim. Acta.* **1974,** *57,* 151.
76. Koenig, T.; Tuttle, M.; Wielesek, R. A. *Tetrahedron Lett.* **1974,** *29,* 2537.
77. Koenig, T.; Wielesek, R. A.; Snell, W.; Balle, T. *J. Am. Chem. Soc.* **1975,** *97,* 3225.
78. Kovač, B.; Mohraz, M.; Heilbronner, E.; Boekelheide, V.; Hopf, H. *J. Am. Chem. Soc.* **1980,** *102,* 4314.
79. Kovač, B.; Klasinc, L.; Güsten, H. Personal communication.
80. Cram, D. J.; Helgeson, R. C.; Lock, D.; Singer, L. A. *J. Am. Chem. Soc.* **1966,** *88,* 1324.
81. Cram, D. J. *Rec. Chem. Progr.* **1959,** *20,* 71.
82. Cram, D. J.; Allinger, N. L. *J. Am. Chem. Soc.* **1955,** *77,* 6289.
83. Cram, D. J.; Day, A. C. *J. Org. Chem.* **1966,** *31,* 1227.
83a. Filler, R., unpublished results.
84. Cram, D. J.; Kierstead, R. W. *J. Am. Chem. Soc.* **1955,** *77,* 1186.
85. Cram, D. J.; Wechter, W. J.; Kierstead, R. W. *J. Am. Chem. Soc.* **1958,** *80,* 3126.
86. Cram, D. J.; Bauer, R. H. *J. Am. Chem. Soc.* **1959,** *81,* 5971.
87. Allen, K. J.; Bolton, R.; Williams, G. H. *J. Chem. Soc., Perkin Trans.* **1983,** *2,* 691.
88. Bolton, R.; Sandall, J. P. B.; Williams, G. H. *J. Chem. Res.* (S) **1977,** *24;* (M), 0323.
89. Bolton, R.; Williams, G. H. *Chem. Soc. Rev.* **1986,** *15,* 261.
90. Banks, R. E.; Madany, I. M. *J. Fluorine Chem.* **1985,** *30,* 211.
91. Banks, R. E.; Prakash, A. *Tetrahedron Letters* **1973,** 99.
92. Banks, R. E.; Prakash, A. *J. Chem. Soc., Perkin Trans I* **1974,** 1365.

93. Abramovitch, R. A.; Challand, S. R.; Scriven, E. F. V. *J. Am. Chem. Soc.* **1972,** *94,* 1374.
94. Cristol, S. J.; Lim, W. Y. *J. Org. Chem.* **1969,** *34,* 1.

General References

95. "Molecular Complexes in Organic Chemistry". Andrews, L. J.; Keefer, R. M.; Holden-Day, Inc.: San Francisco, 1964.
96. "Organic Charge-Transfer Complexes". Foster, R.; Academic Press: New York, 1969.
97. "Molecular Interactions", Vol. 2. Ratajczak, H.; Orville-Thomas, W. J., Eds.; John Wiley and Sons: New York, 1981.
98. "Electrooxidation in Organic Chemistry". Yoshida, K.; John Wiley and Sons: New York, 1984.
99. "Cyclophanes", Vol. 1. Keehn, P. M.; Rosenfeld, S. M., Eds.; Academic Press: New York, 1983.
100. "Strained Organic Molecules". Greenberg, A.; Liebman, J. F.; Academic Press: New York, 1978.

CHAPTER 3

The Effect of Fluorine as a Substituent on Selected Properties of Neutral and Charged Aromatic Systems

Anne Skancke

Department of Mathematical and Physical Sciences, University of Tromsø, N-9001 Tromsø Norway

CONTENTS

1. INTRODUCTION

The vast majority of quantum mechanical studies of molecules are based on the variation method. This implies that the quality of a molecular wavefunction is estimated with reference to the value of the total electronic energy predicted. The energy, being an integrated quantity, is not always a good or even appropriate indicator of the accuracy of other properties characterizing a molecule.

Thus, for molecular structure calculations even a solution at the Hartree-Fock limit may lead to geometries that are in disagreement with the "true" values as determined experimentally.[1] This is due to the well-known shortcomings of the Hartree-Fock model (eg., neglect of electron correlation) and to the role of anharmonicities in molecular vibrations. Molecular ab initio calculations make use of truncated basis sets yielding wave functions that normally are far away from the Hartree-Fock limit. Absolute structure predictions based on such calculations have to be corrected, usually by empirically chosen adjustment or off-set values that are characteristic for given structure parameters and chosen basis sets. Such off-set values have been estimated by several groups.[2-7]

Furthermore, since there is no variation principle for energy differences in molecular calculations, predicted relative stability of different isomers of a species and estimates of energy barriers to chemical reactions must be calculated indirectly. The success of such estimates is critically dependent on cancellation of errors. In many cases cancellations do occur and thus give a reliable basis for isodesmic[8] and homodesmotic[9] models which often are very important in discussions of stabilization energies. For hydrocarbons, it has recently been shown by Wiberg[10a] that the total molecular SCF energies may be converted to heats of formation by using empirical adjustments based on the idea of characteristic group equivalents. Ibrahim and Schleyer have calculated more general atom equivalents[10b] which are useful in discussing molecular stabilities.

The prevailing tendency of the chemist to make references to "molecular families" renders a chemical property interesting only when considered as a relative quantity. For instance, the total energy of fluorovinylidene (C_2HF) is some -175.6 au (or about 100,000 kcal/mol).[11] Whereas this piece of information may be of minor interest to a chemist, the relative importance of the 1,2-H shift and the 1,2-F shift poses a chemically interesting question. Although both the 1,2-H shift and the 1,2-F shift produce the same product, they would proceed via different mechanisms.

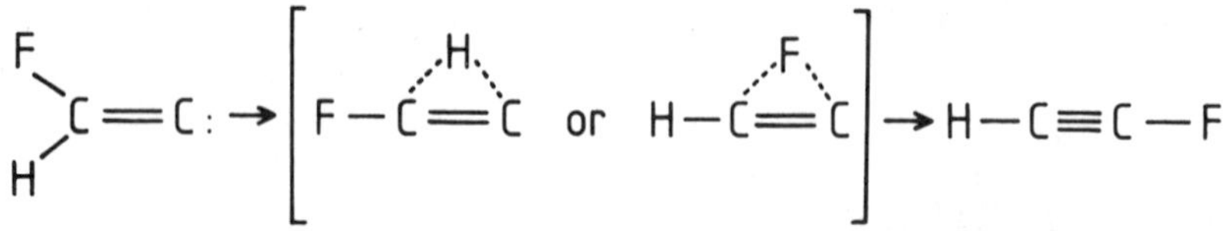

The energy difference between the two different transition states is 0.03 au or about 19 kcal/mol which is sufficiently large to give an unambiguous conclusion regarding the reaction path (the 1,2 H shift being the more facile one).[11] This prediction is based on an accuracy greater than 0.02% in the total energy. Frequently, energy differences between two species of interest are much smaller, requiring even an order of magnitude better energy calculations.

One of the more successful applications of quantum chemistry in recent years has been the calculation of molecular equilibrium geometries and force constants by the gradient method. A large number of studies have been carried out using the gradient technique, the results of which have elucidated trends within chemically related species. The data obtained may sometimes be explained as traditional chemical "effects," like hyperconjugation and aromaticity.

In this chapter, we report on some data from ab initio calculations within the field of carbon-fluorine chemistry. Of particular interest are some cases where consistent calculations were carried out within a series of species at a level of sophistication that should be sufficient for quantitative conclusions. We will, at present, focus on the effect of fluorine upon "aromaticity."

2. BENZENOID AROMATICS

It is well documented that fluorine as a substituent does not perturb the geometry of a benzenoid hydrocarbon to any large extent.[12-18] However, large perturbations of the molecular electronic distribution may be consistent with small geometrical changes. Precise and reliable estimates of small geometric distortions may thus yield valuable information regarding the electronic rearrangements.

The perturbation of the electronic distribution of aromatic systems by fluorine may be thought of as a sum of several different effects, the most important being the σ-inductive effect (which occurs in aliphatic compounds as well) and the back-donation to the π-system.

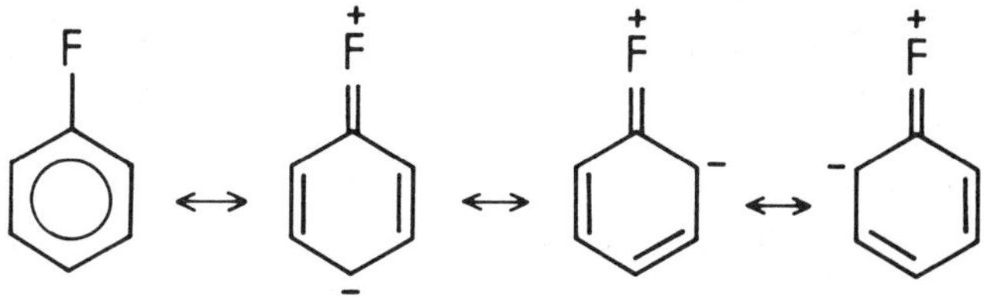

This picture has been used in the interpretation of the ortho–para directing effect of fluorine. The back-donating tendency is often explained in terms of overlap considerations, this effect being favorable for the short C—F bond connecting atoms in the same row.[19] An alternative model for the description of the π-electron distribution is the repulsion model in which the fluorine lone pair repels the π-electrons of the aromatic system.[20]

The merits of the various interaction models will be discussed after a brief look at some experimental and computational results.

In a series of articles Domenicano and co-workers[12-17] have scrutinized the geometrical distortions of the benzene ring caused by different kinds, numbers, and relative positions of substituents. In this context only the influences of fluorine as a substituent will be presented.

A systematic analysis based on experimental results for 199 different compounds led to the conclusion that in monosubstituted benzenes there is a simultaneous change of the internal angles at the ipso, ortho, and para positions of the ring related to the electronegativity of the substituent. An additional distortion involves mainly the internal angles at the meta and para positions and is related to the π-electron donor/acceptor ability of the substituent.[12] Figure 3-1 gives a definition of the applied parameters.

In an ab initio study Boggs and co-workers[18] optimized the geometries of mono-, di-, tri-, penta-, and hexafluorobenzene using the gradient method. In that work, which is carried through at the 4-21G level, empirical corrections have been applied to yield r_e and r_o structures which are compared with experimental counterparts. (The reader is reminded that r_e is the distance between equilibrium nuclear positions, while r_o is the distance between effective nuclear positions derived from rotational constants for the zero-point vibrational level. The value of the latter is therefore slightly larger than the former.) Good agreement is obtained when comparisons are made with experimental moments of inertia. (This is particularly desirable, since these are determined experimentally with a high degree of accuracy.) The following trend was observed: the C—F bond distance decreases in a uniform manner with increasing fluorination of the ring, either on adjacent or on meta sites. Systematic deformations of the ring geometries were also found, notably a decrease of around 0.01Å in the carbon–carbon bond adjacent to the site of substitution and an opening of the ipso angle.

In Table 3-1 we have collected some of the results obtained in the preceding studies. Although experimental structure data exist for several of these compounds, the table includes data only for the cases where ring deformations have been studied.

The data in Table 3-1 demonstrate clearly the opening of the ipso-angle by fluorine substitution. When fluorine atoms are on adjacent carbon atoms, as

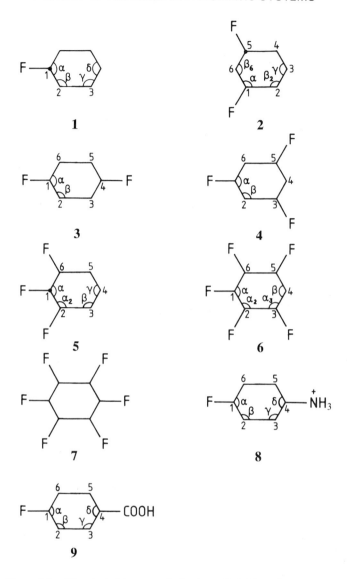

Figure 3-1. Fluorinated benzene derivatives. Definition of parameters.

in **(5)–(7)**, Table 3-1, the effect is diminished due to competing distortions and to the constraints imposed by the hexagon geometry.

By applying a hybridization model we find that an opening of the ipso-angle is consistent with increased p-character in the C—F bond relative to an sp^2 bond, and a concomittant decrease of p-character in the C_1—C_2 bond. This should lead to a lengthening of the C—F bond and a shortening of the C—C bond relative to a model having pure sp^2 hybrids.

As mentioned previously, there are large perturbations of the π-electron

TABLE 3-1. Observed and Predicted (in Parentheses) Geometry Parameters for Fluorine-substituted Benzenes; Distances in Å, Angles in Degrees

Molecule parameter	(1)	(2)	(3)	(4)	(5)	(6)	(7)	(8)	(9)
$R(F—C_1)$	1.356; (1.343)	(1.339)	1.352	1.339; (1.334)	(1.326)	(1.320)	(1.323)	1.363	1.364
$R(F—C_2)$					(1.333)	(1.322)			
$R(F—C_3)$						(1.328)			
$R(C_1—C_2)$	1.387; (1.387)	(1.388)	1.387	1.389; (1.388)	(1.387)	(1.386)	(1.385)	1.373	1.365
$R(C_1—C_6)$		(1.387)							
$R(C_2—C_3)$	1.399; (1.396)	(1.397)	1.399		(1.387)	(1.387)		1.390	1.383
$R(C_3—C_4)$	1.401; (1.397)				(1.396)	(1.386)		1.381	1.389
α	123.4; (122.3)	(122.5)	123.5	123.7; (122.7)	(119.1)	(120.4)	(120.0)	123.2	124.0
α_2					(121.1)	(119.3)			
α_3						(121.0)			
β	118.0; (118.5)		118.5	116.3; (117.3)	(119.1)	(118.9)		118.2	117.8
β_2		(118.4)							
β_6		(117.4)							
γ	120.2; (120.3)	(120.8)			(120.4)			119.2	120.3
δ	120.2; (119.8)							121.8	119.9
Ref.	16 18	18	15	17 18	18	18	18	14	13

system making a discussion based on hybridization alone futile. However, there is a general shortening of the $C_1—C_2$ bond consistent with the hybridization picture.

Politzer and Timberlake have calculated overlap and repulsion integrals for halogenated benzenes.[20] These calculations do not support the assumption that the electron donating part of the electron flow in fluorobenzene is dominated by a favorable carbon–fluorine overlap, but rather by the effect of large repulsion integrals in the C—F region. This repulsion has the effect of pushing the aromatic π-electrons away from the substituted position, leading to a buildup of charge particularly in the para position.

This interpretation is furthermore in agreement with the observed larger para/ortho ratio for electrophilic substitutions in fluorobenzene as compared to chlorobenzene.[21]

A Mulliken population analysis of the computed structure of fluorobenzene[18] gives the π-electron charges reported in Table 3-2. The first step

TABLE 3-2. Gross π-electron Charges for Fluorobenzene[a] 4-21 Basis Set

F	+0.06
C1	0.00
C2	−0.05
C3	+0.03
C4	−0.02

[a] Geometry from Ref. 18.

(which is rate and product determining) in an electrophilic aromatic substitution being the formation of a positive ion, a negatively charged site seems more prone to electrophilic attack. Accordingly, the data in Table 3-2 (although computed from the modes 4-21G basis set) do seem to be in agreement with a preference for ortho/para over meta in electrophilic attack for fluorobenzene. However, the data do not predict any preference for para over ortho attack.

A. Bond Order Considerations In Benzenoid Aromatic And Other Multiple Bonded Systems

An estimate of the C—F bond order[22] may be made from bond distance considerations. Figure 3-2 shows a plot of the C—F bond length for a number of compounds, the extrema being the triply bonded CF^+ and the singly bonded CH_3F. One may argue that the latter point is an arbitrary member of the family CH_3F—CF_4 which displays a significant trend in the CF bond distances.[23,24] This variation may be due to increasing electrostatic attraction on successive fluorination, and the first member of the series is therefore considered as a more realistic reference. With allowance for the decrease of the C—F distance in CF_nH_{4-n} we thus consider a typical C—F single-bond length to be 1.382Å.[23] A comment is needed on the structure of CF^+. Recent velocity modulation laser spectroscopy and millimeter spectroscopy give an amazingly precise determination of the r_e value of 1.1542551(25)Å[25a] for this bond. Implicitly, one may deduce from the data in Reference 25a that they refer to the $^{12}CF^+$ isotopomer. A recent ab initio value (6-31G* level) for this bond is 1.144Å.[25c] We wish to point out that this diatomic species is a member of the 14-electron set (eg., CO, N_2, NO^+, BF, CN). These systems all have closely similar bond lengths and dissociation energies.[26] We thus obtain Figure 3-2 giving the C—F bond orders for double-bonded and aromatic carbons about 1.5, while those of acetylenic carbons can be estimated to be about 1.8. Thus a significant double-bond character is found for fluorobenzene. It may be noted that slightly different values for the C—F bond order for aromatics would have been obtained by using the computed data in Reference 18. In particular, we refer to that paper for a discussion of the C—F bond length in 1,3,5-trifluorobenzene. Also, there is of course an arbitrariness in our choice of end points for the curve. Nevertheless, such curves may be useful for rough estimates of bond orders.

The usefulness of these considerations may be further appreciated when comparison is made with other C—X bonds (X being a second row atom) and corresponding bond orders as shown in Figure 3-3. Although the lower curve in this figure is based upon meager bits of information, apparently C—F bond/bond order relationships follow the general trend for the second row elements. As an application of these curves, consider the isoelectronic ions CO_3^{--} and CF_3^+. The experimental value for the C—O bond length of the former ion (1.29Å)[23] corresponds to a reasonable bond order of about 1.6

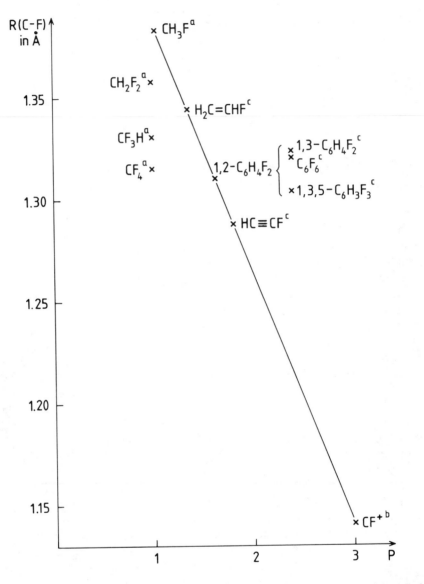

Figure 3-2. Bond order—bond distance relationship for C—F bonds. a) Data taken from Ref. 24. b) Data taken from Ref. 25c. c) Data taken from Ref. 23.

for this system. There is a computed value for the C—F bond of CF_3^+ (1.326Å) in the literature (STO-3G basis set).[27a] This gives an unrealistic small bond order of 1.2 for that system, indicating that the basis set applied does not properly describe the C—F bond. Indeed, a recent computation using a 6-31G* basis set gives a value of 1.218Å for this species, corresponding to a reasonable bond order of 1.6-1.7.[27b]

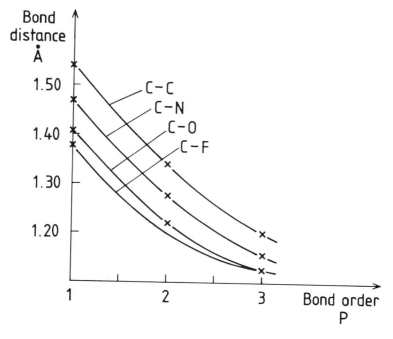

Figure 3-3. Bond order—bond distance relationship for C—X bonds. C—C, C—N, and C—O data taken from Ref. 27. See text for data for C—F curve.

The reason for the high C—F bond order in fluorinated benzenes may be sought in the orbital size and node structure of the second row elements. In a paper by Pyykkö[29] the orbital radii for group 6B atoms are explicitly calculated, showing that for oxygen the nodeless 2p orbital has a smaller radius than the 2s orbital. This is contrary to the results for the heavier elements in the group, where both the ns and the np orbitals are repelled by core orbitals of the same symmetry. The special tendency for second row elements to form double bonds may be related to this effect.

3. NONBENZENOID AROMATICS

Hess and Schaad studied the effect of fluorine on the aromaticity of a large number of carbocyclic systems by means of the Huckel method.[30] The resonance energy per π-electron was computed and compared to chlorine analogs, the key question being π-electron donation vs inductive electron withdrawal. From an inspection of Hammett σ values[31] both fluorine and chlorine should be electron withdrawers relative to hydrogen. The calculations by Hess and Schaad[30] gave an overall electron donation from fluorine, but this was attributed to a cancellation of inductive effects relative to the reference systems. They did, however, conclude that fluorine is a better electron donor than chlorine, a result at variance with what would be expected from pure electronegativity considerations.

Experimental evidence for significant back-donation of electrons from fluorine in aromatic cations may also be inferred from a recent electron ionization study of isomerization of fluorobenzyl and fluorotropylium cations by McLafferty and Amster.[32]

A recent experimental work by Paprott and Seppelt[33a] points in the same direction for a neutral nonbenzenoid aromatic system. In that work, the acidity of C_5F_5H was studied:

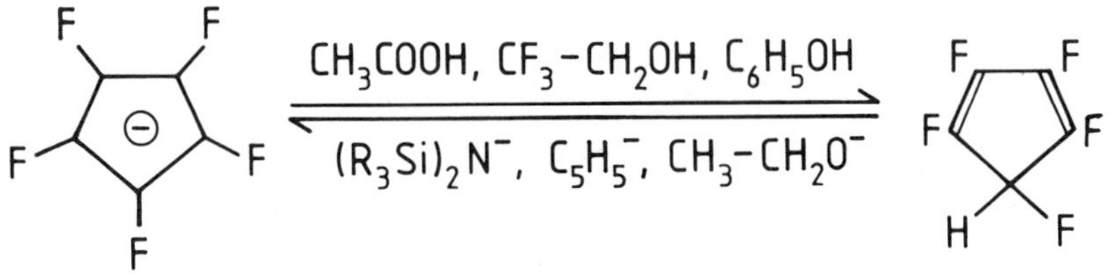

and the results compared to nonfluorinated counterparts. The acid strength (as determined from NMR data) of this system was little different from the reference compound, indicating a balance of electron acceptor and donor tendencies. Support for this was furthermore found in the acid strength of $C_5(CF_3)_5H$ which is reported to be about 18 orders of magnitude larger than for the hydrocarbon.[33b] The CF_3 groups thus have the effect of blocking back-donation.

Related observations have been discussed by Brundle and co-workers and more recently by Liebman, Politzer, and Rosen.[33c] A study of pi and sigma ionization potentials for a large number of fluorinated hydrocarbons containing carbon–carbon multiple bonds shows that the π ionization potential is nearly invariant to the degree of fluorination, whereas there is a considerable variation in the sigma potentials. However, when fluorine substitution occurs on substituent groups (eg, methyl groups) σ—π mixing results in large variation in ionization potentials. The insensitivity for the pure π-systems has given rise to the term π-fluoro effect. Even though this effect refers to the ionization potentials, it is worth noticing that this observable may be correlated to the proton affinities. The work by Liebman and co-workers tries to relate the π-fluoro effect to the electrostatic potentials for hydrogen and fluorine, noting that at about 1.35Å from the nucleus, a free hydrogen atom and a free fluorine atom have exactly the same electrostatic potential.

It is also appropriate in this context to discuss the fluorine substitution effects on a special class of cross-conjugated systems denoted as "Y-aromatic." These systems have been in the focus lately.[34] The archetype Y-aromatic compound is trimethylenmetane dianion (TMM^{2-}) having 6 π-electrons in a Y-shaped structure. Ab initio calculations on this system have been carried out, and the relative stability of its planar form has been studied by comparison to related systems like butadiene dianion, cyclobutadiene dianion, and systems obtained by rotating one, two, or all three CH_2 groups. These calculations were carried out within the Hartree-Fock approximation using a 6-31G basis set. The planar D_3 form of TMM^{2-} was found to be 22 kcal/mol, 61 kcal/mol, and 160 kcal/mol more stable than singly, doubly, and triply rotated forms, respectively. Moreover, the planar form was found to be more stable than the isoelectronic, linearly delocalized butadiene dianions by about 30 kcal/mol. The relative stability of TMM^{2-} as compared to the above mentioned systems was furthermore studied by comparing their charge alternations.[35]

Another system of interest in this connection is cyclobutadiene dianion (CBD^{2-}),[36a,b] which has also been described by quantum chemical methods. The instability of this planar $6\pi e$ system has been attributed to cross-ring or diagonal repulsions,[36a] the global minimum has been found to have C_s symmetry.[36b] The carbon atoms of the planar form were found (from Mulliken population analysis) to be negatively charged. From a naive point of view one might then expect an enhanced stability upon substitution with fluorine, the negative charges being removed from the unfavorable steric positions.

The effect of fluorine substitution on the stabilities of TMM^{2-} and CBD^{2-} is of interest for our discussion. Before reporting results of ab initio calculations, it is necessary to take a brief look at some computational details, especially with regard to the basis set problem including the basis set superposition error. A number of calculations have been carried out on the planar fluorinated CBD^{2-} system in studying these two effects for the problem at hand.

A. Application Of Diffuse Basis Functions

The stabilities of anions are normally expressed in terms of electron affinities or proton affinities. Unfortunately, both of these properties are poorly described by standard calculations within the Hartree-Fock approximation.[37] It is well documented that standard basis sets describe Rydberg states and negatively charged species less well than neutral and positively charged species. Dunning and Hay have demonstrated this in a calculation of electron affinities for atoms using a (9s5p) basis set.[38] Marked improvements have been obtained for electron affinities and proton affinities by augmenting the basis set with functions suitable to describe the extended distribution of the negative charge. It should be noted, however, that certain properties are predicted well by single determinant calculation using simply split-valence basis sets. In particular, energy considerations from a study of homodesmotic reactions and also structure descriptions have been reported to be well reproduced at this level of sophistication.[37a]

In order to test the validity of this for the systems at hand, selected runs have been carried out using diffuse functions, studying the basis set superpo-

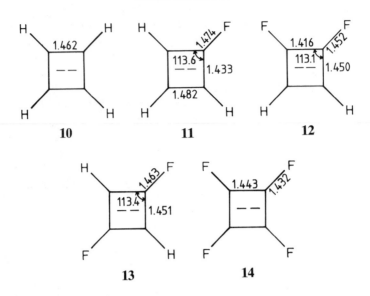

Figure 3-4. Planar cyclobutadiene dianions. Calculated structures.

sition effect and estimating the correlation contribution by Møller-Plesset calculations. The VAX version of the computer program Gaussian 82 has been applied.[39]

Application of diffuse functions to all second row atoms in a fluorine-substituted CBD^{2-} causes a "falling inward" of the diffuse functions. This is because the electronegative fluorines are net acceptors of electrons, leaving the carbons relatively positive. Consequently, diffuse functions were added to only fluorine atoms.

Since the recommended diffuse s and p exponent found in the literature $(0.1076)^{37c}$ has been optimized for the F^- ion and not for fluorine-containing organic systems, a reoptimization was carried out for the CBD^{2-} systems. Systems **11** and **14** (see Figure 3-4) were chosen for this purpose. An exponent value of 0.06 was derived as an average for these two species, and subsequently used in the third set of energy data given in Table 3-3. It is

TABLE 3-3. Total Energies for Dianions (10–14) (For structures, see Figure 3-4.)

	6-31G/6-31G	6-31+G/6-31G[a]	MP2(6-31G)/6-31G	MP2(6-31G*)/6-31G
10	−153.1945		−153.5632	−153.7902
11	−252.0469	−252.0623	−252.5310	−252.8235
12	−350.8907	−350.9170	−351.4903	−351.8505
13	−350.8878	−350.9168	−351.4878	−351.8463
14	−548.5542	−548.5971	−549.3848	

[a] Diffuse sp functions with exponent 0.06 on fluorine only.

worth noticing that this value corresponds to a more diffuse function than the literature value of 0.1076. In establishing the value of 0.06 care was taken (by checking eigenvectors) that the well-known "falling inward" was avoided. This particular point has been considered in some detail in the excellent recent review on basis sets for molecular calculation by Huzinaga.[40] Table 3-3 gives the total energies for the substituted CBD^{2-} systems at different levels of sophistication.

It should be noted that the electron affinity of F is underestimated by about 0.2 eV at the Hartree-Fock level and constitutes one of the classic problems in ab initio quantum chemistry.[41] Extensive correlated treatments and a reliable basis set is required to get correspondence with experimental data. In a recent work by Adamowicz and Bartlett this problem is solved by using size-extensive coupled-cluster or many-body perturbation theory.[42] The wavefunctions obtained in the present work using the basis sets given in Tables 3-3 and 3-4 are not suitable for a study of electron affinities.

B. Basis Set Superposition Error ("BSSE")

As seen in Table 3-3, the 1,2 and 1,3 isomers are very nearly equal in energy, the 1,2 isomer being slightly more stable (from 0.2 to 1.8 kcal/mol depending on the level of sophistication). Is this a result of the basis set superposition error?

This problem has been considered by several authors, and the so-called counterpoise method by Boys and Bernardi[43] has been applied in a series of works. One recent application is given by Vetter and Zulicke[44] in a study of the pentahalocarbonate anions. Noteworthy in this connection is also a paper by Kolos[45] who has studied the effect in a number of molecular fragments. A recent paper by Surjan, Meyer, and Lukovits[46] treats the problem in a direct way through the use of the "chemical Hamiltonian," a method which has the advantage of avoiding the difficulty of calculating a small difference between large numbers.

For the systems at hand, there is no obvious choice of molecular fragments, but an estimate of the effect may be obtained in a simplistic way by mimicking the effect of a fluorine basis adjacent to a fluorine atom. For system **11**, an empty basis (a "ghost") was placed in the extension of the C—F bond and at a distance corresponding to the F—F distance for system **12**. Since this effect should be largest for basis sets containing diffuse functions, this particular search has been restricted to the 6-31+G basis (vide supra). The resulting energy increments were 0.52 kcal/mol for a fluorine-ghost distance of 2.8Å and 0.30 kcal/mol at a distance of 3.2Å. The effect thus seems small, and does not seriously alter the finding that the two difluorinated planar form cyclobutadiene dianions have nearly equal energy.

ANNE SKANCKE

TABLE 3-4. Energy Increments per F Atom. ΔE is Energy (in au) Relative to System I (See Figure 3-4 for structures and structural parameters.)

| System | No of F atoms | 3-21G//6-31G | | 6-31G//6-31G | | 6-31+G//6-31G | | MP2(3-21G)//6-31G | | MP2(6-31G*)//6-31G | |
		$-\Delta E$	$-\Delta E/n$	$-\Delta E$	$-\Delta E/n$	$-\Delta E$	$-\Delta E/n$	$-\Delta E$	$-\Delta E/n$	$-\Delta E$	$-\Delta E/n$
10	0	0		0		0		0		0	
11	1	98.3766	98.3766	98.8524	98.8524	98.8679	98.8679	98.4937	98.4937	99.0335	99.0335
12	2	196.6857	98.3429	197.6962	98.8481	197.7225	98.8613	196.9060	98.4530	198.0603	99.0302
13	2	196.6818	98.3409	197.6933	98.8467	197.7223	98.8614	196.9025	98.4512	198.0561	99.0281
14	4	393.3430	98.3358	395.3597	98.8399	395.4026	98.8507	393.7856	98.4464		

C. The Correlation Energy

The effect of correlation energy has been tested by calculating second order Møller-Plesset energies for a number of the systems, as shown in Table 3-3. For system **10** the MP2 improvements were 0.3625 au and 0.3687 au, respectively, for the 3-21G and 6-31G basis sets. For system **11**, the same calculation yielded values of 0.4723 au and 0.4841 au, respectively. Although the data give an indication that the larger basis set would be preferable for these systems, the lower level calculations do seem to include the major part of the MP2 energy correction. A further investigation of the basis set dependence of relevant energy relationships is shown in Table 3-4 where incremental energies as a function of the number of fluorine atoms are given for a number of basis sets. The trend is seen to be the same at different levels of sophistication: a decreasing energy increment per fluorine atom with increasing number of fluorines.

   In short, the experiences from using different levels of sophistication are as follows: The correlation energy, although large in absolute value, is not important for the stabilization energy as calculated from homodesmotic reactions. The same is true for diffuse functions, although these would be necessary for a discussion of absolute stability from proton affinities. There is a discrepancy between the 6-31G and 3-21G results. Since the larger set gives a better description of the inner core, the 6-31G set is the basis of choice for these systems.

D. Stability Of Fluorinated TMM^{2-} and CBD^{2-} Systems

Table 3-5 gives the energies for the fluorinated TMM^{2-} systems (see Figure 3-5). The present study is confined to planar forms; rotation barriers of fluorinated TMM^{2-} will be discussed elsewhere. Comparison between data for *gem*-difluorinated TMM^{2-} and fluorinated CBD^{2-} species shows the same general trend for both sets of dianions: fluorine destabilizes both series, but the effect is small. System **19**, a symmetrically substituted TMM^{2-}, is slightly more stabilized than the *gem*-substituted systems. The incremen-

TABLE 3-5. Energies of TMM^{2-} (in au) and its Symmetrical Fluorine Derivatives (For structures, see Figure 3-5.)

System		$E(6\text{-}31G/16\text{-}31G)$	No. of F atoms (n)	$-\Delta E$	$-\Delta E/n$
TMM^{2-}	**(15)**	−154.5005	0	0	0
F$_2$−TMM^{2-}	**(16)**	−352.1796	2	197.6791	98.8396
F$_4$−TMM^{2-}	**(17)**	−549.8449	4	395.3439	98.8360
F$_6$−TMM^{2-}	**(18)**	−747.4949	6	592.9944	98.8324
F$_3$−TMM^{2-}	**(19)**	−451.0253	3	296.5248	98.8416

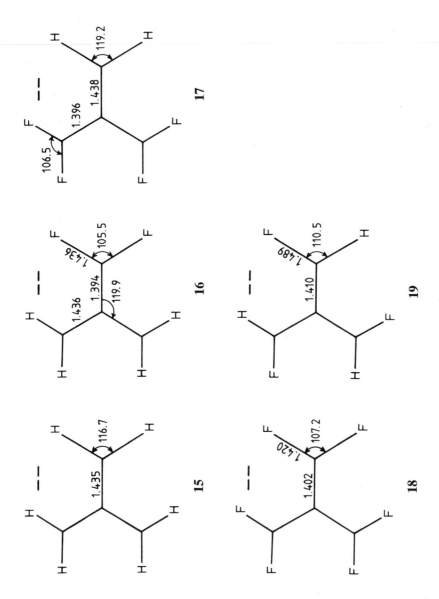

Figure 3-5. Trimethylenemethane dianions. Calculated structures.

tal difference between **16** and **19**, for instance, is only about 1 kcal/mol. However, the BSSE effect should favor **16, 17**, and **18** by about 0.5 kcal/mol per geminal difluoro group (vide supra), making the energy difference somewhat larger. Still, energy effects from substitution are certainly small, and may be artifacts of the method.

There are a large number of conformers of fluorine-substituted CBD^{2-} systems of C_s symmetry. In the present work calculations have been carried out on a selected number of nonplanar forms (Figure 3-6). The energy incre-

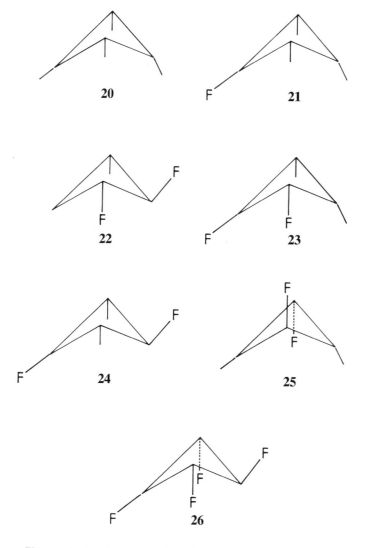

Figure 3-6. Conformations of fluorinated cyclobutadiene dianions studied.

TABLE 3-6. Energies (in au) of Nonplanar CBD^{2-} and Some of its Fluorine-substituted Derivatives (For conformations, see Figure 3-6.)

System	E(6-31G//6-31G)	N. of F atoms	$-\Delta E$	$-\Delta E/n$
20	-153.2092	0	0	
21	-252.0764	1	98.8672	98.8672
22	-350.9666	2	197.7574	98.8787
23	-350.9482	2	197.7390	98.8695
24	-350.9564	2	197.7472	98.8736
25	-350.9555	2	197.7463	98.8732
26	-548.6324	4	395.4232	98.8558

ments are given in Table 3-6. Comparison with Tables 3-4 and 3-5 shows that the nonplanar forms are greatly stabilized compared to their planar counterparts. The effect of successive fluorination is, however, not clear, and comparison between, eg, systems **23** and **24** indicate that the relative position of fluorine atoms is more important than the number of fluorine atoms. Inductive effects thus seem small compared to π-electron effects and hybridization effects.

The destabilizing effect of fluorine in conjugated anions has been reviewed many years ago by Hine, Mahone, and Liotta[47] who attributed the effect to hybridization and increased electronegativity of sp^2 compared to sp^3 carbon. Moreover, the present findings are consistent with the predictions (generalized from a study of acidities) by Streitwieser and Mares[48] that a fluorine substituent stabilizes a pyramidal or nonconjugated anion but destabilizes a conjugated anion. These authors point to increased conjugative destabilization in planar fluorinated systems as the main effect.

Some further insight may be gained by comparing the energy increments which arise from monofluorination of ethane, ethylene, and acetylene. Application of the 6-31G basis set in RHF procedures gives $\Delta E = -98.8243$ au, $\Delta E = -98.8231$ au, and $\Delta E = -98.7901$ au, respectively, for these systems. These data imply a decreasing stabilization when fluorine is attached to multiply bonded carbons, ie, there is a rather strong dependence on the carbon hybridization. See Table 3-6 for energies and geometries. Tables 3-4 and 3-5 (6-31G level) show that the planar dianions are more stabilized upon fluorination than the simple model molecules mentioned above. As to the origin of the hybridization dependence, the geometries given in Table 3-6 point to increasing C—F double bond character in going from sp^3 to sp^2 to sp carbons. This increased bond order is, however, not complemented by stabilization. On the contrary, energy considerations point to increased repulsions at multiple bonding.[20]

E. Geometries Of Fluorinated TMM^{2-} and CBD^{2-} Species

Geometry distortions as a result of fluorine substitution are seen (Figures 3-4 and 3-5) to be quite dramatic, in contrast to the findings for the benzenoid aromatic series. It should be remembered, of course, that the nonbenzenoid aromatic systems studied here are dianions. The limited selection of compounds studied does not allow general conclusions regarding differences between benzenoid aromatics and nonbenzenoid aromatics upon fluorine substitution.

In the planar CBD^{2-} series, the most obvious structural effect of fluorine substitution is an opening of the ipso angle, a concurrent shortening of the C—C bonds adjacent to substitution and a lengthening of the C—F bond, although direct comparison with computed values for the C—F bonds in the fluorobenzene series (Table 3-1) is not justified because the latter have been corrected by off-set values. Still, the dianion C—F bond lengths appear longer than those of the fluorobenzene series and other systems given in Figure 3-5. A likely explanation for these long bond lengths would be electrostatic repulsions in the more electron-rich systems, and conclusions regarding the C—F bond orders are therefore uncertain. It is worth noticing that the opening of the ipso angle and shortening of adjacent C—C bonds were the main features also for the fluorobenzene series (vide supra) where the effects were interpreted in terms of hybridization. In the TMM^{2-} series (Figure 3-5), the C—F bonds appear slightly shorter than in the CBD^{2-} series, but there is an exception for **19**, the only species not *gem*-difluorinated. In **16**, **17**, and **18** the FCF angles are small. A similar effect has been noted for 1,1-difluorocyclopropane[49] and interpreted as a positive lone-pair lone-pair interaction. The small HCF angle in **19** is in agreement with hydrogen bonding for that species.

The structural aspects of nonplanar CBD^{2-} systems are complicated by the low symmetry. For the sake of simplicity, we have listed the average C—F bond distance for the systems **20–26** in Table 3-8. There is a dramatic variation in these distances which range from 1.473Å for system **26** to 1.630Å for system **22**. The latter system may be considered as two fluorine anions weakly interacting with a nonplanar CBD system. This is consistent with the trend within the atomic charges for the fluorine atoms. The average gross atomic charge for the fluorine atoms for system **22** is 9.68 electrons, as compared to, eg, 9.57 for system **26** and 9.61 for system **23**.

Comparison with the energies given in Table 3-7 shows a good correlation between stability and average C—F bond distance. There is a marked con-

TABLE 3-7. Energies (in au) and Heavy Atom
Distances (in Å) (6-31G//6-31G)

Molecule	$-E$	C—F	C—C
H_3C—CH_3	79.19757		1.530
H_3C—CFH_2	178.02187	1.4246	1.509
H_2C=CH_2	78.00446		1.322
H_2C=CFH	176.82754	1.377	1.310
HC≡CH	76.79026		1.194
HC≡CF	175.58032	1.311	1.184

TABLE 3-8. Average
C—F Bond Distance (in
Å) for Nonplanar
Fluorine-substituted
CDB^{2-} (See Figure 3-6 for
conformations.)

System	$(C—F)_{ave}$
21	1.485
22	1.630
23	1.511
24	1.552
25	1.553
26	1.473

trast between the planar and nonplanar systems studied here. Because of the greater flexibility of nonplanar systems, the differences in hybridization may be more pronounced than in the planar ring systems (benzene and CBD^{2-}). Fluorine is seen to stabilize the nonplanar forms, but the amount of stabilization is critically dependent on the mutual orientation of the substituents. Inductive effects in the σ system seem small since there is no additivity upon successive fluorination. Hence, π-electron effects and polarization effects seem to dominate. For the nonplanar systems, an effect of negative hyperconjugation may be operative.[50] It is not easy to point to one specific effect as being the dominating one, since several effects may be interwoven. The planar forms seem to be dominated by destabilizing conjugation.

As a final comment, it may be pointed out that the dianions considered in this work should be regarded as model compounds, and counter ions would stabilize such species in solution.

REFERENCES

1. Boggs, J. E.; Cordell, F. R. *J. Mol. Struct.* **1981,** *76,* 329–347.
2. Pulay, P.; Fogarasi, G.; Pang, F.; Boggs, J. E. *J. Am. Chem. Soc.* **1979,** *101,* 2550–2560.
3. Pulay, P.; Fogarasi, G.; Pongor, G.; Boggs, J. E.; Vargha, A. *J. Am. Chem. Soc.* **1983,** *105,* 7030–7047.
4. Blom, C. E.; Altona, C. *Mol. Phys.* **1976,** *31,* 1377–1391.
5. Blom, C. E.; Slingerland, P. J.; Altona, C. *Mol. Phys.* **1976,** *31,* 1359–1376.
6. Schäfer, L.; Van Alsenoy, C.; Scarsdale, J. N. *J. Chem. Phys.* **1982,** *76,* 1439–1444.
7. Klimkowski, V. J.; Ewbank, J. D.; van Alsenoy, C.; Scarsdale, J. N.; Schäfer, L. *J. Am. Chem. Soc.* **1982,** *104,* 1476–1480.
8. Hehre, W. J.; Ditchfield, R.; Radom, L.; Pople, J. A. *J. Am. Chem. Soc.* **1970,** *92,* 4796–4801.
9. George, P.; Trachtman, M.; Bock, C. W.; Brett, A. M. *Tetrahedron* **1976,** *32,* 317–323.
10. (a) Wiberg, K. B. *J. Comp. Chem.* **1984,** *5,* 197–199. (b) Ibrahim, M. R.; Schleyer, P. V. R. *J. Comp. Chem.* **1985,** *6,* 157–167.
11. Goddard, J. D. *Chem. Phys. Lett.* **1981,** *83,* 312–316.

12. Domenicano, A.; Murray-Rust, P.; Vaciago, A. *Acta Cryst.* **1983**, *839*, 457–468.
13. Colapietro, M.; Domenicano, A.; Ceccarini, G. P. *Acta Cryst.* **1979**, *835*, 890–894.
14. Colapietro, M.; Domenicano, A.; Marciante, C.; Portalone, G. *Acta Cryst.* **1981**, *B37*, 387–394.
15. Domenicano, A.; Schultz, G.; Hargittai, I. *J. Mol. Struct.* **1982**, *78*, 97–111.
16. Portalone, G.; Schultz, G.; Domenicano, A.; Hargittai, I. *J. Mol. Struct.* **1984**, *118*, 53–61.
17. Almenningen, A.; Hargittai, I.; Brunvoll, J.; Domenicano, A.; Samdal, S. *J. Mol. Struct.* **1984**, *116*, 199–206.
18. Boggs, J. E.; Pang, F.; Pulay, P. *J. Comp. Chem.* **1982**, *3*, 344–353.
19. Sheppard, W. A.; Sharts, C. M. In "Organic Fluorine Chemistry"; W. A. Benjamin: New York, 1969.
20. Politzer, P.; Timberlake, J. W. *J. Org. Chem.* **1972**, *37*, 3557–3559.
21. Agranat, I. Personal communication.
22. Coulson, C. A. "Valence", 2nd. ed. Oxford University Press: London, 1961, p. 270.
23. Landolt-Bornstein. "Numerical Data and Functional Relationships in Science and Technology", New Series, Vol. 7. Springer Verlag: Berlin, 1976.
24. Chambers, R. D. "Fluorine in Organic Chemistry"; In *Interscience Monographs on Organic Chemistry*, Olah, G. A. ed.; John Wiley & Sons: New York, 1973, p. 138.
25. (a) Gruebele, M.; Polak, M.; Saykally, R. J. *Chem. Phys. Lett.* **1986**, *125*, 165–169. (b) Plummer, G. M.; Anderson, T.; Herbst, E.; Lucia, F. C. *J. Chem. Phys.* **1986**, *84*, 2427–2428. (c) Hopkinson, A. C.; Lien, M. H. *Can. Journ. of Chem.* **1985**, *63*, 3582–3586.
26. Huber, K. P.; Herzberg, G. "Molecular Spectra and Molecular Structure". Van Nostrand Reinhold: New York, 1979.
27. (a) So, S. P. *Chem. Phys. Lett.* **1979**, *67*, 516–518. (b) O'Keeffe, M. *J. Am. Chem. Soc.* **1986**, *108*, 4341–4143.
28. Wells, A. F. "Structural Inorganic Chemistry", 4th ed. Oxford: University Press, London, 1975, p. 730.
29. Pyykkö, P. *J. Chem. Research (S).* **1979**, 380–381.
30. Hess, B. A.; Schaad, L. J. *Israel Journal of Chemistry* **1978**, *17*, 155–159.
31. Wiberg, K. B. "Physical Organic Chemistry". Wiley: New York, 1964, p. 410.
32. McLafferty, F. W.; Amster, I. J. *J. Fluorine Chem.* **1981**, *18*, 375–381.
33. (a) Paprott, G.; Seppelt, K. *J. Am. Chem. Soc.* **1984**, *106*, 4060. (b) Laganis, E. D.; Lemal, D. M. *J. Am. Chem. Soc.* **1980**, *102*, 6633–6634. (c) Liebman, J. F.; Politzer, P.; Rosen, D. R. In "Chemical Applications of Atomic and Molecular Electrostatic Potentials"; pp. 295–308. Politzer, P.; Trular, D. G., eds.; Plenum Press: New York 1981.
34. (a) Gund, P. *J. Chem. Educ.* **1972**, *49*, 100–103. (b) Agranat, I.; Skancke, A. *J. Am. Chem. Soc.* **1985**, *107*, 867–871, and references therein.
35. Klein, J. *Tetrahedron* **1983**, *39*, 2733–2759.
36. (a) Skancke, A.; Agranat, I. *Nouveau Journ. de Chimie* **1985**, *9*, 577–579. (b) Hess, B. A.; Ewig, C. S.; Schaad, L. J. *Journ. Org. Chem.* **1985**, *50*, 5869–5871.
37. (a) Radom, L. In *Modern Theor. Chem.*, Vol. 4; "Application of Electronic Structure Theory"; Schaefer, H. F., III, Ed.; Plenum: New York, 1977, p. 333. (b) Chandrasekhar, J.; Andrade, J. G.; Schleyer, P. v. R. *J. Am. Chem. Soc.* **1981**, *103*, 5609–5612. (c) Clark, T.; Chandrasekhar, J.; Spitznagel, G. W.; Schleyer, P. v. R. *Journ. Comput. Chem.* **1983**, *4*, 294–301.
38. Dunning, T. H., Jr.; Hay, P. J. In *Modern Theor. Chem.*, Vol. 3; "Methods of Electronic Structure Theory"; Schaefer, H. F., III, Ed.; Plenum: New York, 1977, p. 1.
39. Binkley, J. S.; Frisch, M.; Raghavachari, K.; DeFrees, D.; Schlegel, H. B.; Whiteside, R.; Fluder, E.; Seeger, R.; Pople, J. A.; Gaussian 82, Release A, Carnegie-Mellon University.
40. Huzinaga, S. *Computer Physics Reports* **1985**, *2*, 279–340.
41. (a) Sasaki, F.; Yoshimine, M. *Phys. Rev.* **1974**, *A9*, 26–34. (b) Roos, B. O.; Sadlej, A. J.; Siegbahn, P. E. M. *Phys. Rev.* **1982**, *A26*, 1192–1199. (c) Kucharski, S. A.; Lee, Y. S.; Purvis, G. D.; Bartlett, R. J. *Phys. Rev.* **1984**, *29*, 1619–1626.
42. Adamowicz, L.; Bartlett, R. J. *J. Chem. Phys.* **1986**, *84*, 6837–6839.

43. Boys, S. F.; Bernardi, F. *Mol. Phys.* **1970,** *19,* 553–566.
44. Vetter, R.; Zulicke, L. *Chem. Phys.* **1986,** *101,* 201–209.
45. Kolos, W. *Theor. Chim. Acta* **1979,** *51,* 219–240.
46. Surjan, P. R.; Meyer, I.; Lukovits, I. *Chem. Phys. Letters* **1985,** *119,* 538–542.
47. Hine, J.; Mahone, L. G.; Liotta, C. L. *J. Am. Chem. Soc.* **1967,** *89,* 5911.
48. Streitwieser, A., Jr.; Mares, F. *J. Am. Chem. Soc.* **1968,** *90,* 2444–2445.
49. Skancke, A.; Flood, E.; Boggs, J. E. *J. Mol. Struct.* **1977,** *40,* 263–270.
50. Forsyth, D. A.; Yang, J.-R. *J. Am. Chem. Soc.* **1986,** *108,* 2157–2161, and references therein.

CHAPTER 4

The Electrocyclic Ring Opening of Fluorinated Cyclobutene Derivatives

William R. Dolbier, Jr.

Department of Chemistry, University of Florida, Gainesville, Florida

Henryk Koroniak

Department of Chemistry, University of Florida, Gainesville, Florida
Faculty of Chemistry, Adam Mickiewicz University
Grunwaldzka 6, Poznan, Poland

CONTENTS

1. INTRODUCTION: HYDROCARBON SYSTEMS

Electrocyclic processes are defined as reactions which result in the formation or breaking of a single σ bond between the ends of a linear system of π
electrons. The prototypical system which undergoes such a transformation
is the cyclization of 1,3-butadiene to cyclobutene and the respective opening
of the cyclobutene ring.

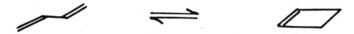

Our general understanding of these processes derives from the concept of
conservation of orbital symmetry[1,2] along with consideration of any steric
interactions in the transition state. The electrocyclic ring openings of the
isomeric 1,2,3,4-tetramethylcyclobutenes are good examples.[3,4,5] *Cis*-1,2,3,4-
tetramethylcyclobutene(1) ring opens via the allowed conrotation process to
give the expected single product *E,Z*-3,4-dimethyl-2,4-hexadiene(2). In contrast, the isomeric trans isomer **(3)** converted exclusively to E,E-3,4-
dimethyl-2,4-hexadiene**(5)** although both isomers **4** and **5** could have been
formed by conrotatory processes.

The stereospecificity of these electrocyclic processes is consistent with
Woodward–Hoffmann theory.[1,2] They pointed out that the conservation of
orbital symmetry of reactants throughout the reaction pathway allows the
reaction to proceed with the lowest energy of activation. The changing of the
symmetry of molecular orbitals in going from starting material to product
requires additional energy, therefore such processes are energetically disfa-
vored and are considered to be symmetry forbidden. The symmetry of a
system's molecular orbitals determines its reaction pathway and, in general,
thermodynamic control (ie, relative stability of products) does not play a
significant role in determining the stereochemical outcome of pericyclic processes.

The electrocyclic ring openings of cyclobutene and its derivatives, being
four-electron processes, and the reactions being dominated by the symmetry
of the diene HOMO (Highest Occupied Molecular Orbital), results in the
prediction that such electrocyclic processes will proceed in a conrotatory

manner. Ring opening of *cis*-3,4-disubstituted cyclobutenes thus always results in the formation of E,Z-derivatives of 1,3-butadiene, whereas *trans*-3,4-disubstituted cyclobutenes potentially can yield either E,E or Z,Z derivatives.

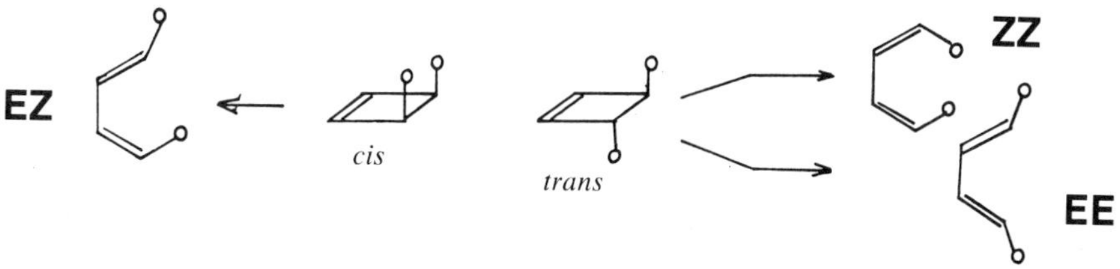

The lack of formation of Z,Z-3,4-dimethyl-2,4-hexadiene(5) in the reaction mixture from ring opening of **3** was initially explained on the basis of detrimental steric interactions in that transition state where both methyl substituents rotate towards each other. This explanation had been widely employed to explain the early experimental studies involving alkyl substituents at the 3 and/or 4 positions. In all pre-1980 examples, the behavior of the cyclobutene derivatives were consistent with this general explanation. Table 4-1 summarizes all kinetic data available for hydrocarbon systems.

A study of Curry and Stevens in 1980, however, provided good reason to suspect such a rationale to be seriously flawed.[23] In a study of 3,3-disubstituted cyclobutenes, Curry and Stevens showed that steric effects cannot be the primary factor responsible for the stereochemical control of product formation in the cyclobutene ring-opening process. In the study, 3-methyl-3-ethylcyclobutene (**6**) was shown to thermally ring-open to give both isomers **7** and **8**, with surprisingly the ostensibly more sterically hindered **8** being formed predominantly.

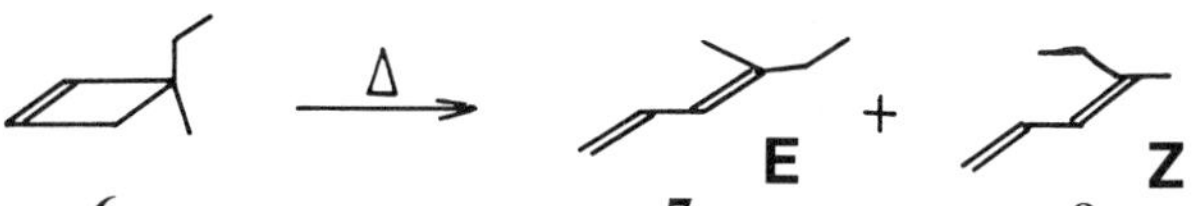

The preferential [2.1 : 1] formation of the Z-isomer **8** wherein the ethyl (not methyl) group rotates inward in the transition state indicated strongly that factors other than steric must be involved in giving rise to these substituent effects. Their studies of other 3,3-disubstituted cyclobutenes also were consistent with such a conclusion. Table 4-2 summarizes their results. Curry and Stevens provided an explanation of these results based upon the relative electronic properties of methyl vs other alkyl groups.

Note the rather substantial apparent isotope effect exhibited by the C_2D_5 case. The result is opposite to that expected for a steric effect and the authors invoked a hyperconjugative interaction of the ethyl methylene group

TABLE 4-1. Activation Parameters for Hydrocarbon Cyclobutene to Butadiene Thermal Conversions

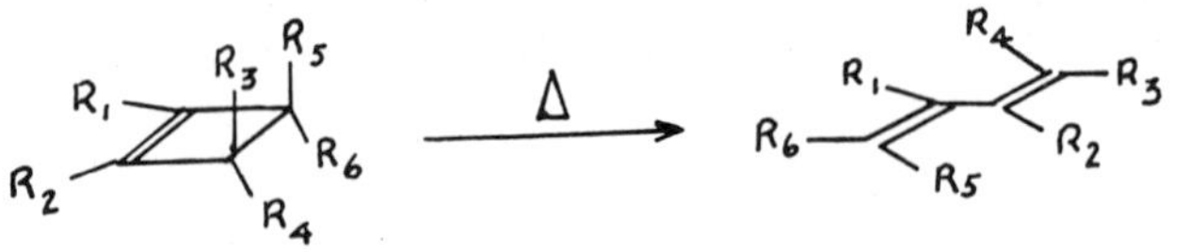

		Substituents						
R_1	R_2	R_3	R_4	R_5	R_6	Log A	$E_a{}^a$	Ref.
H	H	H	H	H	H	13.1	32.5	6
H	H	Me	H	H	H	13.5	31.5	7
H	H	Et	H	H	H	13.5	31.6	8
H	H	Ph	H	H	H	12.4	26.0	9
H	H	Me	H	Me	H	13.9	34.3	10
H	H	Me	H	H	Me	14.0	30.6	11
H	H	Ph	H	Ph	H	13.1	24.5	12
Me	H	H	H	H	H	13.8	35.1	13
Et	H	H	H	H	H	13.8	34.8	8
n-Pr	H	H	H	H	H	13.6	34.6	14
isoPr	H	H	H	H	H	13.6	34.7	14
2-propenyl	H	H	H	H	H	13.5	34.2	14
cyclopropyl	H	H	H	H	H	13.5	34.1	14
Me	Me	H	H	H	H	13.8	36.0	15
Me	H	H	H	Me	H	13.5	33.4	16
Me	H	Me	H	H	H	13.7	33.0	16
Me	H	Me	Me	H	H	13.9	37.0	17
H	H	Me	Me	H	H	13.9	36.1	17
Me	Me	Me	H	H	H	13.3	33.7	18
Me	Me	Me	H	Me	H	14.1	37.4	5
Me	Me	Me	H	H	Me	13.9	33.6	5
H	H	Me	Et	H	H	13.5	35.2	19
H	H	Et	Me	H	H	13.5	35.9	19
H	H	Et	Et	H	H	13.5	34.7	19
H	H	tBu	Me	H	H	14.1	35.8	23
H	H	Me	tBu	H	H	14.1	36.5	23
H	H	Ph	Me	H	H	12.5	29.9	23
H	H	Me	Ph	H	H	12.5	30.6	23
Me	Me	Me	Me	Me	Me	—	40.0	4
Me	Me	Me	Me	Me	H	—	34.0	4
Ph	Ph	H	H	H	H	12.8	32.0	20
Ph	Ph	Ph	H	Ph	H	12.8	25.0	21
Ph	Ph	Ph	H	H	Ph	11.1	21.0	4
Ph	Ph	Me	Ph	Me	Ph	—	24.7	22
Me	CH$_2$=CH	H	H	H	H	13.7	35.7	18

^{a} kcal/mol

with the developing *p*-orbital on C-3 during its rehybridization from sp^3 to sp^2.[23]

In another, related study, cyclobutene **9** ring-opened to form the product wherein the larger group has rotated inward exclusively.[24] This result,

TABLE 4-2. Competitive Ring Openings of 3,3-Dialkylcyclobutenes at 180°C

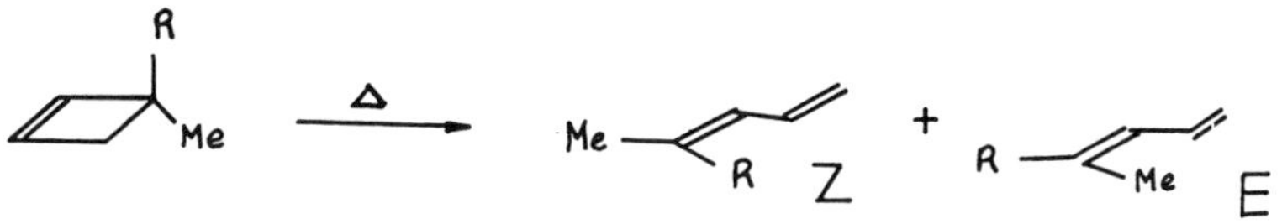

| | Ratio of products | |
Substituent	Z	E
Et	68	32
C_2D_5	61	39
n-Pr	62	38
isopr	65.5	34.5
cyclopropyl	43	57
t-Butyl	32	68
Phenyl	30	70
4-Methoxyphenyl	52	48
3-Methoxyphenyl	32	68[a]
4-Cyanophenyl	45	55[a]

[a] at 161°C

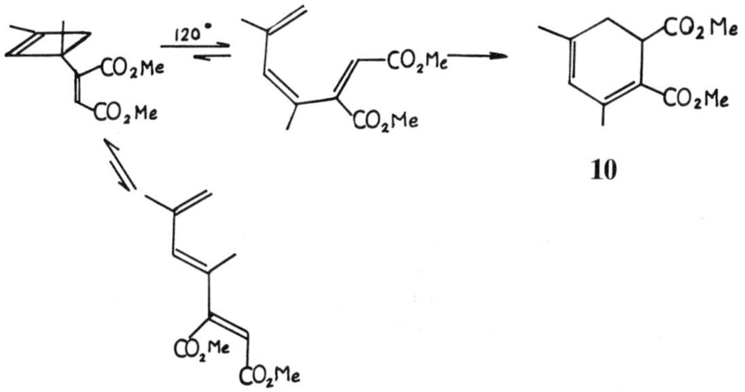

however, did not require preferential kinetically controlled formation of this diene. In all likelihood, the E-diene is in equilibrium with cyclobutene **9** and the Z-diene, which converts irreversibly to the more stable cyclohexadiene **10**. Cyclobutenes trisubstituted at the 3,3,4-positions also have been shown to undergo a conrotatory ring-opening dominated by outward rotation of two (out of three) substituents.[25] For example, the 3-bromo-1,2,3,4-tetraphenyl-cyclobutene **11** in the electrocyclic ring opening gives as the only product **12** with the isomeric **13** not being found.

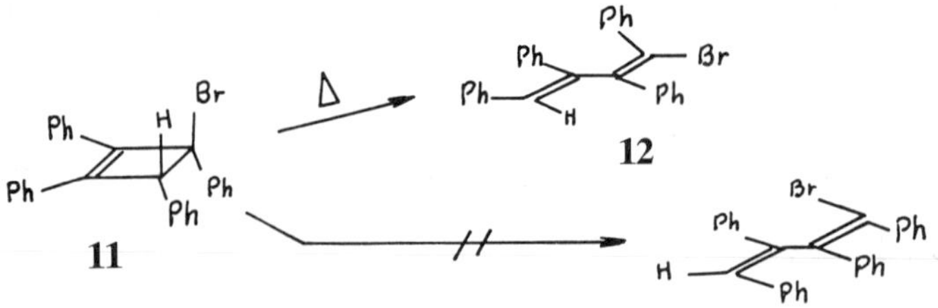

Other such reported examples of the type (ie, **14** and **15**)[26,27,28] also follow this expected pattern in the formation of their respective dienes.

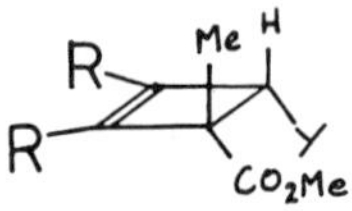

14 R = CH$_3$ Y = COPh
15 R = Ph   Y = CO$_2$CH$_3$

Perhaps the most interesting example was the finding that *cis*-4-chlorocyclobut-2-enecarboxylic acid **(16)** underwent facile thermal rearrangement exclusively to *Z,E*-5-chloropenta-2,4-dienoic acid **(17)**[29]

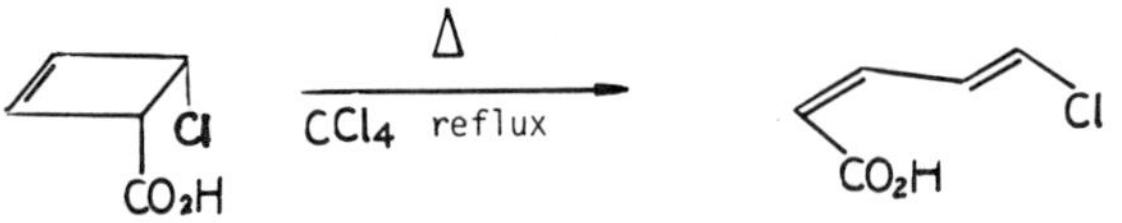

wherein the chlorine has rotated specifically outwards with the carboxyl group rotating inward.

As can be seen from the preceding examples, the presence of virtually any individual substituent at the 3 or 4 position of cyclobutene appears to lower the activation barrier of ring opening. However, 3,3- or *cis*-3,4- identically disubstituted cyclobutenes notably do *not* give rise to rate enhancement. There also seems to be a clear indication that rather than steric effects being the key to determining the mode of conrotatory ring opening, electronic effects dominate. At this point it seemed clear that electron–donor substituents such as alkyl and halo groups had a propensity for outward rotation to a much greater extent than did electron acceptors such as carbonyl functions.

Prior to 1980 there was little attention paid to this fact by theoreticians. Jensen and Kunz, using MNDO Synchronous Transit MINIMAX calculations, estimated the activation energy for the cyclobutene ring opening of 3,4-diamino, diammonio, and amino–ammonio derivatives[30] and compared them with the data calculated for cyclobutene itself.[31]

Since such derivatives are not known, these results could only show the predicted trend for such substituents. As one can see in Table 4-3, each combination of substituents gave rise to a predicted activation energy lower-

TABLE 4-3. Ring Opening of 3,4-Disubstituted Cyclobutenes

X	Y	$E_a{}^{calc}$
H	H	49.9
NH_2	NH_2	34.7
$NH_3{}^+$	$NH_3{}^+$	29.8
NH_2	NH_3	26.0

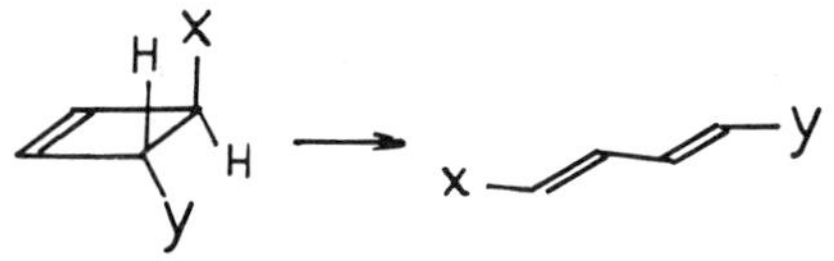

ing for formation of the respective 1,3-dienes. Formation of Z,Z dienes was apparently not considered in this work and as is often the case with such calculations, it is the predicted relative trend in values that is of value. The absolute value for the known case, X = Y = H, (32.5 kcal/mol) is not close to that calculated.

2. PERFLUOROCYCLOBUTENE SYSTEMS

Electrocyclic ring openings of perfluoroderivatives of cyclobutene have long been known to exhibit unusual behavior largely based upon the known propensity of the equilibrium in such cases to be more disposed to lie on the cyclobutene side.[32] Such behavior is generally attributed to a dramatic effect of fluorine's high electronegativity on the relative thermodynamic stability of sp^3-bound fluorine versus sp^2-bound fluorine, and is seen from the limited data in Table 4-4, to be strongly dependent upon both geminal and vicinal fluorine–fluorine and fluorine–other group interactions. (Note that a *negative* $\Delta H°$ indicates the cyclobutene to be less stable then the butadiene.)

TABLE 4-4. Thermodynamic Data on Fluorine-Substituted Cyclobutene-Butadiene Equilibria

Cyclobutene substituents							
1	*2*	*3*	*4*	$E_a{}^a$	$\Delta H°^a$	ΔS^b	*Ref.*
—	—	—	—	32.5	−8	4.5	6
F	F	F_2	F_2	47.1	11.7		32
F	—	—	F_2	—	2.5		33
—	—	F_2	F_2	47.9	−2.4^c		34, 35
CH_3	—	F_2	F_2	—	3.5^c		35
C_2H_5	—	F_2	F_2	—	4.4^c		35
CF_3	CF_3	F_2	F_2	46.0	0.4	4.9	36
CF_3	—	—	F_2	—	1.0		33

^{a} kcal/mol;
^{b} cal/deg;
^{c} calculated from $\Delta G°$ (−5.1, 0.8, and 1.7, resp.), assuming $\Delta S = 4.7$ cal/deg at 300°C.

There is clearly not enough data to define interactive parameters definitively. For example, it has not yet been demonstrated whether the obvious dramatic effect of substituents at the 1-position on cyclobutene-butadiene equilibria of 4,4-difluorocyclobutenes derives from an effect on the cyclobutene or on the butadiene.

Perfluoro-2,4-hexadiene System

While the earlier-mentioned work[23] reported by Curry and Stevens on 3,3-disubstituted cyclobutenes certainly pointed up the shortcomings of a steric rationale for substituent effects on cyclobutene ring openings, it wasn't until the truly dramatic effects exhibited by the perfluoro-3,4-dimethylcyclobutene system were reported by Dolbier, Burton, et al. that serious efforts to understand these substituent effects were undertaken.

Perfluoro *cis*-3,4-dimethylcyclobutene (18) as expected undergoes conrotatory thermal ring opening to yield stereospecifically perfluoro-E,Z-2,4-hexadiene (19),[37,38] with relative normal activation parameters.

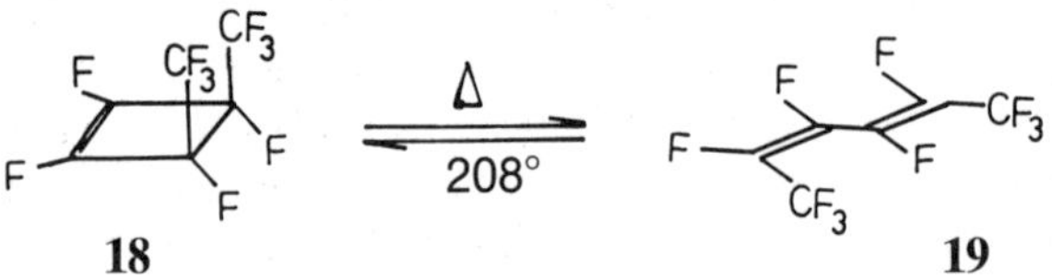

On the other hand, quite unexpectedly, perfluoro-*trans*-3,4-dimethylcyclobutene(20) undergoes thermal electrocyclic ring opening with significantly diminished activation energy requirements to form specifically the Z,Z-diene 21. Because of the reversible nature of the 20 ⇌ 21 interconversion, it was possible by raising the temperature to 250°C to eventually observe the formation of the E,E-isomer 22. Activation parameters for these two competing orbital-symmetry-allowed processes are given along with those for the 18 ⇌ 19 interconversion in Table 4-5. Thermodynamic parameters for these interconversions are given in Table 4-6.

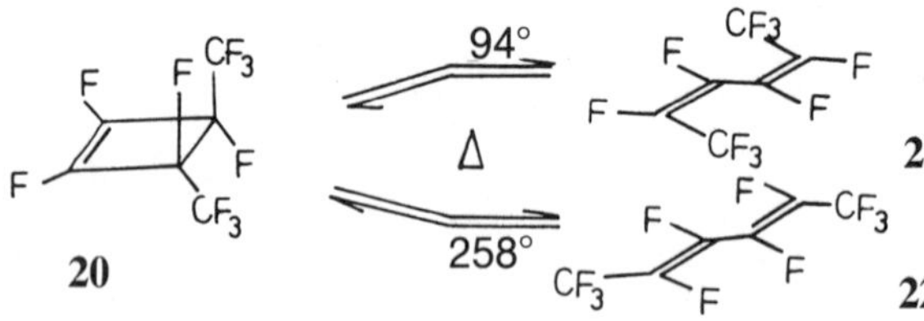

What can be seen from Table 4-5 is a dramatic, kinetic preference for conversion of cyclobutene 20 to the Z,Z-hexadiene 21 rather than the E,E-hexadiene 22. This kinetic preference is reflected by a 21.5 kcal/mol difference in activation energies and a 4.7×10^9 relative rate factor at 100°C. It can be seen from Table 4-6 that there is no significant thermodynamic difference

TABLE 4-5.   Activation Parameters for the Perfluoro-2,4-hexadiene System

	Log A	E_a^e	$\Delta H^{\ddagger a}$	$\Delta S^{\ddagger b}$	$\Delta G^{\ddagger a}$	Mean temp, °C
k_{EE}	15.4 ± 0.4	50.6 ± 0.9	49.5	7.0	45.8	257.8
k_{-EE}	12.5 ± 0.4	43.6 ± 0.9	42.5	−4.5	44.9	257.8
k_{ZZ}	12.1 ± 0.3	29.1 ± 0.5	28.3	−5.4	30.3	93.8
k_{-ZZ}	10.4 ± 0.3	23.5 ± 0.5	22.8	−13.2	27.6	93.8
k_{EZ}	13.9 ± 0.3	38.1 ± 0.7	37.1	2.0	36.2	207.8
k_{-EZ}	11.1 + 0.3	32.7 + 0.7	31.8	−10.5	36.8	207.8

[a] in kcal/mol;
[b] in cal/deg.

between dienes **21** and **22,** and that therefore the observed kinetic difference must be a pure substituent effect on the relative transition state stability for the two competing conrotatory processes.

An additional surprise was finding that the Z,Z-diene(**21**) was actually slightly more stable thermodynamically than its E,E-counterpart **22.** However, since it is not expected that either the E,Z- or Z,Z-dienes (**19** and **21**) are planar, such steric interactions as would be expected to destabilize **19** and **21** relative to **22** may not be significant. The observed λ max's in the UV spectra of **22, 19,** and **21,** 219 nm, 212 nm, and 207 nm, respectively, are indicative of an increase in nonplanarity in going from E,E to E,Z to the Z,Z diene structure. Theoretical calculations also support nonplanarity in these systems.[39]

Perfluoro-3,5-octadiene System

A study of the perfluoro-3,5-octadiene system provided insight as to what effect, if any, an increase in steric bulk of the perfluoroalkyl substituents would have upon the kinetic and thermodynamic profile of the equilibria.

Kinetic data was obtained only for cyclization of the Z,Z- and E,E- dienes, ie, **24** and **25** ⇌ **23,** while equilibrium data was obtained for all four GC-visible components. Perfluoro-cis-3,4-diethylcyclobutene could not be ob-

TABLE 4-6.   Thermodynamic Parameters for the
Perfluoro-2,4-hexadiene System[a]

	20	**18**	**21**	**22**	**19**
$\Delta H^{\circ b}$	0	1.4 ± 0.1	5.6 ± 0.2	6.8 ± 0.3	7.1 ± 0.4
ΔS^c	0	−1.3 ± 0.1	7.8 ± 0.1	11.2 ± 0.1	9.9 ± 0.1

[a] the thermodynamic parameters are given relative to **20**;
[b] in kcal/mol;
[c] in cal/deg.

TABLE 4-7. Activation Parameters for the Perfluoro-3,5-octadiene System

	Log A	$E_a{}^a$	$\Delta H^{\ddagger a}$	$\Delta S^{\ddagger b}$	$\Delta G^{\ddagger a}$	Mean temp, °C
k_{ZZ}	11.9 ± 0.5	27.0 ± 0.8	26.3	−6.3	28.6	92.4
k_{-ZZ}	10.1 ± 0.5	23.4 ± 0.8	22.7	−14.5	28.0	92.4
k_{EE}	13.6 ± 0.5	44.8 ± 1.4	43.6	0.6	43.4	308.5
k_{-EE}	11.9 ± 0.5	42.0 ± 1.4	40.8	−7.2	45.0	308.5

^{a} in kcal/mol;
^{b} in cal/deg.

served in equilibrium with **23, 24, 25, 26**. As Tables 4-7 and 4-8 indicate, there was no dramatic effect from changing the trifluoromethyl groups to pentafluoroethyl groups, other than those changes due to the increase in steric bulk.

One can see from Table 4-8 that, somewhat surprisingly, all three of the bis(pentafluoroethyl) dienes are more stable relative to the *trans*-cyclobutene(**23**) than were their bis(trifluoromethyl) counterparts. Apparently the vicinal C_2F_5/F interactions of **23** are affected more deleteriously than are the cis interactions of C_2F_5 in the dienes **24** and **26**. Note that in the perfluoro-3,4-octadiene system, the *E,E*-diene **25** is slightly more stable than the *Z,Z*-diene **24** while in the perfluoro-2,4-hexadiene system the opposite was true, a fact consistent with a relative lack of steric effect of CF_3 in the hexadiene system combined with the emergence of a steric effect for C_2F_5 in the octadiene system.

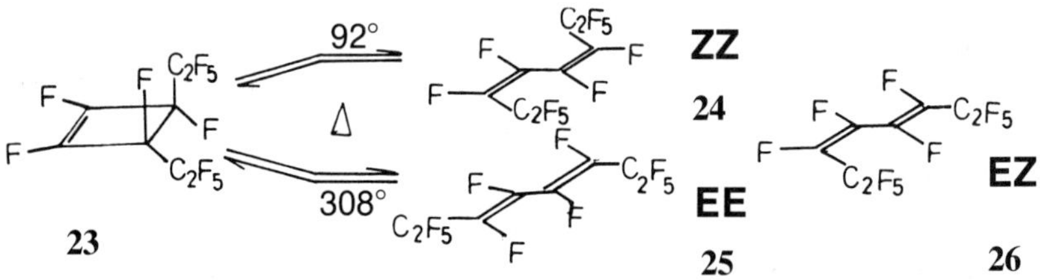

The kinetic picture for the 3,4-octadiene system, as seen in Table 4-7, shows that while the ring-opening E_a's and $\Delta G^{\ddagger}$'s are lowered relative to the

TABLE 4-8. Thermodynamic Parameters for the Perfluoro-3,5-octadiene System^a

	23	26	24	25
$\Delta H^{o b}$	0	3.4 ± 0.2	3.6 ± 0.3	2.8 ± 0.1
ΔS^c	0	9.8 ± 0.1	8.2 ± 0.1	7.7 ± 0.1

^{a} the thermodynamic parameters are all given relative to **23**;
^{b} in kcal/mol;
^{c} in cal/deg.

2,4-hexadiene system (ie, k_{ZZ} and k_{EE}), those for cyclization (k_{-ZZ} and k_{-EE}) are not much different. This is, as noted above, probably a reflection of the increase in stability of dienes **24** and **25** relative to cyclobutene **23.** The cyclization processes appear to be independent of whether the R_f substituent is CF_3 or C_2F_5.

While the interconversion of E,Z-3,5-octadiene, **26,** and *cis*-3,4-diethyl-cyclobutene could not be examined because the latter was so unstable as to be unobservable in equilibrium with **26,** leakage between the E,E/Z,Z system and E,Z system could be measured. A log A of 13.1 ± 0.6 and an E_a of 46.2 ± 1.7 kcal/mol were attributable to this process, whose value was similar to that observed for leakage in the 2,4-hexadiene system.

Perfluoro-1,3-pentadiene System

In a test to determine the kinetic effect of a single $CFCF_3$ site on butadiene cyclization rates, the Z- and E-perfluoro-1,3-pentadienes, **28** and **29,** were examined as to their rates of electrocyclic cyclization to perfluoro-3-methyl-cyclobutene, **27.**[38,40] Indeed it was found that as in the 2,4-hexadiene and 3,5-octadiene systems, the Z-diene underwent cyclization much more readily than the (E)-diene. Table 4-9 provides the activation parameters and Table 4-10 the thermodynamic parameters for these thermal, gas-phase interconversions.

As can be seen from Table 4-10, the equilibrium lies far toward the cyclo-butene product **27** in this system, with only 0.5% of **28** and 0.3% of **29** being present in equilibrium at 200°C. Consistent with the 2,4-hexadiene results, the (Z)-diene **28** proved to be 2.2 kcal/mol more stable than the (E)-diene **29.** While the CF_3 group is considered to be quite sizable (ie, it has an A value

TABLE 4-9. Activation Parameters for the Perfluoro-1,3-pentadiene System

	Log A	$E_a{}^a$	$\Delta H^{\ddagger a}$	$\Delta S^{\ddagger b}$	$\Delta G^{\ddagger a}$	Mean temp, °C
k_Z	12.6 ± 0.4	36.7 ± 0.7	35.9	−3.5	37.4	157.7
k_{-Z}	11.2 ± 0.4	28.5 ± 0.7	27.7	−10.1	32.1	157.7
k_E	14.7 ± 0.1	49.6 ± 0.3	48.5	5.5	45.5	268.3
k_{-E}	12.3 ± 0.1	39.2 ± 0.3	38.1	−5.3	41.0	268.3

^{a} in kcal/mol;
^{b} in cal/deg.

TABLE 4-10. Thermodynamic Parameters
for the Perfluoro-1,3-pentadiene System[a]

	27	28	29
$\Delta H^{\circ b}$	0	8.2 ± 0.5	10.4 ± 0.5
ΔS^c	0	6.6 ± 0.2	10.9 ± 0.2

[a] the thermodynamic parameters are relative to **27**;
[b] in kcal/mol;
[c] in cal/deg.

(2.4–2.6 kcal/mol)[41] similar to that of an isopropyl group (2.20) or a
trimethylsilyl group (2.50)[42]), apparently its size is not the deciding factor in
the Z/E relative stability. Theoretical work on conformational effects in
perfluorobutadiene would seem to indicate that planarity of a perfluorinated
diene is not an important stabilizing factor.[39] (The minimum energy structure
for perfluorobutadiene was a skewed-cis structure with $\theta = 58.4°$, which is
1.8 kcal/mol lower than the s-trans and 5.7 kcal/mol lower than the s-cis
structure. This compares with the hydrocarbon in which the skewed-cis
structure is calculated to be 3.15 kcal above the s-trans and in which the s-cis
barrier is only at 3.5 kcal/mol.)[43] Hence, for **29**, conformations with the two
double bonds orthogonal to each other, such as **30,** could well be involved
such that the cis F/CF_3 interaction of the E isomer might be more destabilizing than the alternate CF_2═CF/CF_3 interaction of the Z isomer. At present
there is no thermodynamic data available which give specific insight into Z
and E CF_3/F interactions.

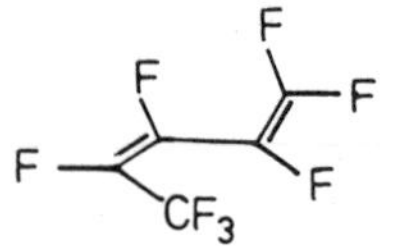

30

Kinetically the preference for outward rotation of F along with inward
rotation of CF_3 in the ring opening of **27** was reflected by a ΔE_a of 12.9 kcal/
mol. This compares with the 21.5 kcal/mol ΔE_a observed for the bis(trifluoromethyl) system **20.**

3. OTHER SUBSTITUENTS

Concomitant with the publication of these results on perfluoro-cyclobu-
tene systems was the publication by Rondan, Houk, and Kirmse[11] of related
kinetic results on the kinetic effect of other substituents on the cyclobutene–

TABLE 4-11. The Kinetic Effect of Other Substituents upon the Ring-Opening of Cyclobutene

Substituents	Log A	$E_a{}^a$	Product stereochemistry	Ref.
None	13.1	32.5	—	6
1-Cl	13.3	33.6	—	44
1-Br	13.5	33.8	—	44
1-CN	13.4	32.8	—	45
3-Cl	13.1	29.4	E	11
cis-3,4-diCl	13.9	35.6	E,Z	11
trans-3,4-diCl	13.0	25.7	E,E	11
3-OEt	12.7	23.5	E	11
3-OAc	12.6	27.8	E	11
cis-3,4-diOMe	13.25	28.6	E,Z	11
cis-3,4-diOEt	12.7	23.5	E,Z	11
cis-3-Cl-4-Me	12.7	31.6	E,Z	11
cis-3-Cl-4-OCH₃	12.9	29.1	Z, E	11
cis-3-OCH₃-4-Me	11.7	25.2	E,Z	11
3,3-diOMe	13.7	32.8	—	46
3-CH₂CH₂OAc	—	29.8	E	47
3-CHO	14.2	27.2	Z	48
3-CF₃	14.3	36.3	95% E	49

^{a} kcal/mol

butadiene interconversion, especially with respect to the relative importance of the two competitive conrotatory processes (Table 4-11). They reported the data for methyl, chloro, and alkoxy groups shown in Table 4-11, which showed that virtually all of these substituents exhibited a kinetic preference for outward rotation relative to inward rotation.[11] These authors hypothesized that indeed all donor substituents should exhibit preferential outward rotation, with substituents having lone–pairs such as alkoxy, and halogen showing a greater effect than alkyl groups.

4. THEORY

Rondan and Houk hypothesized that outward rotation of such groups gave rise to a rate-enhancing bonding interaction with the antibond of the breaking cyclobutene σ bond, while inward rotation would lead to a rate-diminishing anti-bonding interaction of the substituent with the bonding orbital of that same σ bond. (See Figure 4-1).[50] Fluorine and hydroxy substituents were proposed to have the most dramatic effect in this system. Table 4-12 provides a summary of Houk's and Rondan's predictions.

While the calculations carried out by Rondan and Houk clearly showed that electron-donating substituents at the 3 and/or 4 positions of cyclobutene

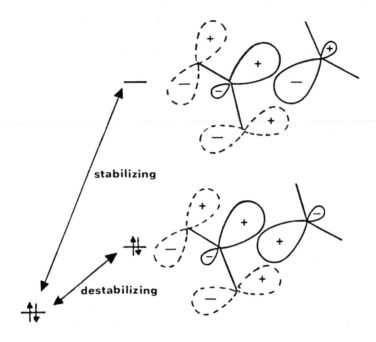

Figure 4-1. The origin of stereoselectivity of the conrotatory electrocyclic reactions of substituted cyclobutenes.[50]

will prefer to rotate outward, the situation was less clear in the case of electron-withdrawing substituents. Parenthetically it should be noted that carboxyl functions have been shown to exhibit little reluctance to rotating inward. (Vide supra for rearrangements of **14, 15,** and **16.**)[26–29] It was suggested that a very strong electron acceptor such as the BH_2 group would rotate inward. (In this calculation the empty p orbital on boron was aligned with the σC_1C_4 orbital.) On the other hand, 3-cyanocyclobutene should

TABLE 4-12. 3-21G Relative Energies of Model Transition States in Ring-Openings of Substituted Cyclobutenes[50]

Substituents	Calculated relative E_a's outward vs inward
trans-3,4-dimethyl	−13
trans-3,4-dihydroxy	−31.6
3-fluoro	−13
trans-3,4-difluoro	−29
trans-3,4-dichloro	−22
trans-3,4-dicyano	−9
trans-3,4-diboryl	+13

preferentially rotate outward by a ΔE_a of 4.3 kcal/mol. CF_3, not calculated, might be expected to show electron-accepting properties and indeed, 3-trifluoromethylcyclobutene(**31**) ring-opened to produce a 95 : 5 ratio of E : Z products (**32** and **33**) at 160°C. Considering that the smaller methyl substituent gave rise to ring-opening without a trace of Z-diene product being formed, the 5% of Z-diene formed from **31** is a significant indication that substituent electron-donor power, not substituent size is the dominant factor in determining the mode of conrotatory motion.

Interestingly, a formyl group (CHO) was predicted to exhibit a 4.5 kcal preference for *inward* rotation, and an E_a 6.9 kcal/lower than that calculated for parent cyclobutene.[48] Experimentally it was confirmed that Z-2,4-pentadienal(**35**) was the only observed product in the electrocyclic ring opening of 3-formyl cyclobutene(**34**), even though it is thermodynamically less stable than its E-counterpart:

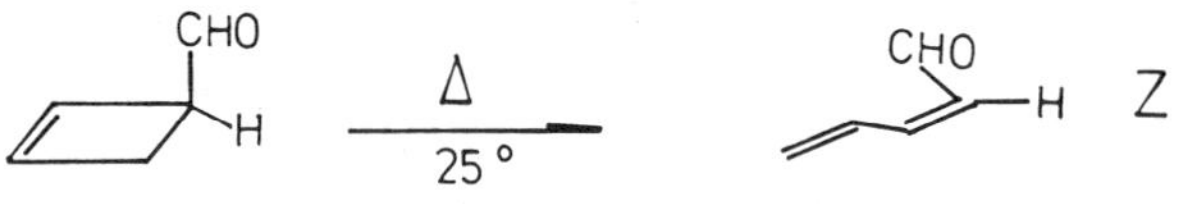

Determined Arrhenius parameters [log A = 14.2 (±1.2), E_a = 27.2 (±1.8) kcal/mol] are in good agreement with their theoretical predictions.[48]

5. CONCLUSION

While work is really just beginning in the area of probing these remarkable substituent effects on the electrocyclic ring-opening of cyclobutene, within the last few years the reality of such effects has been dramatically demonstrated. It is likely that related effects will be observed in other pericyclic reactions, particularly in other electrocyclic processes involving small rings. We have already demonstrated a lack of a significant fluorine-substituent kinetic effect in the Cope rearrangement and the homo-1,5-hydrogen shift rearrangement,[51] but current work indicates that systems in which the rate-determining transition states are significantly far-removed in structure from those of the products are systems where such effects can and will be observed.

REFERENCES

1. Woodward, R. B.; Hoffmann, R. "The Conservation of Orbital Symmetry". Academic Press: New York, 1970.
2. Woodward, R. B.; Hoffmann, R. *J. Am. Chem. Soc.* **1965**, *87*, 395.
3. Criegee, R.; Noll, K. *Liebigs Ann. Chem.* **1959**, *1*, 627.
4. Criegee, R.; Seebach, D.; Winter, R. E.; Borretzen, B.; Burne, H.-A. *Chem. Ber.* **1965**, *98*, 2339.
5. Branton, G. R.; Frey, H M ; Skinner, R. F. *Trans. Faraday Soc.* **1966**, *62*, 1546.
6. Cooper, W.; Walters, W. D. *J. Am. Chem. Soc.* **1958**, *80*, 4220.
7. Frey, H. M. *Trans. Faraday Soc.* **1964**, *60*, 83.
8. Frey, H. M.; Skinner, R. F. *Trans. Faraday Soc.* **1965**, *61*, 1918.
9. Pomerantz, M.; Hartmann, P. H. *Tetrahedron Lett.* **1968**, 991.
10. Srinivasan, R. *J. Am. Chem. Soc.* **1969**, *91*, 7557.
11. Kirmse, W.; Rondan, N. G.; Houk, K. N. *J. Am. Chem. Soc.* **1984**, *106*, 7989.
12. Brauman, J. I.; Archie, W. C., Jr. *Tetrahedron* **1971**, *27*, 1275.
13. Frey, H. M. *Trans. Faraday Soc.* **1962**, *58*, 957.
14. Dickens, D. C.; Frey, H. M.; Skinner, R. F. *Trans. Faraday Soc.* **1969**, *65*, 453.
15. Frey, H. M. *Trans. Faraday Soc.* **1963**, *59*, 1619.
16. Frey, H. M.; Marshall, D. C.; Skinner, R. F. *Trans. Faraday Soc.* **1965**, *61*, 861.
17. Frey, H. M.; Pope, B. M.; Skinner, R. F. *Trans. Faraday Soc.* **1967**, *63*, 1166.
18. Frey, H. M.; Metcalf, J.; Pope, B. M. *Trans. Faraday Soc.* **1971**, *67*, 750.
19. Frey, H. M.; Solly, R. K. *Trans. Faraday Soc.* **1969**, *65*, 448.
20. Battiste, M. A.; Burns, M. E. *Tetrahedron Lett.* **1966**, 523.
21. Freedman, H. H.; Doorakian, G. A.; Sandel, V. R. *J. Am. Chem. Soc.* **1965**, *87*, 3019.
22. Doorakian, C. A.; Freedman, H. M. *J. Am. Chem. Soc.* **1968**, *90*, 5310.
23. Curry, M. J.; Stevens, I. D. R. *J. Chem. Soc. Perkin Trans 2* **1980**, 1391.
24. Pomerantz, M.; Wilke, R. N.; Gruber, G. W.; Roy, U. *J. Am. Chem. Soc.* **1972**, *94*, 2752.
25. Doorakian, G. L.; Freedman, H. M. *J. Am. Chem. Soc.* **1968**, *90*, 3582.
26. Maier, G.; Wiessler, M. *Tetrahedron Lett.* **1975**, 4987.
27. Mundnick, R.; Plieninger, H. *Tetrahedron* **1978**, *34*, 887.
28. Semmelhack, M. F.; De Franco, R. J. *J. Am. Chem. Soc.* **1972**, *94*, 2116.
29. Pirkle, W. H.; McKendry, C. H. *J. Am. Chem. Soc.* **1969**, *91*, 1179.
30. Jensen, A.; Kunz, H. *Theoret. Chim. Acta (Berl.)* **1984**, *65*, 33.
31. Jensen, A. *Theoret. Chim. Acta (Berl.)* **1983**, *63*, 269.
32. Schlag, E. W.; Peatman, W. B. *J. Am. Chem. Soc.* **1964**, *86*, 1676.
33. Dolbier, W. R., Jr.; Medinger, K. S. Unpublished results.
34. Frey, H. M.; Hopkins, R. G.; Vinall, I. C. *J. Chem. Soc. Faraday I* **1972**, *68*, 1874.
35. Dolbier, W. R., Jr.; Al-Fekri, D. M. *Tetrahedron Lett.* **1983**, *24*, 4047.
36. Chesick, J. P. *J. Am. Chem. Soc.* **1966**, *88*, 4800.
37. Dolbier, W. R., Jr.; Koroniak, H.; Burton, D. J.; Bailey, A. R.; Shaw, G. S.; Hansen, S. W. *J. Am. Chem. Soc.* **1984**, *106*, 1871.
38. Dolbier, W. R., Jr.; Koroniak, H.; Burton, D. J.; Heinze, P. L.; Bailey, A. R.; Shaw, G. S.; Hansen, S. W. *J. Am. Chem. Soc.* **1987**, *109*, 219.
39. Dixon, D. A. *J. Phys. Chem.* **1986**, *90*, 2038.
40. Dolbier, W. R., Jr.; Koroniak, H.; Burton, D. J.; Heinze, P. L. *Tetrahedron Lett.* **1986**, *27*, 4387.
41. Della, E. W. *J. Am. Chem. Soc.* **1967**, *89*, 5221.
42. Kitching, W.; Olszowy, H. A.; Drew, G. M.; Adcock, W. *J. Org. Chem.* **1982**, *47*, 5153.
43. Bock, C. W.; George, P.; Trachtman, M. *Theoret. Chem. Acta* **1984**, *64*, 293.
44. Dickens, D.; Frey, H. M.; Mercalf, J. *Trans. Faraday Soc.* **1971**, *67*, 2328.
45. Sarner, S. F.; Gale, D. M.; Hall, H. K., Jr.; Richmond, A. B. *J. Phys. Chem.* **1972**, *76*, 2817.

46. Kirmse, W. Unpublished result cited in Moss, R. J.; White, R. O.; Rickborn, B. *J. Org. Chem.* **1985,** *50,* 5132.
47. Mcdonald, R. N.; Lyznicki, E. P., Jr. *J. Am. Chem. Soc.* **1971,** *93,* 5920.
48. Rudolf, K.; Spellmeyer, D. C.; Houk, K. N. *J. Org. Chem.* **1987,** *52,* 3708.
49. Dolbier, W. R., Jr.; Gray, T. Unpublished results.
50. Rondan, N. G.; Houk, K. N. *J. Am. Chem. Soc.* **1985,** *107,* 2099.
51. Dolbier, W. R., Jr.; Alty, A. C.; Phanstiel, O., III. *J. Am. Chem. Soc.* **1987,** *109,* 3046.

CHAPTER 5

Secondary Orbital Effects in Fluorine Compounds

Carl L. Bumgardner and Myung-Hwan Whangbo

Department of Chemistry, North Carolina State University, Raleigh, North Carolina

CONTENTS

1. INTRODUCTION

A useful approach to structural and conformational problems in organic chemistry involves analysis of the relative overlap between fragment orbitals in different structural arrangements.[1] This overlap is either primary or secondary. The primary overlap refers to one between the two atomic centers of a bond across which molecular fragments are considered to interact, and all other overlap is considered secondary.[1] In most cases, it is straightforward to assign the highest occupied molecular orbital (HOMO) and the lowest unoccupied molecular orbital (LUMO) of a molecular fragment. The HOMO–HOMO and the HOMO–LUMO interactions are destabilizing and stabilizing, respectively, and both interactions increase with overlap. Thus the optimum arrangement of the molecular fragments is the one that leads to the minimum HOMO–HOMO overlap and the maximum HOMO–LUMO overlap. This method of evaluating the relative stability of structures is particularly well suited to compounds containing fluorine, a first row π-donor that is at the same time the most electronegative element. This interplay of π-donation and σ-withdrawal makes the behavior of fluorine-containing compounds sometimes puzzling but always interesting and challenging. In this chapter we focus on a number of enigmatic experimental observations involving fluorine compounds and present rationales which are based, at least in part, on secondary orbital considerations. Cases where analogous effects may be operating are also explored.

2. THE ROLE OF THE CF$_3$ GROUP IN DETERMINING RELATIVE STABILITY OF E AND Z ISOMERS

The nucleophilic addition of PhOH and PhSH to trifluoromethyl acetylenes, L—C≡C—CF$_3$ (L=Ph, PhCO, etc), generalized in Scheme I, shows a striking effect of the CF$_3$ group on the stereochemistry of the reaction.[2,3] As shown in Scheme II, addition of PhSH to phenyl trifluoromethyl acetylene (**1**) in ethanol containing a catalytic amount of t-BuO⁻ gives **2-Z** as the sole kinetic product, but **2-E** as the major thermodynamic product. In contrast, addition of PhOH to **1** under similar conditions yields **3-Z** under both kinetically and thermodynamically controlled reaction conditions. Noteworthy is the fact that in **3-Z** the CF$_3$ and PhO groups are cis while in **2-E** the CF$_3$ and PhS groups have the trans arrangement.[2] Summarized in Scheme III are the

PhXH (X=O, S)

t-BuO⁻ + PhXH → t-BuOH + PhX⁻

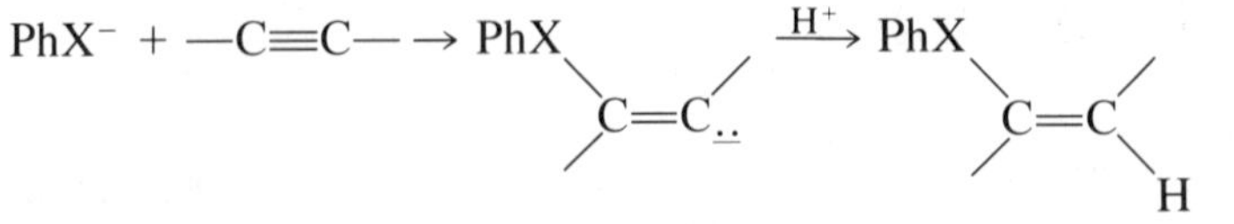

PhX⁻ + —C≡C— → PhX$\diagdown$C=C$\diagup_{..}$ $\xrightarrow{H^+}$ PhX$\diagdown$C=C$\diagup$H

Scheme I

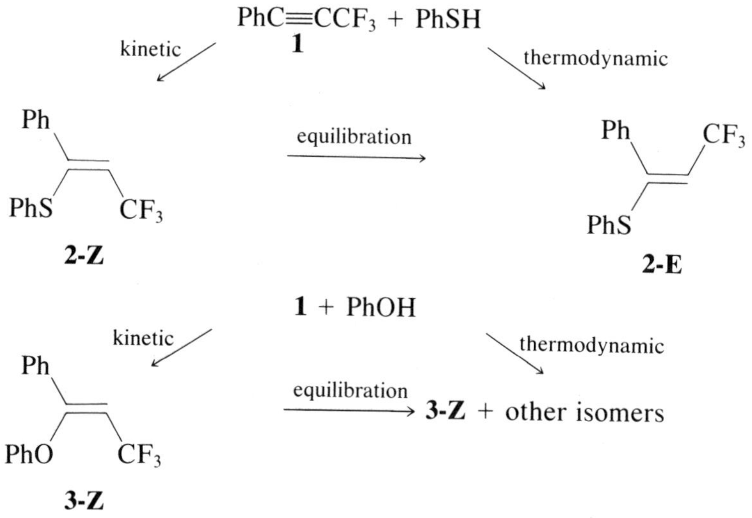

Scheme II

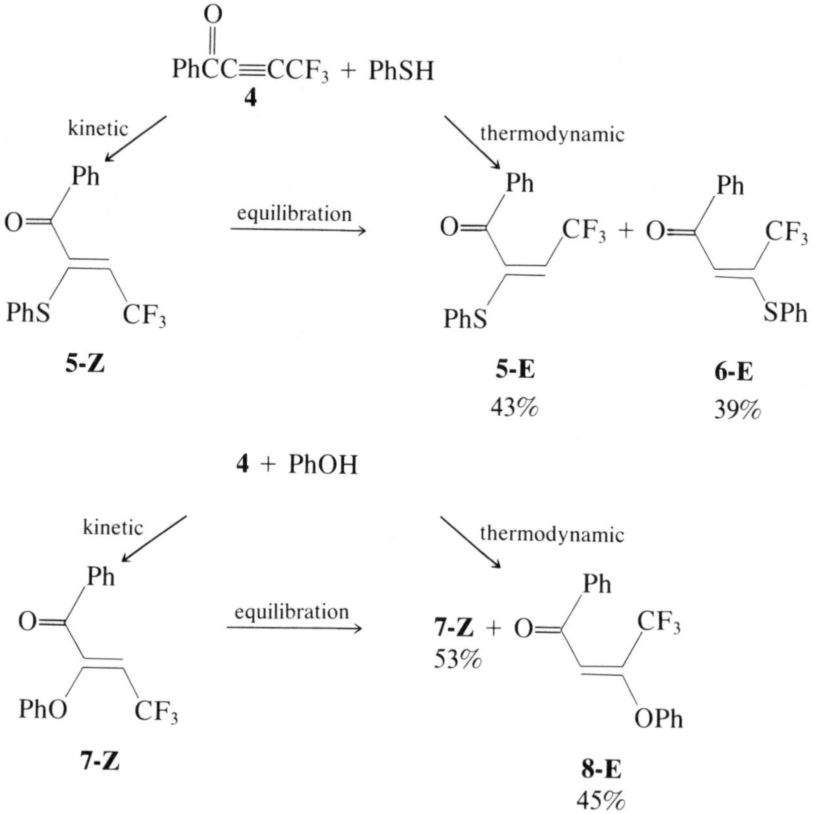

Scheme III

results of the nucleophilic addition of PhOH and PhSH to benzoyl trifluoro-methyl acetylene (4). There is an overwhelming kinetic preference for the anti-Michael adducts,[3] ie, 5-Z and 7-Z for the addition of PhSH and PhOH, respectively. In the thermodynamic products from reaction of 4 with PhSH, both 5-E and 6-E have a cis arrangement of CF_3 and carbonyl groups. In the thermodynamic products of the reaction of 4 with PhOH, however, 7-Z and 8-E have a trans and cis arrangement of CF_3 and carbonyl groups, respectively. Two questions arise from review of these data: (i) why the PhO function and CF_3 groups prefer to have a cis arrangement on a double bond as exemplified by 3-Z and 7-Z, and (ii) why the carbonyl and CF_3 groups favor a cis-arrangement as evidenced by 5-E, 6-E, and 8-E, despite the wide disparity in the steric parameters of these functional groups.[4]

The cis preference of CF_3 and PhO groups on a double bond can be rationalized in terms of the stabilizing π-type orbital interactions between the CF_3 and vinyl ether fragments as depicted in 9. In 9b the vinyl ether moiety is represented by the HOMO π', and the CF_3 group by its LUMO $\pi^*_{CF_3}$. The π'—$\pi^*_{CF_3}$ interaction is expected to be more stabilizing for the cis than the trans arrangement of CF_3 and PhO because of the secondary orbital interaction indicated by the double-headed arrow in 9b. However, as shown in Scheme II, the thermodynamic product derived from the second-row nucleophile PhS$^-$ is 2-E, where the CF_3 and PhS groups are trans. This contrasting tendency between first- and second-row nucleophiles may be ascribed to the long C—S bond length, which makes weak the interaction of the sulfur 3p orbital with the vinyl group. Therefore, the π' orbital of the vinyl thioether moiety will have small contribution from the sulfur 3p orbital, so that the secondary orbital interaction of (π'—$\pi^*_{CF_3}$), indicated by the double-headed arrow, is weak for sulfur and does not override the usual tendency for the trans arrangement of bulky groups on a double bond. To account for the results in Scheme III we note that the interaction of $\pi^*_{CF_3}$ with the HOMO π of an enone framework as shown in 10 should lead to a cis preference for the CF_3 and carbonyl groups. In terms of secondary orbital effects, the exceptional case 7-Z can be accommodated if the (π'—$\pi^*_{CF_3}$) interaction is stronger than the (π—$\pi^*_{CF_3}$) interaction (Cf. 9b and 10b). The secondary orbital argument indicated in 10, where the enone is assumed to be in the s-trans conformation, will also be valid if the enone framework has an s-cis conformation as shown in 11.

The thermodynamic preferences discussed above are reminiscent of the most stable conformations of vinyl ether 12,[5,6] and crotyl anion 13.[7] They

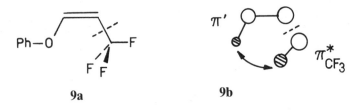

9a   9b

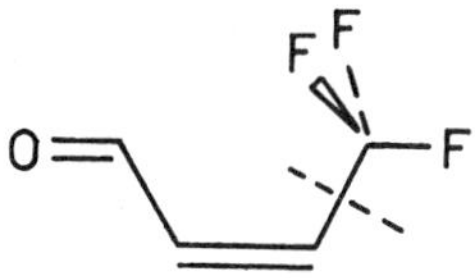

10a 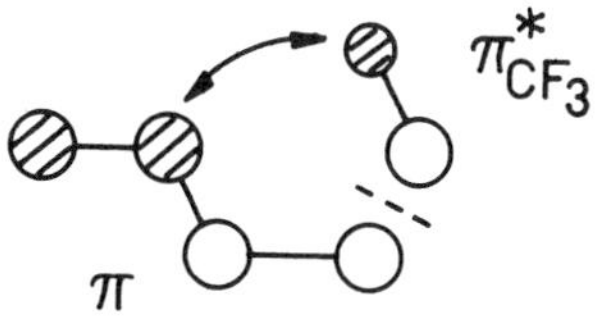   **10b**

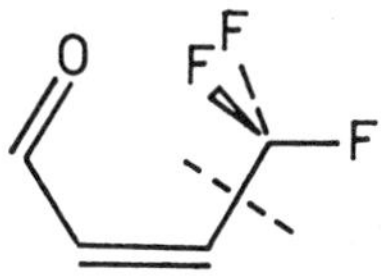

11a 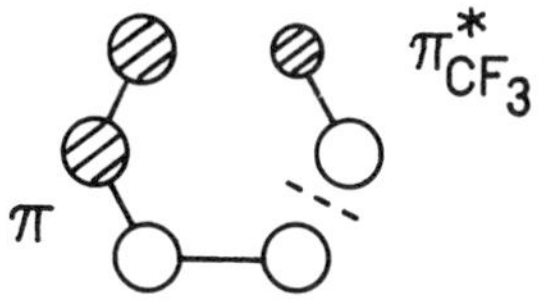   **11b**

adopt the most cyclical arrangement of the π framework, a result rationalized in terms of π-type orbital interactions associated with the CH_3 group in each molecule. Similarly, an equilibrium mixture of 1-methoxy propenes was found to consist mainly of the cis isomer,[8] **(14)**. An example where secondary orbital interactions may play a role in favoring a particular transition state is the observation[9] that application of the vinylogous Ramberg-Bäcklund reaction to **15** gives predominantly the isomer **16**. This "syn" effect appears to be analogous to the crotyl anion case mentioned earlier. What separates these secondary orbital effects with CH_3 from those discussed above involving the CF_3 group is the fact that $\pi^*_{CF_3}$ lies much lower in energy than $\pi^*_{CH_3}$, thereby resulting in a more dramatic stabilization of the cyclic forms containing the CF_3 group.[10]

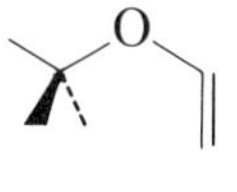

12 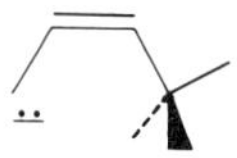   **13** 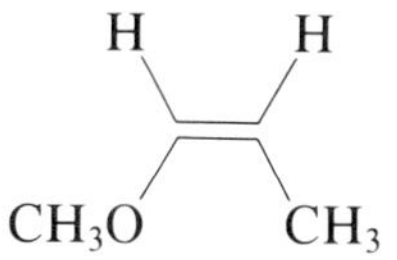   **14**

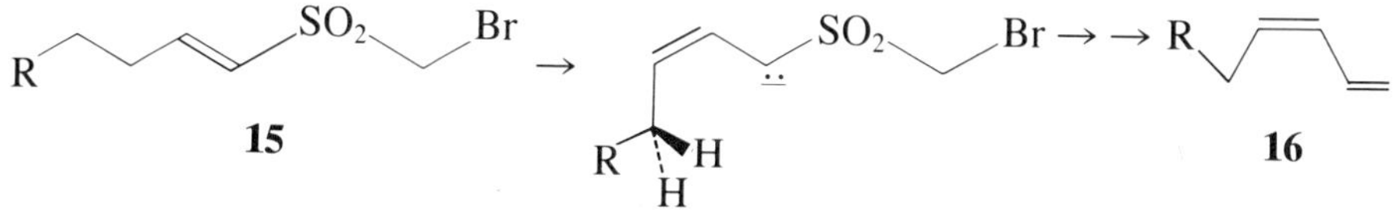

15   **16**

3. THE EFFECTS OF THE FLUORINE ATOM ON KETO-ENOL EQUILIBRIA AND ON ACIDITY OF 1,3-DIKETONES.

While developing a new method for introducing fluorine into the 2-position of 1,3-diketones,[11] we noted a curious dichotomy in the relative keto–enol content of the fluorinated compounds. As shown in Scheme IV, the acyclic compound 17(X − F) prefers the diketo form, 17a (X = F), whereas the cyclic compound 18 (X = F) favors the enolic tautomer 18b(X = F). This difference is underscored by the fact that the hydrogen analogs 17(X = H) and 18(X = H), regardless of whether they are cyclic or acyclic, show a strong bias for the enolic forms 17b (X = H) and 18b (X = H).[12] Thus the exceptional member of this series appears to be 17 (X = F). The questions then arise: does fluorine stabilize the diketone 17a or destabilize the enol 17b? And why does the acyclic series differ from the cyclic one? In addition, the fluorinated diketone 19 (X = F) is a stronger acid by about 2 pKa units than diketone 19 (X = H).[11] A similar observation has been reported for 20.[13] This finding contrasts with the effect of fluorine reported for 21[14,15] and 22[16] where the compounds with X = F are less acidic than those where X = H. On those occasions where fluorine deacidifies, the π-donating ability of F presumably outweighs the σ-withdrawing capacity. The converse would be true in those instances where F is acidifying. Thus the question looms: is there a pattern to these variations which can account for the direction of tilt in the weighing of F π-donation against σ-withdrawal?

The preference of the fluorinated diketone 17a (X = F) over the enolic isomer 17b (X = F) cannot be explained in terms of the relative stability of a fluorine atom attached to an sp^3 vs sp^2 carbon, for $FC(H)=CHCH_3$ is more stable than $CH_2=CHCH_2F$.[17] We note that in the enol 17b (X = F), the

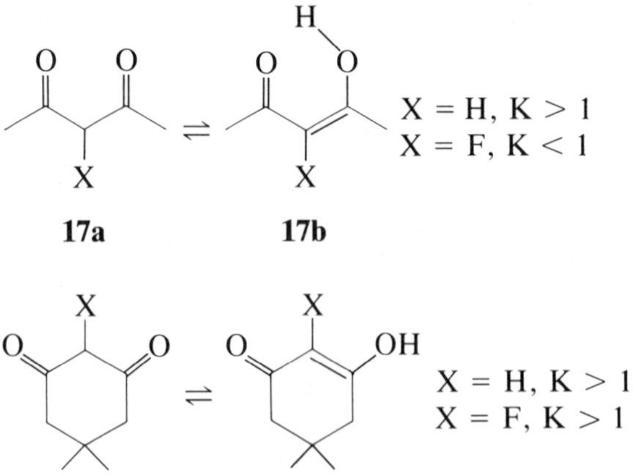

17a   17b   X = H, K > 1
                    X = F, K < 1

18a   18b   X = H, K > 1
                    X = F, K > 1

Scheme IV

fluorine atom is not attached to a simple ethylenic group but to a more extended conjugated framework which is isoconjugate with a pentadienyl anion moiety. To estimate the electronic effects of such an arrangement, we examine the interaction of the pentadienyl anion HOMO with a filled *p*-orbital on F, shown in **23.** This HOMO–HOMO interaction is destabilizing in terms of primary orbital overlap. Because of its node the pentadienyl anion LUMO does not overlap with the filled *p*-orbital of the fluorine atom as shown in **24,** which does not provide any stabilizing effect. Thus, we conclude that F destabilizes **17b** (X = F) with respect to **17a** (X = F) leading to K less than 1 when X = F. If a preferential destabilization of the enol form accounts for the behavior of **17a** (X = F) and **17b** (X = F), then why do the cyclic compounds **18a** (X = F) and **18b** (X = F) follow the norm where the enolic partner is favored? We note in this instance that the shape of the conjugated framework is necessarily altered to accommodate the ring. Combining a filled *p*-orbital from F with the HOMO of this newly shaped pentadienyl anion system still results in a destabilizing primary HOMO–HOMO interaction, but now two secondary orbital interactions indicated by the double headed arrows in **25** decrease the net overlap and thus negate to a large extent the π-destabilization. Since F does not appreciably destabilize the enol in this case, the factors which generally favor the enol[18] prevail, making **18b** (X = F) more stable than **18a** (X = F). Let us now consider our observation that **19** (X = F) is a stronger acid than the nonfluorinated analog. Apparently this fact is contradictory since our explanation for the position of keto–enol equilibria in acyclic 1,3-diketones relied on the fluorine

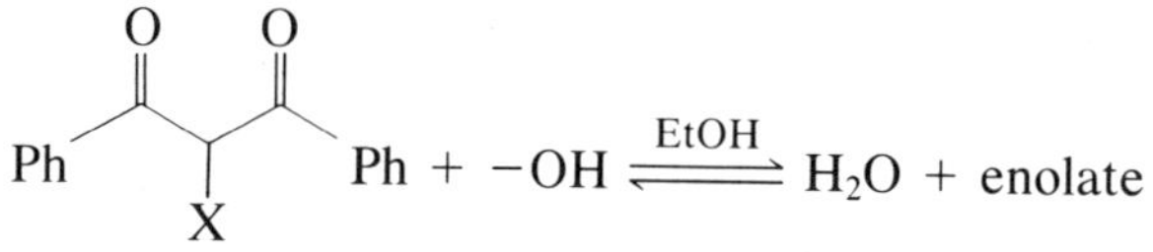

19

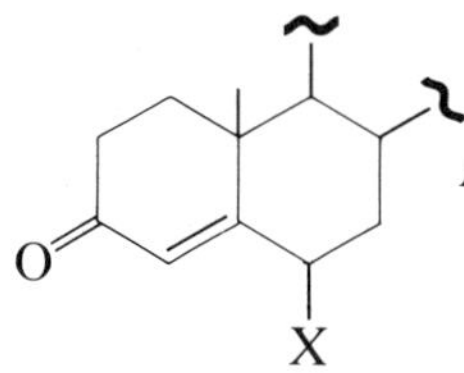

Androst-4-ene-3,17-dione and 6α-fluoro

20

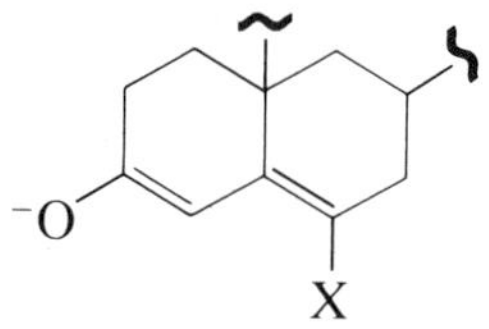

20'

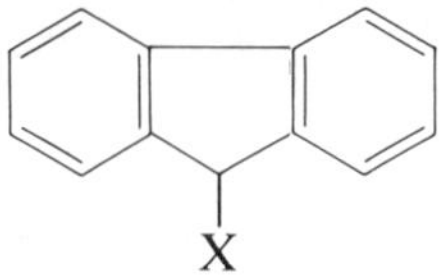

21

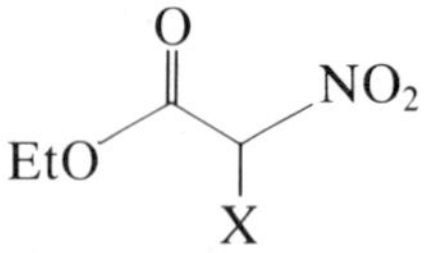

22

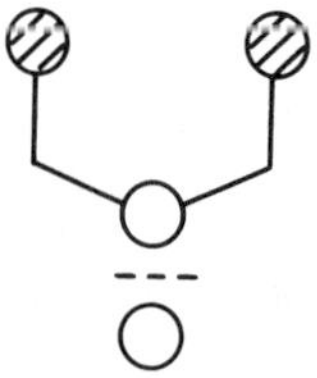

23 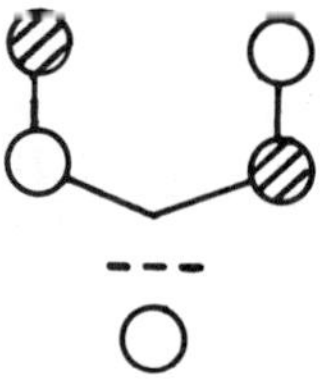   **24** 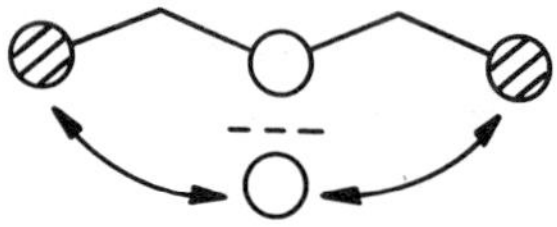   **25**

atom destabilizing the enol form. Why then doesn't fluorine destabilize the enolate as well and make the fluorinated diketone less acidic than the non-fluorinated species? The reason, we believe, is found in Scheme V which shows that the enolate derived from 1,3-diketone does not have the same conformation[19] that the hydrogen-bonded enol has in **17b** (X = F). For the enolate ion **26** our orbital picture would be described as shown in **27.** In this geometry, not only are the two partially charged oxygen atoms remote from each other, but also the destabilizing HOMO–HOMO interaction of fluorine with the pentadienyl anion framework is muted by the secondary orbital interaction shown by the double-headed arrow in **27.** With the π-donating capacity of F thus reduced, the σ-withdrawing (inductive effect) of F becomes relatively more important, resulting in a net acidifying effect on **19.** The conformational adjustments described in Scheme V are not available to **21** and **22** so the π-donation of F operates at full capacity leading to a net deacidifying effect. This reasoning may now be applied to **20** (X = F). The enolate of **20** (i.e., **20′**) can be approximated by the pentadienyl anion. Then the HOMO–HOMO and HOMO–LUMO interactions present in **20′** (X = F) are given by **28** and **29,** respectively. Note that the primary HOMO–HOMO destabilization is partially offset by a secondary orbital interaction shown in

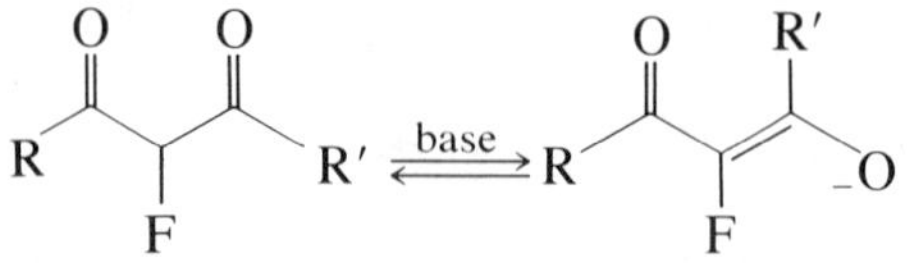

26

Scheme V

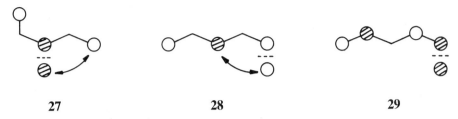

27   28   29

28. Furthermore, since F is now at the end of the framework, the HOMO–LUMO interaction **29** is nonzero and stabilizing. These factors therefore should make the fluorine compound, in this case, **20,** more acidic than the hydrogen counterpart, as observed. We conclude that the effect of F on keto–enol equilibria and on acidity of 1,3-diketones is related to the shape and phase of the conjugating framework to which the F is attached. These π-interactions will determine to what extent the π-donation of F will be effective and where the balance between σ-withdrawal and π-donation will be struck.

The above discussion of keto–enol equilibria and acidity brings to mind the remarkable case of Meldrum's acid,[20] **30.** The $_pK_a$ in water of this cyclic diester is 4.83, comparable to that of acetic acid, and less than that of dimedone, **18** (X = H), discussed earlier. Although more acidic than dimedone, Meldrum's acid exists predominantly in the diketo form whereas dimedone prefers the enol structure. Thus there appears to be no correlation of acidity with enol content. Concerning the relative acidity of **30** Arnett and Harrleson[20] pointed out that in the cyclic arrangement, conformational restraints inhibit the normal π-donation from the oxygen atom into the carbonyl moiety. Thus only the electron-withdrawing capacity of the alkoxy group is operative, producing an acidifying effect. This rationale is similar to that presented in the discussion of **19** where we suggested that π-donation from F is reduced by secondary orbital effects (see **27**), thereby allowing σ-electron withdrawal to become relatively more important resulting in a net acidifying effect.

30

4. THE EFFECT OF FLUORINE ON ELECTROCYCLIC REACTIONS

While studying thermal electrocyclic reactions of perfluorocyclobutene **31,** Dolbier and co-workers observed that symmetry-allowed conrotary ring

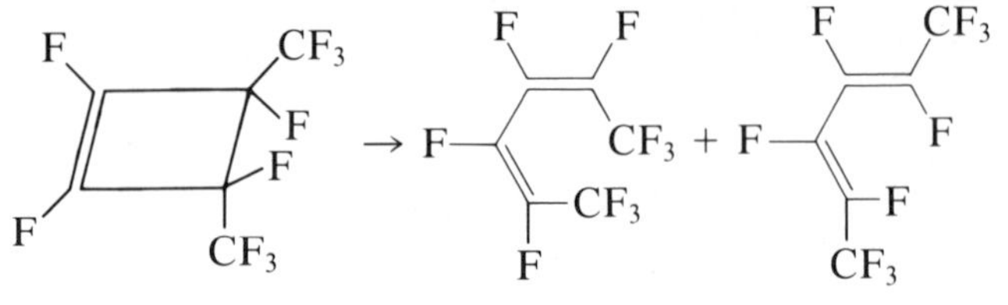

Scheme VI

opening leads preferentially to **32** rather than to **33** (Scheme VI).[21] Surprisingly, formation of **32,** sterically encumbered with the two CF_3 groups, is favored over formation of **33** where the bulky CF_3 groups are in the remote "outside" positions. Rondan and Houk[22] accounted for these results in terms of secondary orbital effects operating in the transition states for the inward ring opening **31** → **32** and the outward ring opening **31** → **33**. The HOMO–HOMO and the HOMO–LUMO interactions in the transition state for **31** → **32** are shown in **34a** and **34b,** respectively, while those for **31** → **33** are shown in **35a** and **35b,** respectively. The destabilizing HOMO–HOMO intereaction associated with the fluorine p orbitals is greater in **35a** than in **34a** due to the secondary overlap, and the stabilizing HOMO–LUMO interaction associated with the fluorine p orbitals is smaller in **35b** than in **34b** due also to the secondary overlap. Thus the transition state for **31** → **32** is lower in energy than that for **31** → **33.** Apparently the secondary orbital effects of the fluorine

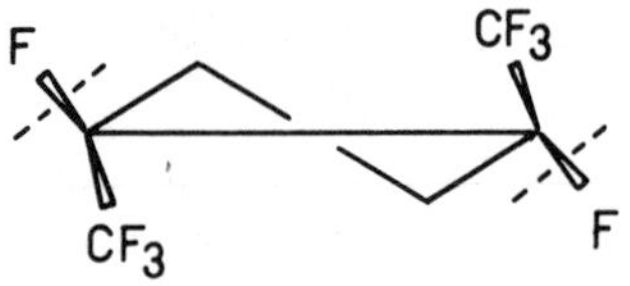

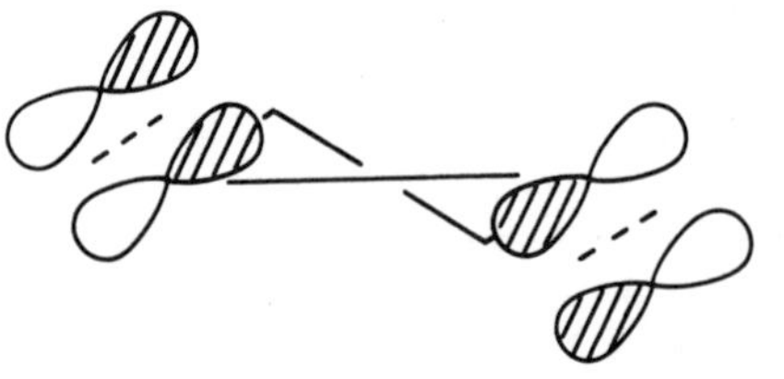

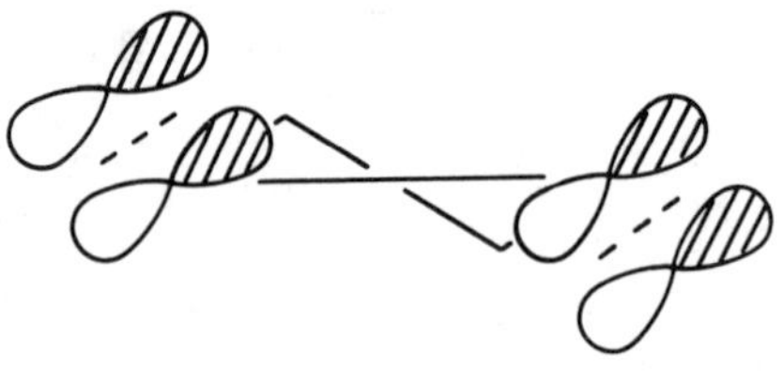

donor are enough to outweigh the steric effects of the bulky CF_3 group. The electrocyclic transformation discussed above as well as the examples examined in Section 3 serve to underscore the enormous destabilization associ-

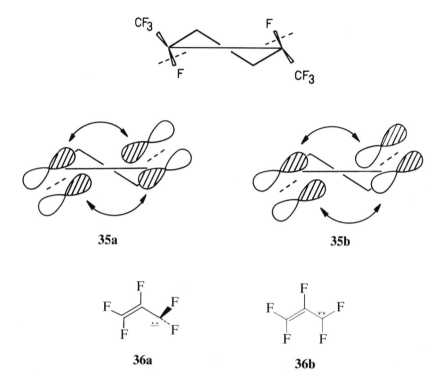

ated with a filled p-orbital on fluorine. Such destabilization accounts for the dramatic variation in pyramidal barriers[23] in going from CH_3^- (15 kcal/mol) to CF_3^- (115 kcal/mol). Another example of fluorine destabilization resulting from a filled p-orbital on fluorine is found in the work of Dixon and Smart[23] whose ab initio calculations indicate that the preferred conformation of the perfluoroallyl anion is **36a** and not the planar **36b**.

5. FLUOROALKENES

The stability trend of the difluoroethylenes is known to be 1,1 greater than cis-1,2 greater than trans-1,2. This intriguing pattern can be rationalized by considering destabilizing HOMO–HOMO interactions between vinyl fluoride and fluorine fragment orbitals as shown in Scheme VII. This scheme shows that the primary destabilizing overlap is smallest in **37** since the orbital coefficient on carbon atom 1 is smaller than that on carbon atom 2.[1] The secondary overlap, which is negative in each isomer thereby reducing the overall overlap, is clearly more important in **38** than in **39**. Thus a combination of primary and secondary orbital interactions account for the stability trend.

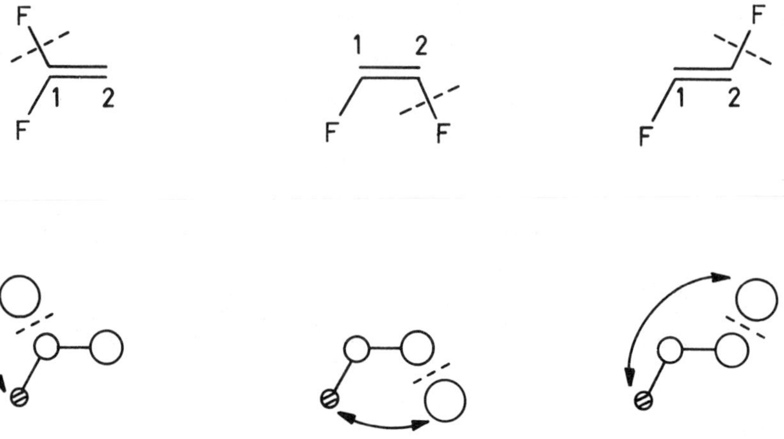

37   38   39

Scheme VII

Cycloaddition to 1,1 -difluoroallene, **40**, provides another striking example of the effect of fluorine. It was observed[24] that 2 + 2 cycloaddition takes
place preferentially at the $C_1=C_2$ double bond of **40** to give **41**, whereas 2 +
4 cycloaddition occurs at the $C_2=C_3$ double bond of **40** to yield **42** as outlined in Scheme VIII. This reversal in the regioselectivity was explained by
noting that the 2 + 4 addition is concerted, but the 2 + 2 addition proceeds
stepwise via free radical intermediates. This latter mode of reaction favors
attack at the fluorine-containing double bond where the sp^2 carbon bearing
the fluorine atoms is converted to an sp^3 carbon atom, which is favored
thermodynamically.[25] In addition, the diradical intermediate **43** is stabilized
by the lone pairs of the fluorine atoms attached to the carbon radical center.

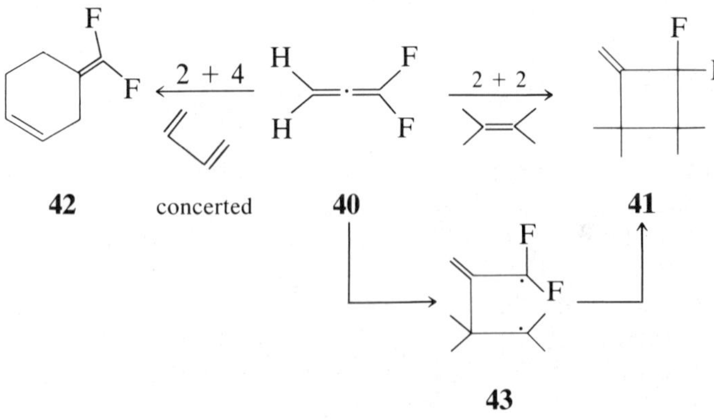

42   concerted   40   41

43

Scheme VIII

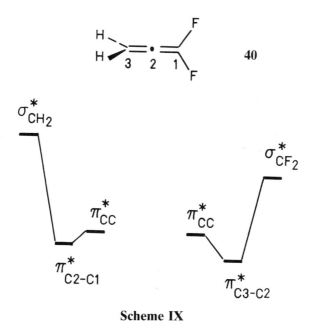

Scheme IX

Why, then, does the preference for C_1=C_2 bond not prevail in the 2 + 4 cycloadditions? Here, a different fluorine effect operates,[26] as shown in Scheme IX. The 1,1-difluoroallene **40** consists of two ethylene units, each of which has an acceptor orbital π^*_{CC}. The CH$_2$ and CF$_2$ units of **40** have $\sigma^*_{CH_2}$ and $\sigma^*_{CF_2}$ orbitals, respectively, which interact with the π^*_{CC} orbitals of the adjacent ethylene units. The $\sigma^*_{CF_2}$ orbital lies lower in energy than the $\sigma^*_{CH_2}$ orbital, so that the π^* orbital of the C_2=C_3 unit, $\pi^*_{C_2-C_3}$, is lowered more in the energy than the π^* orbital of the C_1=C_2 unit, $\pi^*_{C_1-C_2}$, as shown in Scheme IX. Therefore, the LUMO of 1,1-difluoroallene is represented by the $\pi^*_{C_2-C_3}$ orbital, and concerted cyclization reaction of **40** is governed by the C_2=C_3 bond rather than the C_1=C_2 unit. Similarly, 1,3-dipolar cyclo-additions to 1,1-difluoroallene and to monofluoroallene are also found to involve the C_2=C_3 bond.[27]

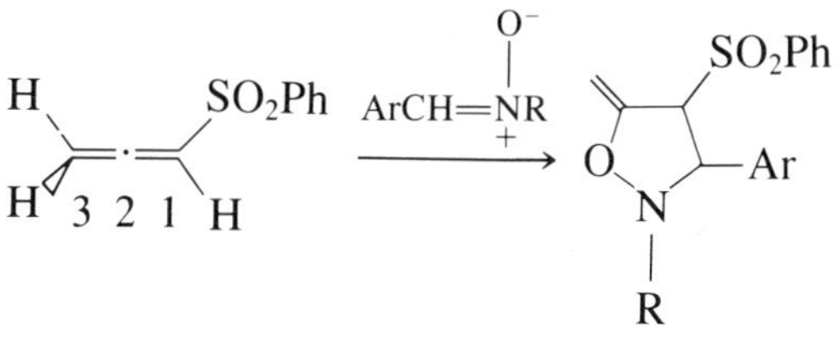

Scheme X

Underscoring the striking effect of fluorine on the site selectivity of cycloadditions to allene **40** are the results of nitrone addition to **44.** As shown in Scheme X, cycloaddition takes place at the $C_1=C_2$ double bond[28] in contrast to those additions described above where the $C_2=C_3$ double bond is the site of attack. In **44** the major effect of the electron-withdrawing sulfonyl group is to lower the $\pi^*_{C_1-C_2}$ level appreciably below that of the $\pi^*_{C_2-C_3}$ level, as shown in Scheme XI, since the σ^* orbital of the S—O bond is lower-lying than the σ^* orbital of the S—C bond. Thus, the $C_1=C_2$ bond is preferentially attacked in cycloaddition reactions of **44.**

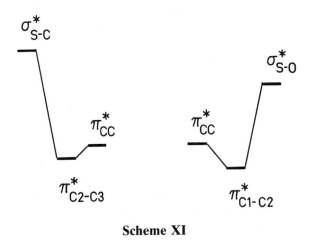

Scheme XI

6. CONCLUDING REMARKS

In this review we delineated a number of puzzling observations involving fluorine-containing compounds and related systems, and presented rationales in terms of secondary orbital effects. The validity and generality concerning such analyses must be answered by additional experiments and accumulation of data. We hope this survey stimulates further studies that enhance our understanding of fluorine chemistry.

ACKNOWLEDGMENTS

We would like to thank Dr. John E. Bunch, whose initial work on trifluoromethylacetylenes led us to explore the applicability of secondary orbital effects in fluorine chemistry.

REFERENCES

1. Whangbo, M.-H.; Wolfe, S. *Isr. J. Chem.* **1980,** *20,* 36.
2. Bumgardner, C. L.; Bunch, J. E.; Whangbo, M.-H. *Tetrahedron Lett.* **1986,** *27,* 1883.
3. Bumgardner, C. L.; Bunch, J. E.; Whangbo, M.-H. *J. Org. Chem.* **1986,** *51,* 4082.
4. Unger, S. H.; Hansch, C. In "Progress in Physical Organic Chemistry," Vol. 12; Taft, R. W., Ed.; Wiley: New York, 1976, p. 91.
5. Bernardi, F.; Epiotis, N. D.; Yates, R. L.; Schlegel, H. B. *J. Am. Chem. Soc.* **1976,** *98,* 2385.
6. Forsyth, D. A.; Osterman, V. M.; DeMember, J. R. *J. Am. Chem. Soc.* **1985,** *107,* 818.
7. Schleyer, P.v. R.; Dill, J. D.; Pople, J. A.; Hehre, W. J. *Tetrahedron* **1977,** *33,* 2497.
8. Salomaa, P.; Nissi, P. *Acta Chem. Scand.* **1967,** *21,* 1386.
9. Bloch, E.; Aslam, M. *J. Am. Chem. Soc.* **1983,** *105,* 6164.
10. Albright, T. A.; Burdett, J. K.; Whangbo, M.-H. "Orbital Interactions in Chemistry." Wiley: New York, 1985, pp. 82, 178.
11. Purrington, S. T.; Bumgardner, C. L.; Lazaridis, N. V.; Singh, P. *J. Org. Chem.* **1987,** *52,* 4307.
12. Burdett, J. L.; Rogers, M. T. *J. Am. Chem. Soc.* **1964,** *86,* 2105.
13. Subrahamanyam, G.; Malhota, S. K.; Ringold, H. J. *J. Am. Chem. Soc.* **1966,** *88,* 1332.
14. Vlasov, V. M.; Yakabson, G. G. *Zh. Org. Khim* **1981,** *71,* 242.
15. Streitwieser, A.; Mares, F. *J. Am. Chem. Soc.* **1968,** *90,* 2444.
16. Adolph, H. G.; Kamlet, M. J. *J. Am. Chem. Soc.* **1966,** *88,* 4761.
17. Dolbier, W. R., Jr.; Medinger, K. S.; Greenberg, A.; Liebman, J. F. *Tetrahedron* **1982,** *32,* 2415.
18. Schwarzenbach, G.; Suter, H.; Lutz, K. *Helv. Chim. Acta* **1940,** *23,* 1191.
19. Olmstead, M. N.; Bordwell, F. G. *J. Org. Chem.* **1980,** *45,* 3302.
20. Takegoshi, K.; McDowell, C. A. *J. Am. Chem. Soc.* **1986,** *108,* 6852. Arnett, E. M.; Harrelson, J. A. *J. Am. Chem. Soc.* **1987,** *109,* 809. For recent studies of Meldrum's acid see Wang, X. and Houk, K. N. *J. Am. Chem. Soc.* **1988,** *110,* 1870, and Wiley, K. B. and Laidig, K. E. *J. Am. Chem. Soc.* **1988,** *110,* 1872.
21. Dolbier, W. R., Jr.; Koroniak, H.; Burton, D. J.; Heinze, P. L.; Bailey, A. R.; Shaw, G. S.; Hansen, S. W. *J. Am. Chem. Soc.* **1987,** *109,* 219.
22. Rondan, N. G.; Houk, K. N. *J. Am. Chem. Soc.* **1985,** *107,* 2099.
23. Dixon, D. A.; Smart, B. E. Eighth Winter Fluorine Conference, **1987,** St. Petersburg, FL, paper 50.
24. (a) Dolbier, W. R., Jr.; Piedrahita, C. A.; Houk, K. N.; Strosier, R. W.; Gandour, R. W. *Tetrahedron Lett.* **1978,** 2231. (b) Dolbier, W. R., Jr.; Burkholder, C. R. *J. Org. Chem.* **1984,** *49,* 2381.
25. Dolbier, W. R., Jr.; Piedrahita, C. A.; Al-Sader, B. H. *Tetrahedron Lett.* **1979,** 2957.
26. Domelsmith, L. N.; Houk, K. N.; Piedrahita, C. A.; Dolbier, W. R., Jr. *J. Am. Chem. Soc.* **1978,** *100,* 6908.
27. (a) Dolbier, W. R., Jr.; Burksholder, C. R.; Wicks, G. E.; Palenik, G. J.; Gawron, M. *J. Am. Chem. Soc.* **1985,** *107,* 7183. (b) Dolbier, W. R., Jr.; Wicks, G. W.; Burkholder, C. R. *J. Org. Chem.* **1987,** *52,* 2196.
28. Parpani, P.; Zecchi, G. *J. Org. Chem.* **1987,** *52,* 1417.

CHAPTER 6

Partitioning of Carbanion Intermediates Generated in Alcoholic Media

Heinz F. Koch and Judith G. Koch

Department of Chemistry, Ithaca College, Ithaca, New York

CONTENTS

1. INTRODUCTION

For over twenty years, our research efforts have included investigations of nucleophilic reactions of alkenes, dehydrohalogenation and proton exchange reactions, and primary kinetic isotope effects (PKIE) associated with these reactions.[1] The unifying theme is processes that generate carbanions in protic solvents and the goal is to understand the reactions of these intermediates. To determine the chemistry of carbanions in protic solvents, an appropriate method must be used to generate them in situ. It is common to form carbanions by a proton transfer from carbon to base; however, if this reaction is accompanied with excessive amounts of internal return, it is impossible to determine the conditions required to initially generate the carbanion intermediate. To avoid such complications alternate reactions must be utilized. Miller and co-workers[2] reported that reaction of $CF_2{=}CCl_2$ with ethanolic ethoxide results in formation of only the saturated ether, $C_2H_5OCF_2CHCl_2$. Hine's group[3] later reported methanolic methoxide-catalyzed hydrogen exchange of $CH_3OCF_2CHCl_2$ with methanol occurs at a rate 10^4 faster than the dehydrofluorination. This confirmed that $(C_2H_5OCF_2CCl_2)^-$, formed by the reaction of $C_2H_5O^-$ with $CF_2{=}CCl_2$, accepts a proton more readily than the ejection of the fluoride to give a vinyl ether. Therefore we have the ideal situation: a carbanion can be generated from either direction. However the chemistry is rather dull since proton transfer between carbon and oxygen appears to be the only reaction.

Reaction of ethanolic ethoxide with 2-phenyl perfluoropropene (1) is more interesting since Z-1-ethoxy-2-phenyl perfluoropropene (2) is the major product, 76%, while the isomeric vinyl ether (3) accounts for 9%, and only 15% of the saturated ether (4) is formed.[4] In principle, **INT-1** in Scheme I can also be generated by reaction of ethoxide with 4. This is not a competitive process below room temperature, and the back reaction from 4 to form **INT-1** under the reaction conditions of $-78°C$ is not shown. Loss of fluoride from the trifluoromethyl group to yield the allylic ether (5) is not observed, but 5 is the only product formed from reaction of 3-chloro-2-phenyl perfluoropropene (6) under similar conditions. When reaction of 1 occurs

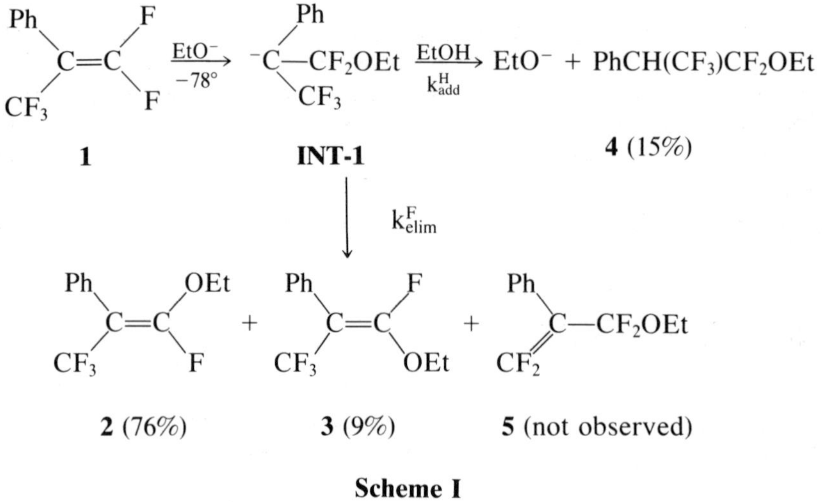

Scheme I

with methanolic methoxide, the corresponding carbanion partitions to give more saturated ether, 34%, at the expense of both vinyl ethers.

Dehydrofluorination of **4** occurs readily at 40°C and the question can be asked: *Is the same carbanion generated from the reaction of ethoxide with starting alkene as the one formed with reaction of saturated ether?* When the methoxy saturated ether, $PhCH(CF_3)CF_2OMe$ **(7)**, reacts with ethoxide in EtOD (40°C) until 20% $PhC(CF_3)=CFOMe$ **(8)** is formed, recovered **7** contains 3–4% deuterium.[5] Under similar conditions reaction of **1** results in 15% **4-d** and 85% **2** and **3**. Both reactions result in a similar partitioning between fluoride loss and deuteron addition, which suggests product distribution is independent of the method of intermediate generation from either **1** or **7**. Since analysis of the reaction mixture of **7** in EtOD indicated no **2**, **3**, or **4**, methoxide loss from $\{PhC(CF_3)CF_2OCH_3\}^-$ **(INT-2)** to form **1** will not compete with either fluoride ejection or proton transfer. Quantitative numbers are not available for the leaving ability of halides compared to alkoxides, but experimental evidence suggests that even fluoride is a better leaving group than an aliphatic alkoxide.[6]

Transfer of proton from oxygen to carbon is fast,[7] but other reactions can readily compete. Loss of fluoride is competitive with proton transfer to neutralize carbanion intermediates formed by the reaction of alkoxide with alkene. The proton transfer is not able to compete with ejection of chloride since **5** is the only product from the reaction of alkoxide with **6**:

$$PhC(CF_2Cl)=CF_2 + EtO^- \xrightarrow[-78°C]{EtOH} PhC(=CF_2)CF_2OMe$$
$$\textbf{6} \qquad\qquad\qquad\qquad\qquad\qquad \textbf{5}$$

Loss of beta chloride or bromide occurs faster than protonation in many nucleophilic vinyl halide substitutions; however, one must first determine if a carbanion is formed along the reaction pathway or if the mechanism is concerted.[8] Allylic substitution could occur by an S_N2' mechanism. Bach and Wolber report calculations on synchronous vinyl[9] and allyl[10] displacements. Kinetic arguments suggested reaction of **6** occurs by the formation of a carbanion, $\{PhC(CF_2Cl)CF_2OEt\}^-$ **(INT-3)**, in the rate-limiting step.[4]

Leaving ability of fluoride from various environments was studied by using a series of compounds, $PhCR_f=CF_2$, where $R_f = CF_3$ **(1)**, CF_2CF_3 **(9)**, and CHF_2 **(10)**.[5] Since alkoxide preferentially attacks $=CF_2$, a fluoride leaving from R_f in $\{PhCR_fCF_2OR\}^-$ can be compared to proton transfer from ROH to the carbanion or ejection of fluoride from $-CF_2OR$. Reaction of **9** results in some displacement of allylic fluoride, but vinyl ethers are still the major products; however, reaction of **10** results in greater than 98% E-$PhC(CF_2OR)=CHF$ as the kinetic product. Table 6-1 summarizes the product distributions. The leaving ability of fluoride from different groups is $CF_2H > CF_2CF_3 > CF_2OR >> CF_3$. Proton transfer from alcohol to the carbanion is faster than loss from CF_3, and this is confirmed by exchange and elimination studies on $PhCH(CF_3)_2$.[11]

TABLE 6-1. Product Distribution, Activation Parameters, and Rates for the Reactions:[a]
$C_6H_5CR_f=CF_2 + MeONa \rightarrow C_6H_5CHR_fCF_2OMe$(satd) +
$C_6H_5CR_f=CFOMe$(vinylic) + $C_6H_5C(CF_2OMe)=R_f$(allylic)

R^f		Temp, °C	%Satd	%Vinylic	%Allylic	$kx10^3$ $M^{-1}s^{-1}$ $-50°C$	$\Delta H^{\ddagger}$ kcal/mol	$\Delta S^{\ddagger}$ eu
CF_2Cl	(6)	−78	$(0)^b$	$(0)^b$	$(100)^b$	135	9.6 ± 0.1	−19 ± 1
CF_3	(1)	−78	$34(15)^h$	$66(85)^h$	$0(0)^h$	105	8.7 ± 0.3	23 ± 1
CF_3	(1)	20	$44(21)^b$	$56(79)^b$	$0(0)^b$			
C_2F_5	(9)	−78	$(4)^b$	$(74)^b$	$(22)^b$	40.6	9.2 ± 0.2	−23 ± 1
CF_2H	(10)	−50		$<2^c$	>98	3.63	11.2 ± 0.3	−19 ± 1

[a] Data from Ref 5.
[b] Values in parentheses are for reactions with EtONa in EtOH.
[c] Thought to be $C_6H_5C(CF_2H)=CFOMe$.

2. HYDROGEN-BONDED CARBANION INTERMEDIATES

The near unity PKIE, $k^H/k^D = 1.3$ at 40°C, associated with ethanolic ethoxide-promoted dehydrofluorination of **4** and **7**, suggests reaction occurs with substantial internal return, $k_{-1} \gg k_2$, Scheme II.

$$\overset{|}{\underset{|}{C}}-H + {}^-OR \underset{k_{-1}}{\overset{k_1}{\rightleftharpoons}} \overset{|}{\underset{|}{C}}{}^- \cdots H-OR \overset{k_2}{\longrightarrow} products$$

Scheme II

Therefore, proton return to carbon is faster than loss of fluoride from the initially formed carbanion when alkoxide reacts with saturated ether. Ergo this carbanion cannot be the same as **INT-1** or **INT-2**, and a second carbanion must be formed as shown in Scheme III.[5] One intermediate, **(H)**, is stabilized by a hydrogen bond, while the other, **(F)**, has no contact stabilization from either a cation or solvent molecule. Internal return, k_{-1}, is the low free energy process for the initial intermediate when alkoxide abstracts a proton from **4** or **7**. The hydrogen bond is broken, k_2, in the rate-limiting step to form intermediate F, which can partition to yield mixtures of vinyl or saturated ethers. The intermediate F is formed in a rate-limiting step for reaction of alkoxide with alkene, k_N. Ejection of fluoride, k^F_{elim}, is faster than the formation of **H**, k_{-2}, which is rate-limiting on the reaction pathway from F to **4** or **7**. When reaction proceeds by Scheme III near unity experimental PKIE are expected for the hydron transfer in both directions.

If proton transfer between carbon and oxygen has a highly asymmetric transition structure, there is no need to suggest a mechanism that features extensive internal return. Scheme I could explain experimental results, and Scheme III would not be necessary. Theoretical calculations of isotope effects associated with elimination reactions suggest that greater than 95% C—H bond breaking is required in a transition structure to obtain k^H/k^D

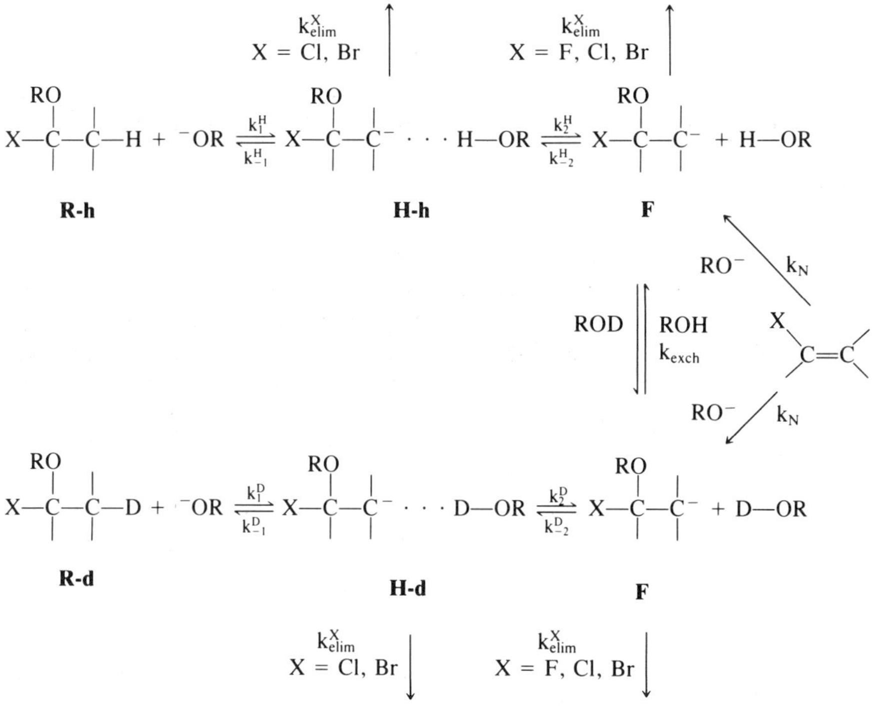

$$\begin{array}{ccc}
& k_{elim}^X \uparrow & k_{elim}^X \uparrow \\
X = Cl, Br & X = F, Cl, Br
\end{array}$$

RO   RO   RO

$$X-C-C-H + {}^-OR \underset{k_{-1}^H}{\overset{k_1^H}{\rightleftharpoons}} X-C-C^- \cdots H-OR \underset{k_{-2}^H}{\overset{k_2^H}{\rightleftharpoons}} X-C-C^- + H-OR$$

R-h   **H-h**   **F**

$$ROD \Updownarrow ROH \quad k_{exch}$$

$$RO^- \searrow k_N$$
$$RO^- \nearrow k_N$$

$$\mathop{X}\limits^{} \mathop{C=C}\limits^{}$$

RO   RO   RO

$$X-C-C-D + {}^-OR \underset{k_{-1}^D}{\overset{k_1^D}{\rightleftharpoons}} X-C-C^- \cdots D-OR \underset{k_{-2}^D}{\overset{k_2^D}{\rightleftharpoons}} X-C-C^- + D-OR$$

R-d   **H-d**   **F**

$$\begin{array}{cc}
k_{elim}^X \downarrow & k_{elim}^X \downarrow \\
X = Cl, Br & X = F, Cl, Br
\end{array}$$

Scheme III

values as low as 1.5.[12] Streitwieser[13] suggested a method to differentiate between these two explanations for low observed isotope effects that makes use of the Swain-Schaad relationship:[14]

$$k^H/k^D = (k^D/k^T)^y \tag{6-1}$$

where y can be 2.26 to 2.344 depending on the assumptions made in the derivation.[15] Albery and Knowles[16] warn that rate constants must be determined to a high degree of accuracy to make use of deviations from a rather insensitive Swain-Schaad relationship to calculate internal return. Since exchange of $PhC^iH(CF_3)CF_2OMe$ competes with dehydrofluorination, an accurate measurement of k_{elim}^T would be difficult.[17] An alternate method was used to address this question. Reaction of **1** in 1:1 mixtures of EtOH:EtOD resulted in isotope effects of $k^H/k^D = 1.5$ ($-78°C$) and 1.9 (20°) associated with the hydron transfer from ethanol to **INT-1**. The slight increase of PKIE with increasing temperature has been observed for reactions of other alkenes,[18] and is not consistent with an asymmetric transition structure. Therefore the use of Scheme III appears to be justified.

Since hydron transfer does not occur in the rate-limiting step, Scheme III predicts experimental isotope effects would be near unity for hydrogen exchange reactions. *Is this a common behavior for carbon acids?*

3. PKIE ASSOCIATED WITH HYDROGEN EXCHANGE REACTIONS

Table 6-2 summarizes results for the exchange of eleven carbon acids with methanolic methoxide. Only four result in experimental PKIE values that are of a normal magnitude: 9-trifluoromethylfluorene (**9-TFMFl**), 9-phenylfluorene (**9-PhFl**), fluorene (**Fl**), and 9-methylfluorene (**9-MeFl**). The range of reactivity of the entire group is ca. 10^{16}, with **9-TFMFl** the fastest and toluene the slowest. Reaction of **9-TFMFl** is 5.5×10^3 faster than **9-PhFl**, which is equivalent to 2-hydro-2-phenylperfluoropropane (**11**), and 2-hydroperfluoropropane (**12**). Although **9-PhFl, 11,** and **12** have the same reactivity at 25°C, isotope effects associated with these reactions differ significantly.

TABLE 6-2. Rates of Protodetritiation, Activation Parameters, and PKIE Values for Exchange Reactions in Methanolic Sodium Methoxide

Compound	k, $M^{-1}s^{-1}$ 25°(45°)C	$\Delta H^{\ddagger}$ kcal M^{-1}	$\Delta S^{\ddagger}$ eu	PKIE (°C) k^H/k^D	k^D/k^T
9-TFMFl[a]	9.3	15	−4	k^H/k^T=28 ± 3 (−50)[i]	
$C_7F_{10}H_2$(14)[b]	1.4×10^{-1}	19	2	1.2(−15)	
C_6F_5H[c]	2.5×10^{-2}	20	2		1.0(25)
$PhCH(CF_3)_2$[d](11)	2.1×10^{-3} (2.8×10^{-2})	24.1	10	1.1	1.0(25)
$(CF_3)_2CFD$(12)[e]	2.3×10^{-3}	21	−1	1.4(20)[j]	
9-PhFl[f]	1.7×10^{-3} (3.3×10^{-2})	21.0	−1	6.4	2.5(25)
Fl[f]	3.1×10^{-5} (4.0×10^{-4})	23.4	−1		2.3(25) 2.1(45) 1.8(100)
9-MeFl[f]	(2.2×10^{-4})			5.2	2.3(45)
n-$C_7F_{15}D$(13)[e]	3.8×10^{-8}	30	7	1.4(50)[j]	
Ph_3CH[g]	2×10^{-11}	32	1	1.3	1.3(100)
$PhCH_3$[h]	2×10^{-16}	38	−5		1.0(178)

[a] 9-Trifluoromethylfluorene. M. F. McEntee, unpublished results.
[b] 1,4-dihydroperfluorobicyclo[2.2.1]-heptane. Ref. 21.
[c] Ref. 22.
[d] Ref. 11.
[e] Ref. 19.
[f] Ref. 15. 9-Ph-Fl = 9-phenylfluorene, Fl = fluorene, and 9-Me-Fl = 9-methylfluorene.
[g] Ref. 13a.
[h] Ref. 23.
[i] Both H and T exchange were carried out in MeOD at temperatures between −41° and −51°C.
[j] Corrected for KSIE = 2.6, see text for explanation.

Experimental PKIE for all three isotopes of hydrogen have been reported for **9-PhFl,** $k^H/k^T = 16.0$ and $k^D/k^T = 2.50$ (25°C), and **9-MeFl**, $k^H/k^T = 11.9$ and $k^D/k^T = 2.30$ (45°C).[15] Values of k^H/k^T are for MeOD and those of k^D/k^T are for MeOH. This allowed the calculation of a kinetic solvent isotope effect (KSIE), $k^{OD}/k^{OH} = 1.8$ **(9-PhFl)** and $= 2.2$ **(9-MeFl)**, and the internal return factors, $a^iH = k_{-1}/k_2$: $a^H = 0.48$, $a^D = 0.048$, and $a^T = 0.015$ **(9-PhFl)**; $a^H = 0.56$, $a^D = 0.068$, and $a^T = 0.022$ **(9-MeFl)**. The values obtained for **11** are: $k^H/k^T = 1.10$ (MeOD) and $k^D/k^T = 1.05$ (MeOH), with $k^{OD}/k^{OH} = 2.63$ (25°C).[11] Since this reaction has extensive internal return, calculation of internal return factors is meaningless, and the PKIE associated with reactions of **11** are used as values of equilibrium hydrogen isotope effects for this system.

The published data[19] for $(CF_3)_2CF^iH$ **(12)** should be analyzed before using the reported rates to calculate PKIE associated with these reactions. The rates for $C_7F_{15}D$ and $C_7F_{15}T$ in MeOH resulted in $k^D/k^T = 1.4$ at 70°C, and Equation 6-1 was used to calculate a rate for $C_7F_{15}H$ **(13)** in MeOH. This calculated rate was compared to the experimental k^H (in MeOD) to obtain a value of the KSIE, $k^{OD}/k^{OH} = 1.5$, which was used to calculate k^H/k^D for **12**.[20] When the KSIE measured for **11** (2.63) is used, the corrected value for k^H/k^D is 1.4 instead of the reported value of 2.4 for **12** at 20°C. This suggests extensive internal return accompanies the exchange reactions of both **12** and **13**.

Table 6-2 has four entries with near unity PKIE: three aliphatic compounds {1,4-dihydroperfluorobicyclo[2.2.1]heptane **(14)**[21], **12**, and **13**} and an aromatic compound {pentafluorobenzene **(PFB)**[22]}. *Why are the PKIE values associated with the fluorenes different than those for these compounds?* One major difference between the two sets is that the fluorenyl anions are capable of extensive pi-electron delocalization while the aliphatic anions and $C_6F_5^-$ are not. Three other compounds {$PhCH(CF_3)_2$, Ph_3CH, and $PhCH_3$} that result in near unity PKIE have benzylic hydrogens that undergo exchange with methanol. The benzyl anions are capable of delocalization through a pi-network; however, the experimental results obtained from this set of compounds are similar to those for localized carbanions. The experimental results for Ph_3CH, $k^H/k^T = 1.77$ (MeOD) and $k^D/k^T = 1.34$ (MeOH), with $k^{OD}/k^{OH} = 2.29$ (97.7°C), allow the calculation of $k_1^H/k_1^D = 4.2$.[13a] The PKIE for the exchange of toluene with methanolic methoxide at 178°C ($k^D/k^T = 1.0$)[23] suggests there is more internal return for that reaction than for the exchange of Ph_3CH.

Carbonyl and nitro compounds are not included in Table 6-2 since charge will be localized on oxygen. Highly halogenated hydrocarbons can rival the fluorenes in reactivity because the corresponding carbanions will be stabilized by field effects as well as interaction of lone pair electrons with the σ^* orbital of a β—X bond.[24] The two trifluoromethyl groups in **11** increase reactivity of the benzylic hydrogen by a factor of more than 10^{12}, while the two additional phenyl groups of Ph_3CH increase reactivity by only 10^4. Our working hypothesis is that benzylic carbanions behave as if they have little pi-delocalization in alcohol solvents. The next two sections will develop this concept further.

4. CARBANIONS FROM REACTION OF METHANOLIC METHOXIDE AND SILANES

Much work has been reported on proton-transfer reactions from a neutral carbon to oxide ions; however, few quantitative studies of the reverse process, transfer from hydroxyl to carbanion, have been documented. Bockrath and Dorfman[25] used pulse radiolysis techniques to obtain rates of reaction, $k \times 10^{-8}$ $M^{-1}s^{-1}$ (297°K), for benzyl anion generated in situ with several alcohols and water in tetrahydrofuran (THF): MeOH, 2.3 ± 0.3; EtOH, 1.4 ± 0.2 and EtOD, 1.2 ± 0.2; t-BuOH, 0.16 ± 0.02; H_2O, 0.53 ± 0.17. The near unity PKIE, $k^H/k^D = 1.2$, for reaction with ethanol was in agreement with those reported for reactions of organometallic compounds,[26] where a known mixture of ROH and ROD was added to a solution of organometallic in an inert solvent. The H : D ratio in the resulting hydrocarbon was used to calculate k^H/k^D. An assumption was made that organometallic compounds are good models for carbanions. The kinetic studies seemed to support this as PKIE measured for the free ion with ethanol was similar to that for ion pairs, $k^H/k^D = 1.4$.[25] To study the behavior of free carbanions in protic solvents is more difficult.

The excellent experimental work reported by Eaborn and co-workers[27] was the first systematic investigation addressing this problem. Anions were generated in situ by the reaction of methoxide and an appropriate alkyltrimethylsilane in mixtures of MeOH and MeOD:

$$MeO^- + Me_3SiR \rightarrow MeOSiMe_3 + R^- \tag{6-2}$$

$$MeO^iH + R^- \rightarrow MeO^- + R^iH \tag{6-3}$$

Kinetics were studied using a number of ring-substituted benzyltrimethylsilanes, and this resulted in a $\rho = 4.7$.[28] The measured isotope effects, (k^H/k^D), associated with protonation of most benzylic anions were low (1.1–1.3) but increased steadily from p-CN (2.0), p-PhSO$_2$ (2.9), p-PhCO (7.0) to p-NO$_2$ (10).[27ac] The 9-fluorenyl and 9-methyl-9-fluorenyl anions as well as o-NO$_2$C$_6$H$_4$CH$_2^-$ resulted in values of 10; however, Ph$_3$C$^-$ and Ph$_2$CH$^-$ gave low values of 1.3.[27c] These isotope effects for the proton transfer from an alcohol to a carbanion are in agreement with those found in Table 6-2 which represent the opposite reaction from hydrocarbon to alkoxide.

Eaborn[27b] noted an apparent discrepancy between results for PKIE associated with the methanolic methoxide-catalyzed exchange of Ph_3C^iH (k_1^H/k_1^D = 4.2 at 100°C)[13a] compared to their near unity results for protonation of the triphenylmethyl anion. Based on the above value, they predicted an isotope effect, k_{-1}^H/k_{-1}^D, of about 8 (25°) instead of the experimental value of 1.3. The measured PKIE for Ph_3C^iH, $(k^H/k^D)^{obs}$ = 1.32 at 97.7°C (k^H/k^T = 1.77 ± 0.01 and k^D/k^T = 1.34 ± 0.03) is low due to internal return: a^H = 4.5, a^D = 1.8 and a^T = 0.49. Scheme III can account for these results: the proton-transfer step is not rate limiting. This agrees with the chemistry of carbanions generated from our alkene reactions where the formation of a hydrogen-bonded carbanion is also rate limiting. Results for protonation of 9-methylfluorenyl anion and methoxide-catalyzed exchange of 9-methyl-fluorene-9-t are in good agreement. For the fluorene reactions, the transfer of hydron between carbon and oxygen is rate limiting.

Initial studies of PKIE associated with proton transfer from hydroxyl groups to organometallic compounds were carried out by mixing solutions of the two reactants. The near unity hydrogen isotope effects were thought to arise from diffusion controlled reactions. The pulse radiolysis studies on benzyl anion gave rates below the diffusion limit; however, near unity PKIE were obtained. The Eaborn group[27b] reacted fluoren-9-yl lithium in diethyl ether with an excess of 1 : 1 MeOH : MeOD, and this resulted in k^H/k^D of 1.5. This is much lower than the value of 10 for reaction of 9-fluorenyl anion generated in methanol from the silane. They concluded that methanol in the vicinity of the organometallic is used up prior to diffusion, and results are due to mixing rather than diffusion controlled reaction rates.

Their interpretation of differences in isotope effects associated with proton transfer from methanol was that anions resulting in large PKIE were highly delocalized by conjugation and those with low values were stabilized by delocalization via sigma bonds.[29] Reaction of $PhC(CF_3){=}CF_2$ with RO^- in alcohol also generates a carbanion in situ, and results in a low isotope effect for the neutralization reaction by proton transfer. This suggested a similar study for the reaction of $XC_6H_4CH{=}CF_2$ with methanolic methoxide.[30]

5. PI-DELOCALIZED VS LOCALIZED CARBANIONS

There are advantages to generating carbanions in situ by the reaction of methanolic methoxide with β,β-difluorostyrenes. The nucleophilic attack of methoxide on the alkene is rate limiting, occurs at a lower temperature than is required for methoxide-catalyzed proton abstraction, and eliminates the problem of internal return. Methanol is an excellent trapping agent for carbanions, and only an intramolecular trap, such as β-halide ejection, is more efficient. Reactions of $YC_6H_4CH{=}CF_2$ (15-Y) have the added advantage of competing processes: The intermediates INT(Y)-4 are able to partition between proton transfer to yield $YC_6H_4CH_2CF_2OMe$ **(16-Y)**, and the ejection of a β-fluoride to give $YC_6H_4CH=CFOMe$ **(17-Y)**.

$$YC_6H_4CH=CF_2 \xrightarrow{\ MeO^-\ }$$

15-Y

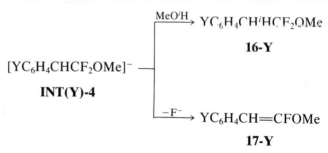

Since reaction of methoxide with **15** is much faster than proton removal from **16** and reaction of **17** with methoxide or fluoride, products are stable under experimental conditions, Scheme IV, and relative amounts of **16** and **17** are formed according to k_{elim}, and

$$k_{add} = k_f k_{prot}/(k_r + k_{prot}) \tag{6-4}$$

Calculation of the PKIE using Equation 6-5 assumes k_{elim} is the same in MeOH and MeOD.[31,32]

$$\frac{k^H}{k^D} = \frac{[\%16/\%17]^{MeOH}}{[\%16/\%17]^{MeOD}} \tag{6-5}$$

The measured PKIE (k^H/k^D) associated with formation of both **16-m-NO$_2$** and **16-m-CF$_3$** increase slightly with increasing temperature: m-NO$_2$ is 1.20 ($-50°$), 1.24 ($-25°$), 1.29 ($0°$), 1.35 ($25°$), and 1.39 ($50°$) while m-CF$_3$ is 1.28 ($0°$), 1.34 ($25°$), and 1.40 ($50°$). If $k_{prot} \gg k_r$ ($k_{-1} \gg k_2$ in Scheme III), Equation 6-4 reduces to $k_{add} = k_f$, and experimental PKIE are associated with the formation of **INT-4-H** from **INT-4-F**. A model for these low observed PKIE could be the proposal by Gold and Grist that KSIE are due to multiple solvated methoxide ions that have a deuterium fractionation factor equal to 0.74.[33] If the magnitude of the fractionation factor is about the same for **INT-4-H**, the measured PKIE values of 1.2 to 1.5 associated with protonation (k_{add}) are reasonable. It is not known if the fractionation factor will increase with temperature.

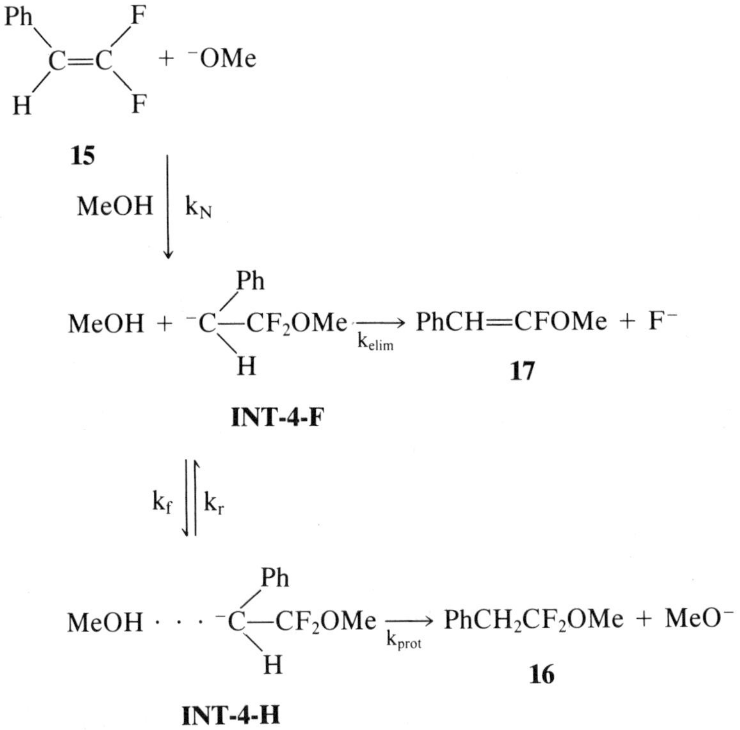

$$\underset{\underset{\displaystyle \textbf{15}}{}}{\overset{Ph}{\underset{H}{}}C{=}C\overset{F}{\underset{F}{}}} \quad + \quad {}^-OMe$$

MeOH $\Big\downarrow$ k_N

$$MeOH + {}^-\overset{Ph}{\underset{H}{C}}{-}CF_2OMe \xrightarrow[k_{elim}]{} PhCH{=}CFOMe + F^-$$
$$\qquad\qquad\qquad\qquad\qquad\qquad\qquad\textbf{17}$$

INT-4-F

$k_f \big\| k_r$

$$MeOH \cdots {}^-\overset{Ph}{\underset{H}{C}}{-}CF_2OMe \xrightarrow[k_{prot}]{} PhCH_2CF_2OMe + MeO^-$$
$$\qquad\qquad\qquad\qquad\qquad\qquad\qquad\textbf{16}$$

INT-4-H

Scheme IV

The PKIE associated with the protonation of **INT(p-NO$_2$)-4** shows a more normal temperature behavior: 11.3 ($-70°$), 9.62 ($-50°$), 8.14 ($-25°$), 7.12 ($0°$), and 6.44 ($25°$);[18] however, $A^H/A^D = 1.9$ is not normal and suggests an anomalous Arrhenius behavior of the type associated with internal-return mechanisms[34] where neither k_f nor k_{prot} is clearly rate limiting. Behavior of k^H/k^D measured for the protonation of **INT(p-CN)-4** is similar to those for **INT(m-NO$_2$)-4** and **INT(m-CF$_3$)-4**: 1.33 ($-70°$), 1.36 ($-50°$), 1.53 ($-25°$), 1.69 ($0°$), 2.04 ($25°$), and 2.11 ($50°$). Isotope effects that increase with temperature have negative values for ΔE_a^{D-H}, and theory predicts that k^H/k^D should be less than unity. Therefore, A^H must be greater than A^D to result in measured k^H/k^D greater than unity. In our systems, A^H/A^D values are 2.2 (m-CF$_3$), 2.0 (m-NO$_2$), and 5.1 (p-CN). Similar unusual behavior for hydrogen isotope effects was recently reported for the deprotonation of methylarene cation radicals in acetonitrile.[35]

Our isotope effects compare to the single temperature values reported by the Eaborn group, and suggest that [YC$_6$H$_4$CH$_2$]$^-$ and [YC$_6$H$_4$CHCF$_2$OMe]$^-$ have a similar reactivity toward methanol. Both p-NO$_2$ intermediates result in large isotope effects. These anions will delocalize through the pi network to place the negative charge mainly on oxygen. Orbitals need to have the proper alignment in an encounter complex, **INT(*p*-NO₂)-4-H**, prior to proton transfer. If the free carbanion is more stable than the hydrogen-bonded species, the barrier to reform **INT(*p*-NO₂)-4-F** is lower than that for the proton-transfer step, and significant isotope effects result.

Intermediates from reaction of methoxide with **15-*m*-NO₂**, **15-*p*-CN**, and **15-*p*-NO₂** behave quite differently toward protonation by methanol compared to the ejection of β-fluoride. Values of **16-*m*-NO₂ : 17-*m*-NO₂** show a slight increase from 1.1 (−50°C) to 1.5 (50°C). A slight increase in the amount of saturated ether formed at the expense of vinyl ethers is normal for all alkenes studied to date with the exception of **15-*p*-CN** and **15-*p*-NO₂**. Ratios of **16-*p*-CN : 17-*p*-CN** decrease from 6.81 (−70°) to 2.87 (50°) while those for **16-*p*-NO₂ : 17-*p*-NO₂** show a more dramatic decrease from 16.0 (−70°) to 1.0 (25°). The calculated $\Delta H^{\ddagger}$, 3.3 ± 0.1 kcal/mol, between addition to form **16-*p*-NO₂** and elimination to give **17-*p*-NO₂** suggests that there should be a larger percentage of the addition product; however, the $\Delta S^{\ddagger}$ of ca 10 eu (A^{add}/A^{elim} = 0.004) which favors the formation of **17-*p*-NO₂** offsets this difference in the enthalpy term. The relative ratios of E/Z vinyl ether observed for reaction of **15-*p*-NO₂** differ significantly to those obtained for **15-*m*-NO₂** and other alkenes studied. This will be discussed in the following section.

6. SUBSTITUTION OF VINYL HALIDES BY ALKOXIDES

Do nucleophilic vinyl substitutions form carbanion intermediates or proceed by a concerted mechanism? There is ample evidence for carbanion formation when fluoride is displaced: (1) carbanions are trapped by proton transfer from solvent;[5,36] (2) reaction of pure E- and Z-alkenes give the same product;[37] (3) rates of fluoride displacement are several orders of magnitude faster than that for chloride.[37] Favoring a single step process is complete retention of configuration observed in the substitution of many vinyl chlorides or bromides.[8]

Modena et al. studied reactions of methoxide and benzenethiolate with *p*-NO₂C₆H₄CH=CHX **(18-X)**, X = F or Cl.[37,38] The reaction of **Z-18-F** or **E-18-F** resulted in *E*-*p*-NO₂C₆H₄CH=CHOMe **(E-19)** or *E*-*p*-NO₂C₆H₄CH=CHSPh **(E-20)**; however, substitution of chloride by benzenethiolate occurred with complete retention.

To explain this discrepancy, they suggested chloride, a good leaving group, is displaced with kinetic control while the substitution of fluoride, a poorer leaving group, results in thermodynamic control allowing rotation of

the bond, $\overset{\diagdown}{\underset{\diagup}{\text{C}}}$-CHFNuc, prior to loss of fluoride.

Reactions of difluoroalkenes and methanolic methoxide occur with kinetic control of products. The exception is *p*-NO₂C₆H₄CH=CF₂ **(15-*p*-NO₂)**

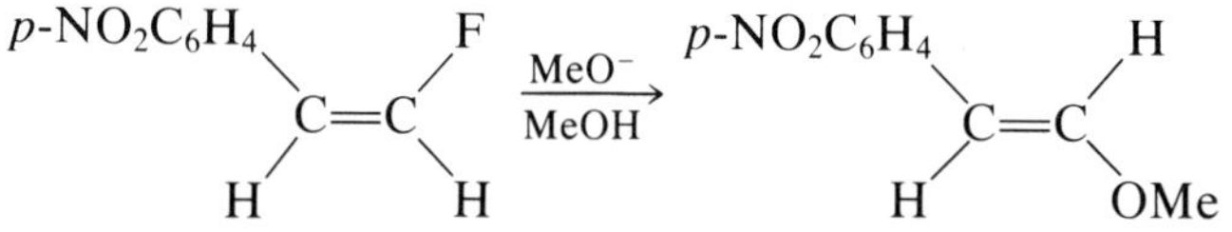

$$p\text{-NO}_2\text{C}_6\text{H}_4\diagdown\text{C}{=}\text{C}\diagup^{\text{F}}_{\text{H}} \quad (\text{H below left C}) \xrightarrow[\text{MeOH}]{\text{MeO}^-} \quad p\text{-NO}_2\text{C}_6\text{H}_4\diagdown\text{C}{=}\text{C}\diagup^{\text{H}}_{\text{OMe}}$$

(Z-18-F)   **(E-19)**

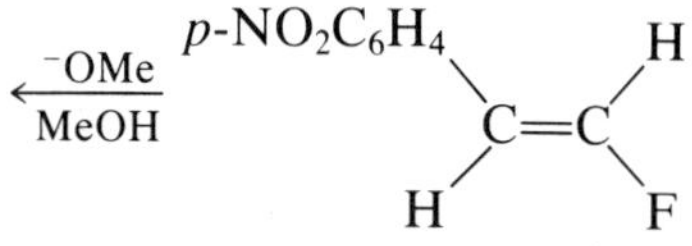

$$\xleftarrow[\text{MeOH}]{^-\text{OMe}} \quad p\text{-NO}_2\text{C}_6\text{H}_4\diagdown\text{C}{=}\text{C}\diagup^{\text{H}}_{\text{F}}$$

(E-18-F)

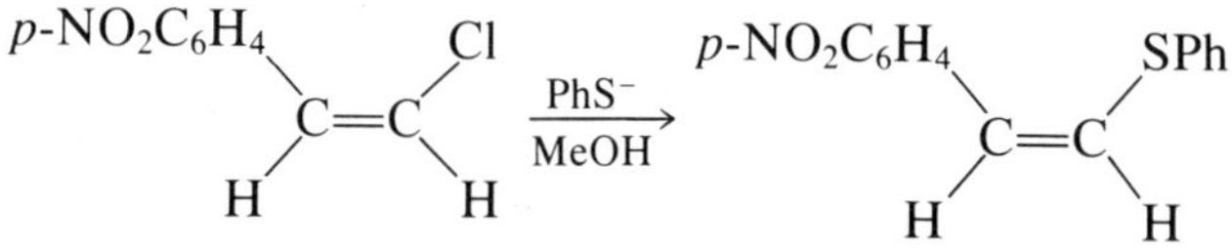

$$p\text{-NO}_2\text{C}_6\text{H}_4\diagdown\text{C}{=}\text{C}\diagup^{\text{Cl}}_{\text{H}} \xrightarrow[\text{MeOH}]{\text{PhS}^-} p\text{-NO}_2\text{C}_6\text{H}_4\diagdown\text{C}{=}\text{C}\diagup^{\text{SPh}}_{\text{H}}$$

(Z-18-Cl)   **(Z-20)**

$$p\text{-NO}_2\text{C}_6\text{H}_4\diagdown\text{C}{=}\text{C}\diagup^{\text{H}}_{\text{Cl}} \xrightarrow[\text{MeOH}]{\text{PhS}^-} p\text{-NO}_2\text{C}_6\text{H}_4\diagdown\text{C}{=}\text{C}\diagup^{\text{H}}_{\text{SPh}}$$

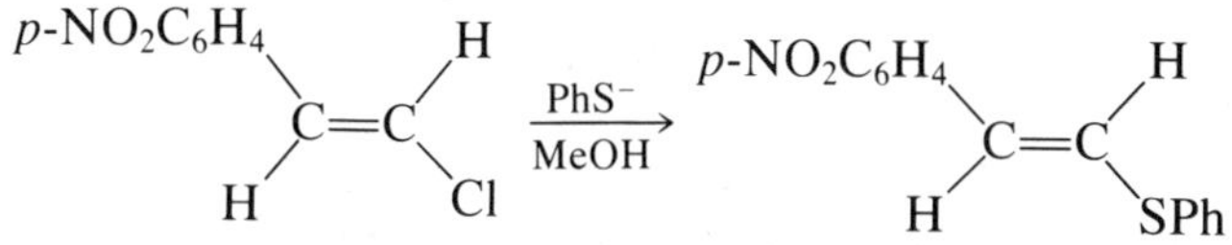

(E-18-Cl)   **(E-20)**

$$\xleftarrow[\text{MeOH}]{^-\text{SPh}} \textbf{Z-18-F or E-18-F}$$

where the ratios of E : Z vinyl ethers are equal from $-70°$ to $0°$.[30] If negative charge in $\{p\text{-NO}_2\text{C}_6\text{H}_4\text{CHCF}_2\text{OMe}\}^-$ is on oxygen, rotation around the

$${=}\text{C}\diagdown_{\text{CF}_2\text{OMe}}^{\diagup}$$ bond can occur prior to the elimination of fluoride. Clearly,

chloride is a better leaving group than fluoride, but the loss of fluoride can compete readily with the proton-transfer reaction. To test the hypothesis that the results from reactions of **18-F** were caused by pi-delocalization rather than the leaving ability of fluoride, we studied the reactions of PhC-$(\text{CF}_3){=}\text{CHF}$ **(21)** with methanolic methoxide.

Reaction of both isomers of **21** resulted in displacement of fluoride with complete retention, and addition of methanol (3–11%) was also observed.[39] Formation of **23** argues against a concerted mechanism, and favors a two-step mechanism similar to that for reaction of PhC(CF₃)=CF₂ **(1)** with methanolic methoxide. That more saturated ether **7** (44%) results from **1**, is reasonable since fluoride leaving from -CHFOMe should be better than from

$$\underset{\text{CF}_3}{\overset{\text{Ph}}{>}}C=C\underset{\text{H}}{\overset{\text{F}}{<}} \quad \xrightarrow[\text{MeOH, 0°}]{^-\text{OMe}} \quad \underset{\text{CF}_3}{\overset{\text{Ph}}{>}}C=C\underset{\text{H}}{\overset{\text{OMe}}{<}} \quad + \text{ PhCH(CF}_3)\text{CHFOMe}$$

(E-21)   97% **(E-22)**   3% **(23)**

$$\underset{\text{CF}_3}{\overset{\text{Ph}}{>}}C=C\underset{\text{F}}{\overset{\text{H}}{<}} \quad \xrightarrow[\text{MeOH, 0°}]{^-\text{OMe}} \quad \underset{\text{CF}_3}{\overset{\text{Ph}}{>}}C=C\underset{\text{OMe}}{\overset{\text{H}}{<}} \quad + \text{ PhCH(CF}_3)\text{CHFOMe}$$

(Z-21)   89% **(Z-22)**   11% **(23)**

-CF$_2$OMe.[40] Therefore, protonation of {PhC(CF$_3$)CF$_2$OMe}$^{\ominus}$ is more favorable compared to fluoride loss than is the case for (PhC(CF$_3$)CHFOMe}$^{\ominus}$ **(INT-5-F)**. The ratio of disastereomers **23** formed from **E-21** or **Z-21** is the same (45:55). Figure 6-1 attempts to rationalize the stereochemistry. The initial intermediate, **INT-5-F**, has three options: (A) formation of a hydrogen bond to methanol that is anti-periplanar to C—OMe, (B) rotation of 60° and the formation of a similar hydrogen bond anti-periplanar to the C—F bond, or (C) rotation of 60° without the formation of a hydrogen bond. The latter conformer can lead to formation of the vinyl ether with retention. Since measured PKIE associated with reaction of **21** and **23** are similar to those for reactions of **1, 4,** or **7,** Scheme III should also define the reaction mechanism for **21**.

Reaction of **E-21** is 370 times faster than that for (*E*)-PhC(CF$_3$)=CHCl with methanolic methoxide. The latter occurs with 100% retention to form **E-22** as the only product. The inverse element effect, $k^{\text{Cl}}/k^{\text{F}} \ll 1.0$, is similar to that reported for aromatic[41] and other vinyl substitutions.[37] Therefore an earlier report[42] that the reaction of PhC(CF$_3$)=CFCl **(24)** with methanolic methoxide resulted in only 96% retention to give PhC(CF$_3$)=CFOMe **(8)** seemed odd. Some substitution of fluoride was also thought to occur, but this was not experimentally documented.[43]

Since the ethoxide-promoted dehydrochlorination of PhCHClCF$_2$Cl is 5.3 × 10^5 faster than elimination of HF from PhCHClCF$_3$,[44] the loss of fluoride from {PhC(CF$_3$)CFClOMe}$^-$ **(INT-6-F)** should not be competitive with chloride ejection. The relative rates quoted are for elimination of HCl that occurs via a hydrogen-bonded intermediate while those for dehydrofluorination are from a free carbanion.[11] *Can fluoride compete with chloride leaving from a carbanion not stabilized by a hydrogen bond?* This question was important enough to reinvestigate the reactions of **24**. Elimination of some fluoride (4%) does occur during the reaction of **24** with methanolic methoxide, Figure 6-2.[45] The initial intermediate, **INT-6-F**, can undergo a 60° clockwise rotation that results in the more stable rotamer, **INT-6-Fa**,[9,46] with the C—Cl bond periplanar to the lone pair electrons or a counterclockwise

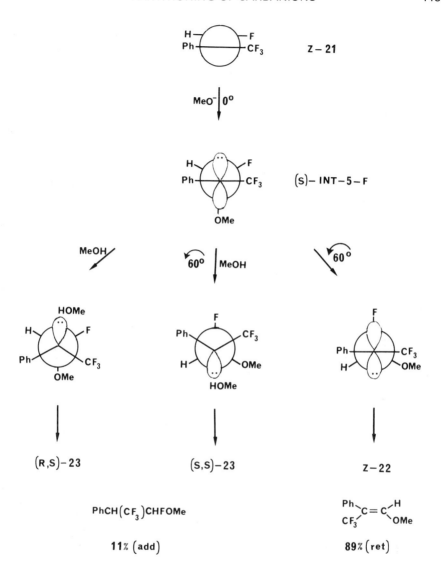

Figure 6-1. Reaction of Z-PhC(CF$_3$) = CHF with methanolic sodium methoxide.

rotation of 60° placing the C—F bond periplanar to the lone pair, **INT-6-Fb**. Loss of chloride from **INT-6-Fa** yields the retention product **Z-8**. Two options are available for **INT-6-Fb**: loss of fluoride forming E-PhC(CF$_3$)= CClOMe **(E-25)**, or a second 60° counterclockwise rotation generating another rotomer with a C—Cl bond periplanar to the lone pair, **INT-6-Fc**. The loss of chloride from **INT-6-Fc** accounts for the 6% inversion product, **E-8**. Similar results were obtained from the reaction of **Z-24** and methanolic methoxide. Assignment of isomers for **25** was based on chemical shifts for

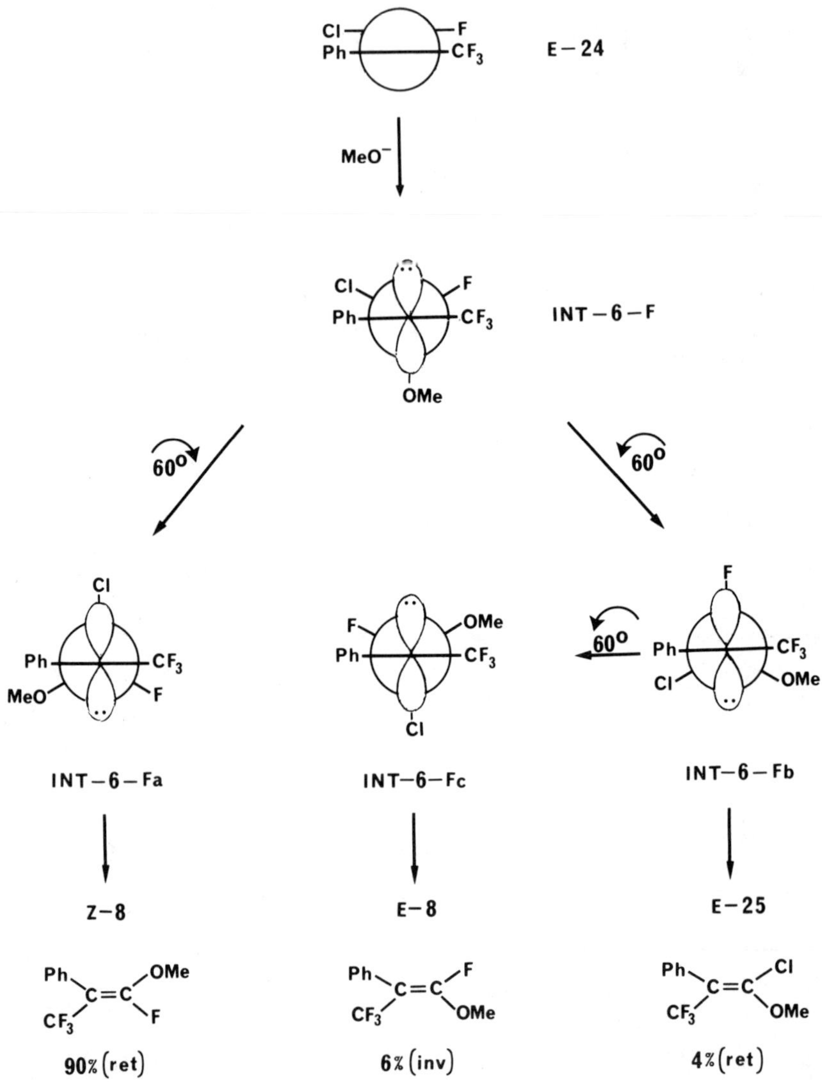

Figure 6-2. Reaction of (E)-PhC(CF$_3$) = CFCl with methoxide.

the trifluoromethyl group since there is no coupling to help determine which is the E- or Z-isomer.

7. MECHANISMS OF ALKOXIDE-PROMOTED DEHYDROHALOGENATIONS

The consequences of forming hydrogen-bonded as well as free carbanions along a reaction pathway have important mechanistic implications, and care must be taken in choosing models to elucidate the mechanism of a reaction.

Scheme III indicates that beta chloride or bromide can leave from the hydrogen-bonded intermediate **H**, while intermediate **F** is required for the loss of fluoride. This conclusion was reached since the ethoxide-promoted dehydrochlorination of PhC^iHClCF_2Cl (25) occurs 5×10^5 faster than the elimination of HF from PhC^iHClCF_3 (26).[44] These results can be compared to those of the corresponding β-phenethyl halides,[47] where elimination of HCl from $PhCH_2CH_2Cl$ (27) is 81 times faster than the loss of HF from $PhCH_2CH_2F$ (28), Table 6-3. Since the α-chlorine and two β-fluorines increase benzylic hydrogen reactivity, it is not surprising that dehydrochlorination of 25 occurs 10^4 faster than that for 27. Dehydrofluorinations are the apparent anomaly. The $\Delta H^{\ddagger} = 25.5$ kcal M^{-1} measured for the dehydrofluorination of 28 compared to $\Delta H^{\ddagger} = 29.7$ kcal M^{-1} for 26 would suggest that reaction of 28 should occur much faster than that for 26; however, a $\Delta S^{\ddagger}$ of 18 eu favors the reaction of 26, which reacts 10 times faster than 28. The C—F bonds in CF_3 groups are far stronger than those in CH_2F and this would explain the enthalpy difference favoring reaction of 28. We have suggested that a more favorable entropy term comes from extensive internal return associated with the reaction of 26.[40]

Near unity PKIE indicates that internal return substantially decreases the observed reactivity for both dehydrofluorination of 26 and the exchange of 26-d.[11] There is no exchange of the benzylic proton of 25 prior to elimination when reaction takes place in EtOD. *Since there is no exchange of proton prior to loss of HCl, does reaction occur by a concerted process?* Incorporation of deuterium can be taken as positive evidence for carbanion formation; however, the lack of exchange cannot rule out the formation of a carbanion. It means the loss of a leaving group can occur much faster than the proton-transfer reaction to neutralize the carbanion.[48]

TABLE 6-3. Rates, Activation Parameters, and PKIE for Alkoxide-Promoted Dehydrohalogenation and Exchange Reactions in Alcohols at 25°C

Compound		ROH	k, $M^{-1}s^{-1}$	$\Delta H^{\ddagger}$, kcal	$\Delta S^{\ddagger}$, eu	k^H/k^D (°C)
$PhCH_2CH_2F^a$	(28)	EtOH	4.34×10^{-8}	25.5	−6	4.50 (50)^b
$PhCHClCF_3^c$	(26)	EtOH	4.40×10^{-7}	29.7	12	1.04 (75)
$PhCDClCF_3^d$	(26-d)	EtOH	8.73×10^{-6}	26.8	8	1.06 (75)
$PhCH_2CH_2Cl^a$	(27)	EtOH	3.52×10^{-6}	23.2	−6	7.42 (25)^e
$PhCHClCF_2Cl^f$	(25)	EtOH	2.33×10^{-1}	19.5	4	2.73 (25)

^{a} Data from Ref. 47a.
^{b} Measured in t-BuOH, Ref. 47b.
^{c} Data from Ref. 11.
^{d} Results for an exchange reaction.
^{e} Unpublished, K. S. Sweinberg.
^{f} Data from Ref. 52.

Although ethoxide-promoted dehydrofluorination of **26** results in near unity PKIE, dehydrochlorination of **25** gives k^H/k^D values of 2.73 (ethoxide) and 2.29 (methoxide) at 25°. The methoxide-promoted dehydrohalogenation of $PhCHBrCF_2Cl$ vs $PhCHBrCF_2Br$ (**29**) has a large element effect, k^{Br}/k^{Cl} = 49 at 25°,[49] and the PKIE associated with **29** is k^H/k^D = 4.00 at 25°. Elimination from $PhCHBrCFClBr$ is anti- and stereospecific.[50] This is a textbook example of experimental evidence that accompanies an elimination occurring by an E2 mechanism;[51] however, a more detailed investigation of the isotope effects associated with these dehydrohalogenations suggests that the reaction is not concerted. The hydrogen isotope effects are almost temperature independent: k^H/k^D = 2.71 (50°), 2.73 (25°), and 2.77 (0°) in EtOH and 2.30 (50°), 2.29 (25°), and 2.27 (0°) in MeOH.[49] This anomalous Arrhenius behavior of PKIE, coupled with different k^{35}/k^{37} for loss of HCl vs. DCl, suggests that a carbanion is formed along the reaction pathway.[52]

Chlorine isotope effects (k^{35}/k^{37}) have been compared to the element effects (k^{Br}/k^{Cl}) for eliminations of $PhC(CH_3)CH_2Cl$ (1.00590 and 52, EtOH at 75°), $PhCH_2CH_2Cl$ (1.00580 and 44, EtOH at 75°), $PhCHClCH_2Cl$ (1.00908 and 27, EtOH; 1.00978 and 31, MeOH at 25°), and $PhCHClCF_2Cl$ (1.01255 and 49, MeOH at 25°).[52] There is no correlation between the element effect and chlorine isotope effects associated with these reactions. A more detailed study using $YC_6H_4C^iHClCH_2X$ (where X = Cl, Br, or I) suggests a major portion of an experimental k^{Br}/k^{Cl} is due to the effect of β-halide on k_1, Scheme III.[53] Since measurement of heavy atom isotope effects requires special equipment, Bunnett's proposal of the "element effect" (the poor man's heavy isotope effect) was ingenious, and works well for nucleophilic aromatic and vinyl substitutions. Based on our results, we question its use to assign mechanisms to elimination reactions.

Another argument against a two-step mechanism for dehydrohalogenation reactions is that the rate of elimination is faster than anticipated proton transfer as predicted by a pK_a difference between the carbon acid and the conjugate acid of the reacting base.[54]

8. CORRELATION OF pK_a FOR CARBON ACIDS AND EXCHANGE RATES

Interest in pK_a values for carbon acids is to help predict relative reactivities for reactions promoted or catalyzed by a base. In many cases, rates of proton transfer from carbon to the base are of greater value than equilibrium measurement. Ritchie[55] concluded: "those carbon acids whose conjugate bases have localized charge are predicted to have proton transfer rates considerably greater than acids of the same thermodynamic strength whose conjugate bases have delocalized charges. That is, saturated hydrocarbons, alkenes, alkynes, and cycloalkanes whose conjugate bases are localized are expected to show 'kinetic acidities' greater than their thermodynamic acidities." An example is methoxide-catalyzed protodetritiation of pentafluoro-benzene (PFB) is 15 times faster than that for 9-phenylfluorene (9-PhFl) at 25°C, Table 6-2. Relative pK_a values measured in cyclohexylamine (CHA) are 18.5 for 9-PhFl and 25.8 for PFB.[56]

The weaker carbon acid, by 7.3 pK_a units, exchanges its acidic proton an order of magnitude faster than the stronger carbon acid. The 9-phenyl-9-fluorenyl anion (9-PhFl⁻) is highly pi-delocalized, and negative charge of the pentafluorophenyl anion (PFB⁻) is localized in an sp^2 orbital that is ortho-ganal to the pi-system. *What about rates of the proton-transfer reactions for these compounds?* PKIE data for 9-PhFl suggests a negligible amount of internal return for that exchange reaction; however, the isotope effect, $k^D/k^T = 1.0$, associated with the exchange of PFB implies extensive internal return. The hydron transfer from PFB to methoxide actually occurs at least 50–100 times faster than the observed rates of exchange. So much for pK_a data to predict relative rates of a base-catalyzed reaction.

The pK_a obtained in CHA are ion-pair values. *Will values measured in Me₂SO give a better correlation?* Since PFB⁻ is not stable, the pK_a for PFB cannot be measured in Me₂SO.[57] The expected pK_a in Me₂SO would be greater than that found in CHA. A ΔpK_a of ca. 5 units between phenylacety-lene and 9-PhFl in CHA increases to about 11 pK_a units in Me₂SO.[58] The ΔpK_a of 7.3 units measured in CHA should be larger in Me₂SO and be a worse indicator.

Values of pK_a cited for measurements in CHA and most of those reported for Me₂SO are obtained using an indicator method that utilizes carbon acids within several pK_a units of the compound to be determined. Therefore, the indicator method compares energy levels of "free" carbanions. The reason phenylacetylene has a lower pK_a in CHA than in Me₂SO can be attributed to the stabilizing effect of the ion-pair gegen-ion.

The highly delocalized 9-PhFl⁻ is more stable than the hydrogen-bonded carbanion, 9-PhFl⁻ · · · HOMe. As hydron is transferred from 9-PhFl to MeO⁻, the incipient intermediate is at a higher energy state than the deloca-lized species. Prior to delocalization, the carbon must rehybridize and break the hydrogen-bond, which is competitive with the internal-return reaction. Proton transfer from PFB to methoxide can occur at a lower energy than that for 9-PhFl⁻; however, PFB⁻ losses stability as the free ion and requires a significant energy to break the hydrogen bond. Internal return is the low free energy path, and washes out the isotope effects associated with proton transfer.*

The exchange of 1,4-dihydroperfluorobicyclo[2.2.1]heptane (14), pK_a = 22.3 (CHA),[21] undergoes protodetritiation 82 times faster than 9-PhFl at 25°C, Table 6-2, and is another example of this apparent anomaly. Scheme III offers an explanation. The pK_a are obtained from comparing energetics of "free ions," and kinetics are an indication of energetics for the transition structures leading to the hydrogen-bonded carbanion for 9-PhFl vs. the free carbanion for PFB and 14. Ions generated from PFB and 14 can have charge delocalized, but not through a pi network. Regardless of the mechanism for this stabilization, the effect is felt in the hydrogen-bonded intermediate.[59] Again the effect of carbon-fluorine bonds dramatically increases the rates of exchange. This allows the study of carbanions that do not require pi-delocalization to aid in the stabilization of the negative charge.

Bordwell and Boyle[60] suggested a localized carbanion is the initial intermediate resulting from a base-catalyzed proton removal from nitroalkanes, and electron delocalization to form a more stable intermediate lags behind proton transfer. The PKIE associated with methoxide-catalyzed exchange of **9-PhFl** and **9-MeFl** are consistent with an internal return mechanism, and suggests that delocalization of the electrons lags behind proton transfer. Bernasconi[61] deals with this problem and refers to it as the principle of imperfect synchronization. Our interpretation would be to use Scheme III to explain the results. The initial intermediate formed by the transfer of proton from a C—H bond to a base will be a hydrogen-bonded species. Prior to delocalization, the carbon must rehybridize and this generates the more stable species.

9. CONCLUDING REMARKS

A major goal for investigations of chemical reaction mechanisms is to understand not only the detailed pathway of a reaction, but also the timing of various steps along that pathway. Proton transfer is a basic reaction of fundamental importance, yet many details of that process are still not known. The kinetic studies of Ahlberg and Thibblin report enhanced values of PKIE due to ion-pair intermediates in base-promoted eliminations,[62] and temperature-independent PKIE.[63] Thibblin has also reported some excellent work on hydrogen-bonded carbanion intermediates that include ArOH and quaternary ammonium ions as the hydrogen source.[64]

A lack of exchange with solvent prior to elimination does not rule out the occurrence of a carbanion intermediate. Buncel and Bourns discuss this in depth in their paper on the mechanism of the ethoxide-promoted carbonyl elimination reaction of benzyl nitrate,[48] yet one still reads statements like: the presence of a carbanion is ruled out due to a lack of exchange prior to elimination. They measured both hydrogen ($k^H/k^D = 5.0$ at 60°) and nitrogen ($k^{14}/k^{15} = 1.0196$ at 30°) isotope effects associated with that elimination, and concluded that reaction occurred by a concerted mechanism rather than a two-step pathway. At that time (1960) another interpretation of the data would not have gotten past a reputable referee.

Similar results were obtained for methoxide-promoted elimination of PhC^{i}HClCH$_2$Cl ($k^H/k^D = 3.83$ and $k^{35}/k^{37} = 1.00978$ at 25°).[52] The Arrhenius behavior of hydrogen PKIE associated with this reaction ($\Delta E_a^{D-H} = 0.79$ and $A^H/A^D = 0.99$) are consistent with residual zero point energy in an asymmetric transition structure; however, this type of behavior can also occur in a reaction that features internal return.[34] A difference in chlorine isotope effects for the elimination of DCl, $k^{35}/k^{37} = 1.00776$, favors the two-step mechanism with internal return. Accurate rates for the loss of TCl were not possible due to a competing elimination that forms PhCH=CHCl, but the rates for all three isomers was possible for m-ClC$_6$H$_4$C^{i}HCH$_2$Cl, and resulted in the calculation of: $a^H = 0.4$, $a^D = 0.1$, and $a^T = 0.03$ at 25°.[53] The ethoxide-promoted elimination reaction of benzyl nitrate may occur by a concerted mechanism, but formation of a carbanion cannot be ruled out.

Methods for studying reaction mechanisms are more sophisticated now than in 1960. The polar C—F bond can stabilize localized carbanions, which allows their investigation.[65] The reactions of fluoroalkenes can generate carbanions in situ, and permits their study under mild conditions in protic solvents. One should not predict the behavior of localized carbanions from the results of investigations on extensive pi-delocalized carbanions.

ACKNOWLEDGMENTS

We wish to acknowledge support by National Science Foundation grant no. CHE-8316219.

REFERENCES AND NOTES

1. Koch, H. F. *Acc. Chem. Res.* **1984,** *17,* 137.
2. Miller, W. T., Jr.; Fager, E. W.; Griswold, P. H. *J. Am. Chem. Soc.* **1948,** *70,* 431.
3. (a) Hine, J.; Wiesboek, R.; Ghirardelli, R. G. *J. Am. Chem. Soc.* **1961,** *83,* 1219. (b) Hine, J.; Wiesboek, R.; Ramsay, O. B. *ibid.* **1961,** *83,* 1222.
4. Koch, H. F.; Kielbania, A. J., Jr. *J. Am. Chem. Soc.* **1970,** *92,* 729.
5. Koch, H. F.; Koch, J. G.; Donovan, D. B.; Toczko, A. G.; Kielbania, A. J., Jr. *J. Am. Chem. Soc.* **1981,** *103,* 5417.
6. Stirling, C. J. M. *Acc. Chem. Res.* **1979,** *12,* 198.
7. Kresge, A. J. *Acc. Chem. Res.* **1975,** *9,* 354.
8. Rappoport, Z. *Acc. Chem. Res.* **1981,** *14,* 7.
9. Bach, R. D.; Wolber, G. J. *J. Am. Chem. Soc.* **1984,** *106,* 1401.
10. Bach, R. D.; Wolber, G. J. *J. Am. Chem. Soc.* **1985,** *107,* 1352.
11. Koch, H. F.; Dahlberg, D. B.; Lodder, G.; Root, K. S.; Touchette, N. A.; Solsky, R. L.; Zuck, R. M.; Wagner, L. J.; Koch, N. H.; Kuzemko, M. A. *J. Am. Chem. Soc.* **1983,** *105,* 2394.
12. Saunders, W. H., Jr. *Chem. Scr.* **1975,** *8,* 27.
13. (a) Streitwieser, A., Jr.; Hollyhead, W. B.; Sonnichsen, G.; Pudjaatmaka, A. H.; Chang, C. J.; Kruger, T. C. *J. Am. Chem. Soc.* **1971,** *93,* 5096. (b) Boerth, D. W.; Streitwieser, A., Jr. *J. Am. Chem. Soc.* **1981,** *103,* 6443.
14. Swain, C. G.; Stivers, E. C.; Reuwer, J. F.; Schaad, L. J. *J. Am. Chem. Soc.* **1958,** *80,* 5885.
15. Streitwieser, A., Jr.; Hollyhead, W. B.; Pudjaatmaka, A. H.; Owens, P. H.; Kruger, T. L.;

Rubenstein, P. A.; MacQuarrie, R. A.; Brokaw, M. L.; Chu, W. K. C.; Niemeyer, H. M. *J. Am. Chem. Soc.* **1971,** *93,* 5088.

16. Albery, W. J.; Knowles, J. R. *J. Am. Chem. Soc.* **1977,** *99,* 637.

17. Since k^H was measured in EtOH and k^D in EtOD, correction was made for the kinetic solvent isotope effect. This effect can be between k^{OD}/k^{OH} = 2.0 to 2.5 for systems that we have studied. An accurate value can only be assigned if k^T is known for both solvents.

18. Koch, H. F.; Koch, A. S. *J. Am. Chem. Soc.* **1984,** *106,* 4536.

19. Andreades, S. *J. Am. Chem. Soc.* **1964,** *86,* 2003.

20. This approach to calculate the KSIE was used by others in the early 1960s. The concept of internal return was proposed at that time and not too widely accepted. As a result, experimental sections must be read carefully before relying on values of KSIE reported in the text of papers.

21. Streitwieser, A., Jr.; Holtz, D.; Ziegler, G. R.; Stoffer, J. O.; Brokaw, M. L.; Guibe, F. *J. Am. Chem. Soc.* **1976,** *98,* 5229.

22. Streitwieser, A., Jr.; Hudson, J. A.; Mares, F. *J. Am. Chem. Soc.* **1968,** *90,* 648.

23. Keevil, T. A. Doctoral Dissertation, University of California, Berkeley, 1972.

24. This is treated using theoretical calculations: (a) Streitwieser, A.; Berke, C. M.; Schriver, G. W.; Grier, D.; Collins, J. B. *Tetrahedron Suppl.* **1981,** *37,* 345. (b) Pross, A.; DeFrees, D. J.; Levi, B. A.; Pollack, S. K.; Radom, L. Hehre, W. J. *J. Org. Chem.* **1981,** *46,* 1693. (c) Schleyer, P. v. R.; Kos, A. J. *Tetrahedron* **1983,** *39,* 1141.

25. Bockrath, B.; Dorfman, L. M. *J. Am. Chem. Soc.* **1974,** *96,* 5708.

26. Wiberg, R. *J. Am. Chem. Soc.* **1955,** *77,* 5987; Assarson, L. O. *Acta Chem. Scand.* **1958,** *12,* 1545; Pocker, Y.; Exner, J. H. *J. Am. Chem. Soc.* **1969,** *90,* 6764.

27. (a) Alexander, R.; Asomaning, W. A.; Eaborn, C.; Jenkins, I. D.; Walton, D. R. M. *J. Chem. Soc. , Perkin Trans.* 2, **1974,** 490. (b) Eaborn, C.; Walton, D. R. M.; Seconi, G. *J. ibid,* **1976,** 1857. (c) Macciantelli, D.; Seconi, G.; Eaborn, C. *ibid,* **1978,** 834.

28. Bott, R. W.; Eaborn, C.; Swaddle, T. W. *J. Chem. Soc.* **1963,** 2342.

29. (a) Seconi, G.; Eaborn, C.; Fischer, A. J. *J. Organomet. Chem.* **1979,** *177,* 129. (b) Seconi, G.; Eaborn, C.; Stamper, J. G. *ibid,* **1981,** *204,* 153.

30. Koch, H. F.; Koch, J. G.; Koch, N. H.; Koch, A. J. *J. Am. Chem. Soc.* **1983,** *105,* 2388.

31. This assumption has validity since a fractionation factor of 1.0 has been reported for fluoride and H_2O and D_2O.[32]

32. Albery, W. J. In "Proton-Transfer Reactions"; Caldin, E. F.; Gold, V. Eds.; Chapman and Hall: London, 1975, p 283.

33. Gold, V.; Grist, S. *J. Chem. Soc.* (B) **1971,** 2282.

34. Koch, H. F.; Dahlberg, D. B. *J. Am. Chem. Soc.* **1980,** *102,* 6102.

35. Parker, V. D.; Tilset, M. *J. Am. Chem. Soc.* **1986,** *108,* 6371.

36. Marchese, G.; Naso, F. *La. Chim. E. Ind.* (Milano) **1971,** *53,* 760.

37. Marchese, G.; Naso, F.; Modena, G. *J. Chem. Soc.* (B) **1969,** 290.

38. Marchese, G.; Naso, F.; Modena, G. *J. Chem. Soc.* (B) **1968,** 958.

39. Koch, H. F.; Koch, J. G.; Kim, S. W. Abst. of Papers for VI IUPAC Conference on Physical Organic Chemistry, *Bull. Soc. Chim. Belg.* **1982,** *91,* 431.

40. Koch, H. F.; Tumas, W.; Knoll, R. *J. Am. Chem. Soc.* **1981,** *103,* 5423.

41. Bunnett, J. F.; Garbisch, E. W.; Pruitt, K. M. *J. Am. Chem. Soc.* **1957,** *79,* 385.

42. Burton, D. J.; Krutzch, H. C. *J. Org. Chem.* **1971,** *36,* 2351.

43. Krutzch, H. C. Doctoral Dissertation, University of Iowa, Iowa City, 1971.

44. Koch, H. F.; Dahlberg, D. B.; Toczko, A. G.; Solsky, R. L. *J. Am. Chem. Soc.* **1973,** *95,* 2029.

45. Koch, H. F.; Koch, J. G.; Barnes, M. J. Abst. of Papers for VII IUPAC Conference on Physical Organic Chemistry, Auckland, NZ, August 20–24, 1984.

46. Apeloig, Y.; Rappoport, Z. *J. Am. Chem. Soc.* **1979,** *101,* 5095.

47. (a) DePuy, C. H.; Bishop, C. A. *J. Am. Chem. Soc.* **1960,** *82,* 2535. (b) DePuy, C. H.; Schultz, A. C. *J. Org. Chem.* **1974,** *39,* 878.

48. Buncel, E.; Bourns, A. N. *Can. J. Chem.* **1960,** *38,* 2457.

49. Koch, H. F.; Dahlberg, D. B.; McEntee, M. F.; Klecha, C. J. *J. Am. Chem. Soc.* **1976,** *98,* 1060.

50. Root, K. S.; Touchette, N. A.; Koch, J. G.; Koch, H. F. Abst. of Papers, Euchem Conference on Mechanisms of Elimination Reactions, Assisi, Italy, Sept., 1977.

51. (a) Streitwieser, A., Jr.; Heathcock, C. H. "Introduction to Organic Chemistry", Third Ed., Macmillan, New York, 1985, p. 242–248. (b) Morrison, R. T.; Boyd, R. N. "Organic Chemistry", Fifth Ed. Allyn and Bacon: Boston, 1987, p. 272–278.

52. Koch, H. F.; McLennan, D. J.; Koch, J. G.; Tumas, W.; Dobson, B.; Koch, N. H. *J. Am. Chem. Soc.* **1983,** *105,* 1930.

53. Koch, H. F.; Koch, J. G.; Lodder, G.; Hage, R.; Bogdan, D. J. Abst. of Papers, International Symposium on Organic Reactivity, Paris, France, July 1987.

54. (a) Bordwell, F. G. *Acc. Chem. Res.* **1972,** *5,* 377. (b) Saunders, W. H., Jr., *Ac. Chem. Res.* **1976,** *9,* 21.

55. Ritchie, C. D. *J. Am. Chem. Soc.* **1969,** *91,* 6479.

56. Streitwieser, A., Jr.; Scannon, P. J.; Neimeyer, N. M. *J. Am. Chem. Soc.* **1972,** *94,* 7936.

57. Bordwell, F. G. Personal communication.

58. Matthews, M. S.; Bares, J. E.; Bartmess, J. E.; Bordwell, F. G.; Cornforth, F. J.; Drucker, G. E.; Margolin, Z.; McCallum, R. J.; McCollum, G. J.; Vanier, N. R. *J. Am. Chem. Soc.* **1975,** *97,* 7006.

59. Schleyer and Kos[24c] discuss the importance of anionic hyperconjugation from a theoretical point, and disagree with the interpretation of calculations by Streitwieser et al.[24a] Both groups agree that there is a large stabilization due to perfluoro substituents, but disagree on the mechanism of stabilization.

60. Bordwell, F. G.; Boyle, W. J., Jr. *J. Am. Chem. Soc.* **1975,** *97,* 3447.

61. Bernasconi, C. F. *Acc. Chem. Res.* **1987,** *20,* 301.

62. Thibblin, A.; Bengtsson, S.; Ahlberg, P. *J. Chem. Soc.* Perkin Trans. II **1977,** 1569.

63. Thibblin, A.; Onyido, I.; Ahlberg, P. *Chem. Scr.* **1982,** *19,* 145.

64. (a) Thibblin, A. *J. Am. Chem. Soc.* **1983,** *105,* 853. (b) Thibblin, A. *J. Chem. Soc. Chem. Comm.* **1984,** 92.

65. Chambers R. D. In "Comprehensive Carbanion Chemistry", Part C; Buncel E.; and Durst, T., Eds.; Elsevier, Amsterdam, 1987.

CHAPTER 7

The Effect of Fluorination on Enolates and Enolate Equivalents

John T. Welch and Seetha Eswarakrishnan

Department of Chemistry, State University of New York at Albany, Albany, New York

CONTENTS

1. INTRODUCTION

The importance of enolate and enol reactivity is easily illustrated by the aldol condensation. The aldol condensation is of fundamental importance in the biosynthesis of a broad range of biologically significant natural products.[1] The recent development of stereoregulated variants of aldol addition reaction, which was first reported in 1838, has proven extremely useful to synthetic chemists.

Enol and enolate reactivity are key features of the aldol reaction. In the crossed aldol reaction, four possible product stereoisomers, 1–4, each containing two asymmetric centers may be created.

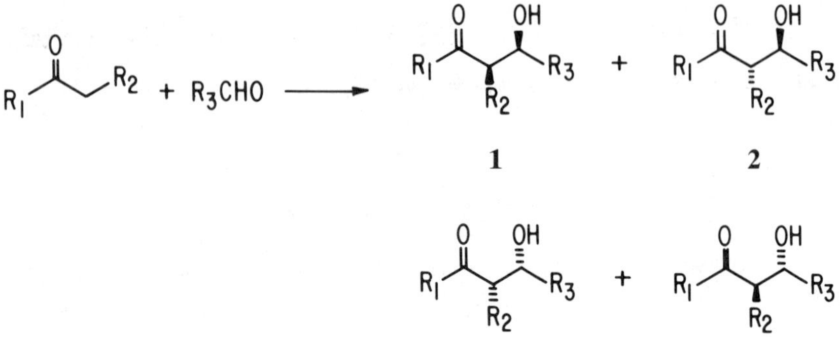

$$(7\text{-}1)$$

The control of this stereochemistry results from control of the geometry and reactivity of the enolate. Kinetic diastereoselection is strongly influenced by enolate geometry, either 5 or 6.[2] The most important variables appear to be the nature of R_1 and R_2 and the nature of the enolate counterion.

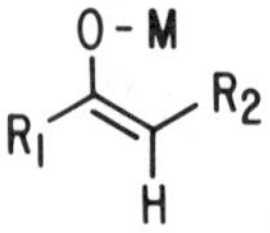

In this report we describe those cases where R_2 is fluorine. The normal correlation that is found when an alkali metal is the counterion is that Z enolates (5) tend to form syn aldols 1 and 3 and E enolates (6) tend to form anti aldols 2 and 4.

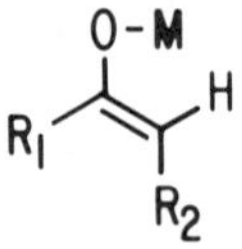

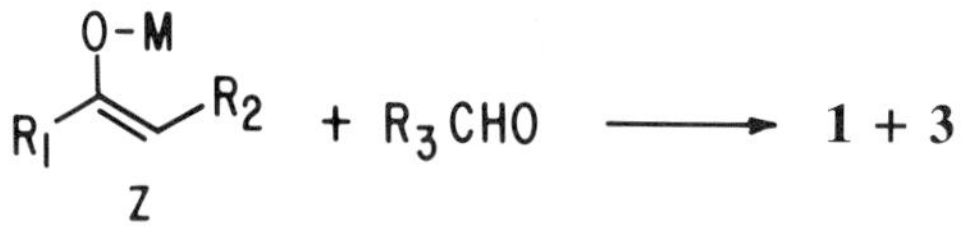

$$\underset{\underset{Z}{R_1}}{\overset{\overset{O-M}{\|}}{\diagup}}\!\!\diagdown R_2 \quad + \quad R_3CHO \quad \longrightarrow \quad 1 + 3 \qquad (7\text{-}2)$$

$$R_1\diagdown\!\!\underset{\underset{E}{R_2}}{\overset{\overset{O-M}{\|}}{\diagup}} \quad + \quad R_3CHO \quad \longrightarrow \quad 2 + 4 \qquad (7\text{-}3)$$

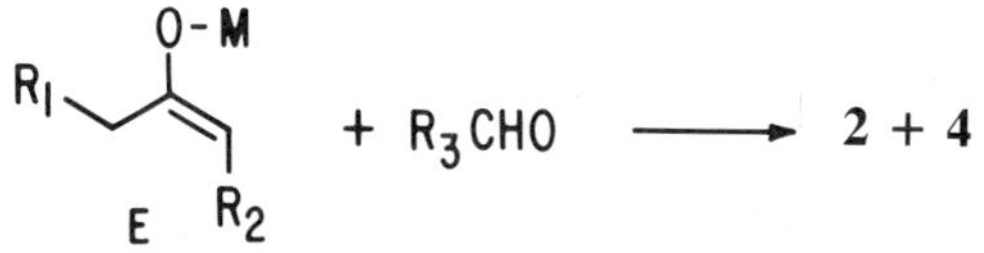

Both the nature of the base and the structure of the carbonyl component affect the geometry of the enolate. With a sterically demanding base, generally the E isomer is formed. The base approaches the C—H bond over the face of the carbonyl group, resulting in steric interactions between the base and the R_2 group, raising the energy of transition state $A^{\ddagger}$ relative to that of $B^{\ddagger}$ (Equations 7-4 and 7-5).

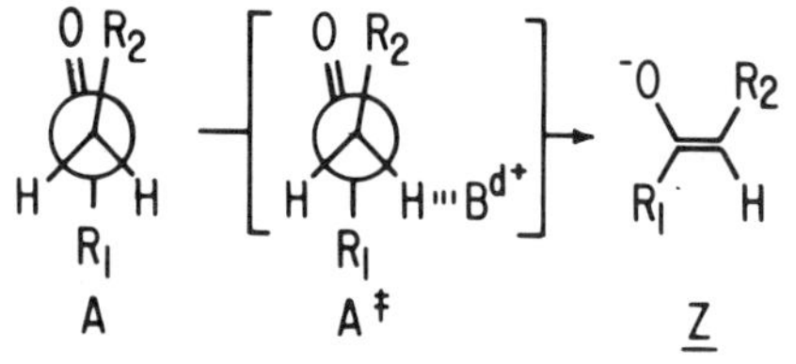

$$\text{(Newman projection A)} \quad \longrightarrow \quad \left[\text{A}^{\ddagger}\right] \quad \longrightarrow \quad Z \qquad (7\text{-}4)$$

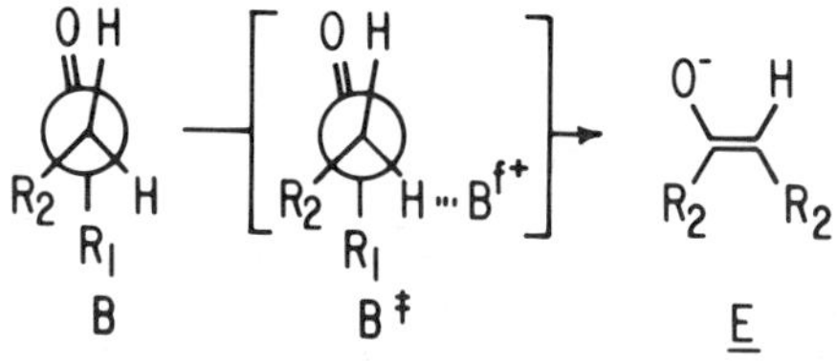

$$\text{(Newman projection B)} \quad \longrightarrow \quad \left[\text{B}^{\ddagger}\right] \quad \longrightarrow \quad E \qquad (7\text{-}5)$$

Both lithium hexamethyldisilazide (LHMDS) and solvent addends such as hexamethylphosphoric triamide (HMPA) promote equilibration of the kinetically formed enolate to the thermodynamically more favored isomer.[3] The influence of carbonyl structure on enolization stereoselection using dialkylamide bases can be rationalized by the Ireland enolization model.[1]

Deprotonation may occur via either of two metal centered pericyclic chairlike transition states, $T^{\ddagger}$ and/or $C^{\ddagger}$, in which synchronous proton transfer and metal ion transfer occur in a bimolecular process. Inspection of developing nonbonded interactions in these diastereomeric transition states reveals that developing R_1---R_2 allylic strain factors in transition state $T^{\ddagger}$ must be weighed against R_2---L nonbonded interactions in transition state $C^{\ddagger}$.

Correlation of enolate geometry to aldol product stereochemistry is possible employing the Zimmerman-Traxler hypothesis.[4] The π systems of the

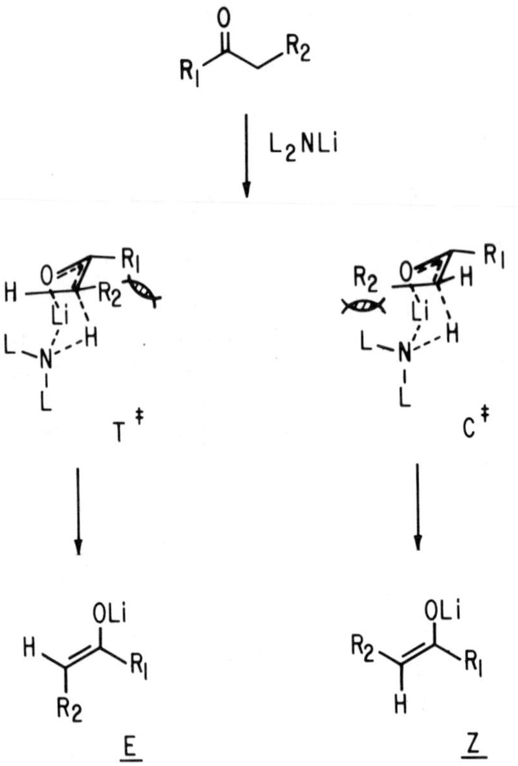

Figure 7-1. Ireland pericyclic model for enolate formation.

enolate and the carbonyl compound overlap in such a manner that the metal atom is chelated by the oxygen of the enolate and the carbonyl oxygen. Inspection of the possible diastereomeric transition states reveal that the lowest energy transition state will be the one in which 1,3-diaxial interactions are minimized. The "closed" or "chelated" transition state nicely accounts for most of the reported structure-stereoselectivity data. For E enolates, transition state E_e (E_e is so named for the reaction of the E enolate to form an erythro product; likewise Z_t describes the transition state leading to a threo product by a Z enolate) is predicted to be destabilized relative to E_t because of the R_1---R_3 interaction, thus forming the anti aldol predominantly. Similarly, transition state Z_t is destabilized relative to Z_e for Z enolates, leading to the syn aldol. Thermodynamic control of the aldol reaction leads to an enhancement of the anti diastereomer population, with the rate of equilibration largely dependent upon R_1.

Consideration of these factors proposed to control stereoselectivity of enolate formation and enolate reactivity in the aldol condensation suggests that steric effects are predominant. Considering the similarity in van der

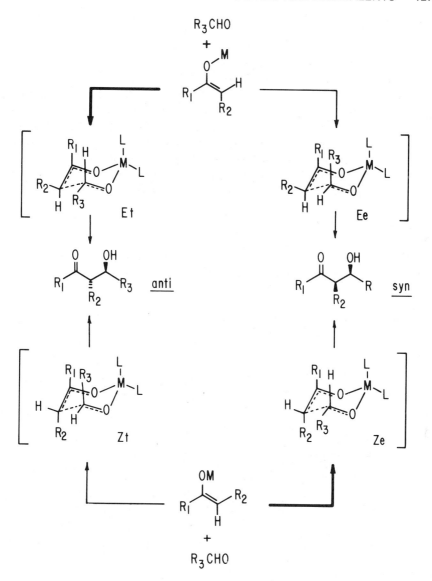

Figure 7-2. Zimmerman-Traxler aldol transition state model.

Waals radii, 1.35Å for C—CH$_3$, it is apparent that fluorine is not a sterically demanding substituent.[5] Stereoselectivity with α-fluorinated carbonyl compounds can then be viewed as a test of the limits to which steric effects can be pushed to control the reactions. However, electronic effects of fluorination, resulting from the high electronegativity of fluorine, may influence the course of the reaction in other ways.

2. α-FLUORINATED ENOLATES

Bergmann and co-workers prepared α-fluoro enolates many years ago but with no attempt to control the enolate geometry.[6]

In the Reformatsky reaction of ethyl α-bromo-α-fluoroacetate high diastereoselectivity was reported.[7] In a parallel work the lithium enolate of t-butyl fluoroacetate was generated, however both of these reports describe difficulty in preparing the lithium enolate of ethyl fluoroacetate.[8a] In more recent work a functionalized ester of 2-fluorohexanoate was successfully deprotonated with LDA and was employed in a non-diastereoselective directed aldol reaction.[8b]

A. Results

We have found that the lithium enolate of ethyl fluoroacetate, **7**, may be readily prepared and efficiently utilized in the directed aldol reaction.[9]

$$CH_2F\,CO_2CH_2CH_3 \xrightarrow{\text{LHMDS}} \overset{\displaystyle \overset{OLi}{|}}{CHF=C-OCH_2CH_3} \xrightarrow{\text{RR'CO}}$$

7   (7-6)

$$RR'C(OH)CHF\,CO_2CH_2CH_3$$

8

TABLE 7-1.   Products of Directed Aldol Reaction of Lithium Enolate of Ethyl Fluoroacetate LiCHFCO$_2$CH$_2$CH$_3$ + RR'CO → RR'COHCHFCO$_2$CH$_2$CH$_3$

R	R'	Yield	Internal asymmetric induction[a]
CH$_3$	(CH$_3$)$_3$C	95%	1:3.8
CH$_3$	CH$_2$CH$_3$	82%	1:1
CH$_3$	Ph	96%	1:1.6
CH$_3$	C$_5$H$_{11}$	93%	1:1.1
Ph	Ph	70%	
	2-adamantyl	75%	
	2-norbornyl	91%	1:1.6
	2-cyclohexenyl	91%	2.6:1
H	(CH$_3$)$_3$C	85%	1:3
H	C$_6$H$_{13}$	20%	1:2
H	Ph	93%	1:2
H	3,3-dimethyl-2,4-dioxol-l-yl	90%	1:1

[a] Stereocontrol, as determined by ^{13}C NMR in this case, where the newly created centers bear a specific relationship only between themselves is termed internal asymmetric induction. Bartlett, P. A. *Tetrahedron* **1980**, *36*, 2–73.

The lithium enolate may be successfully generated with LHMDS at $-78°$ in the presence of HMPA. Success in the generation of the enolate with LHMDS lead to reexamination of the use of LDA. If the temperature was held to $-105°$, the enolate could be prepared in the presence of HMPA. For yields of the aldol products and diastereoselectivity see Table 7-1.

Having successfully prepared the lithium enolate of ethyl fluoroacetate, the preparation of fluoroacetate ester enolates with bulky ester substituents was undertaken, ie, the steric demand of R_1 was increased. Deprotonation of benzyl fluoroacetate, **9**, was effected with LHMDS but the resulting enolate showed no improvement in diastereoselectivity in the directed aldol reaction[10] (Table 7-2).

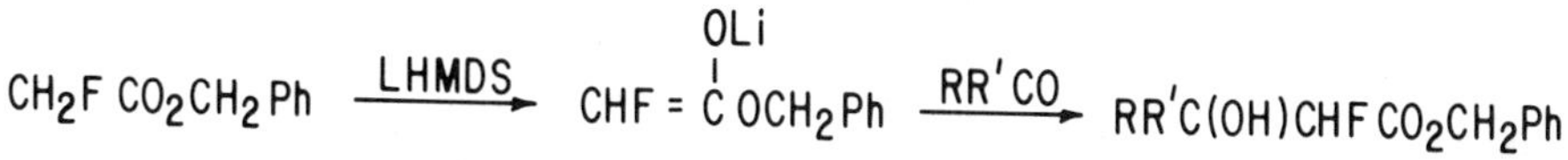

$$CH_2F\ CO_2CH_2\ Ph \xrightarrow{LHMDS} CHF=\overset{\overset{\displaystyle OLi}{|}}{C}OCH_2Ph \xrightarrow{RR'CO} RR'C(OH)CHF\ CO_2CH_2Ph$$

9   **10**

$$(7\text{-}7)$$

In contrast to the ease with which the ethyl and benzyl esters were deprotonated with LHMDS, 2,6-di-*t*-butyl-4-methyl-phenyl(BHT) fluoroacetate, **11**, required treatment with LDA to form the enolate.[11] BHT esters had previously been reported to be significantly more diastereoselective in the aldol reaction presumably as a result of increased steric bulk.[12] As can be seen in Table 7-3, the enolate was more diastereoselective. Ester enolates

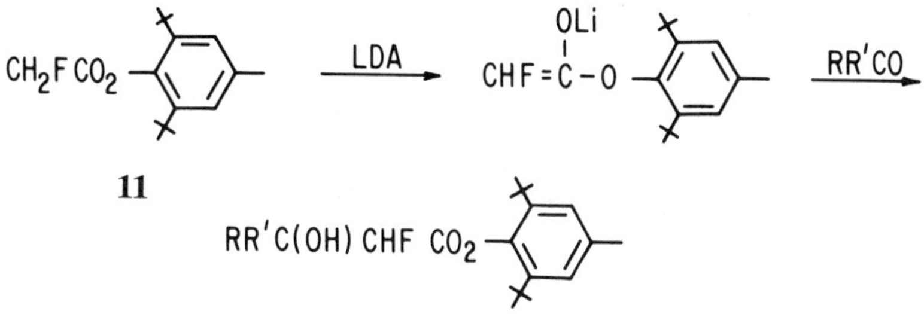

$$(7\text{-}8)$$

11   **12**

TABLE 7-2. Products of Directed Aldol Reaction of Lithium Enolate of Ethyl Fluoroacetate LiCHFCO₂CH₂Ph + RR'CO → RR'COHCHFCO₂CH₂Ph

R	R'	Yield	Internal asymmetric induction[a]
CH₃	Ph	39%	1:1.2
CH₃	C₅H₁₁Ph	50%	1:1.8
H	CH₂CH₃	66%	1:3
H	Ph	61%	1:1

[a] As determined by ¹³C NMR.

TABLE 7-3. Products of Directed Aldol Reaction of Lithium Enolate of 2,6-di-*t*-Butyl-4-Methyl-Phenyl (BHT) Fluoroacetate LiCHFCO$_2$CH$_2$CH$_3$ + RR'CO → RR'COHCHFCO$_2$BHT

R	R'	Yield	Internal asymmetric induction[a]
H	CH$_2$CH$_3$	70%	1 : 1.3
H	CH$_2$CH$_3$[c]	66%	1 : 2.8
H	CH$_2$CH$_3$[d]	92%	1 : 2.5
H	Ph	88%	1 : 7.5
H	(CH$_3$)$_2$C	79%	1 : 1
	(CH$_3$)$_3$C	83%	1 : 19

[a] As determined by ^{13}C NMR.

have been generally found to form E enolates which would therefore be expected to favor formation of the anti aldolate.[1a] Examination of the Zimmerman-Traxler transition state model would predict that increased bulk at R$_1$ by introduction of 2,6-di-*t*-butyl-4-methyl-phenyl group should favor greater anti selectivity.

As the diastereoselectivity of the fluoroacetate enolates was definitely improving as the steric bulk of ester increased, comparison of these findings with the diastereoselectivity observed with fluoroacetamides is useful. It was previously reported that amides can generally be deprotonated with great stereoselectivity for formation of the Z enolates. It was also known, however, that the diastereoselectivity of amide enolates was strongly dependent upon the nature of R$_2$.[13]

N,N-Dimethylfluoroacetamide, 13, was deprotonated with LDA and the amide enolate condensed with several aldehydes and ketones[14] (Table 7-4).

$$CH_2F\,CON(CH_3)_2 \xrightarrow{\text{LDA}} LiCHF\,CON(CH_3)_2 \xrightarrow{RR'CO}$$
13

$$RR'\,C(OH)CHF\,CON(CH_3)_2$$
14

(7-9)

Further increases in the bulk of the alkyl substituents were also ineffective in causing significant improvement of the diastereoselectivity of the fluoroacetamide enolates as can be seen in Tables 7-5 and 7-6.

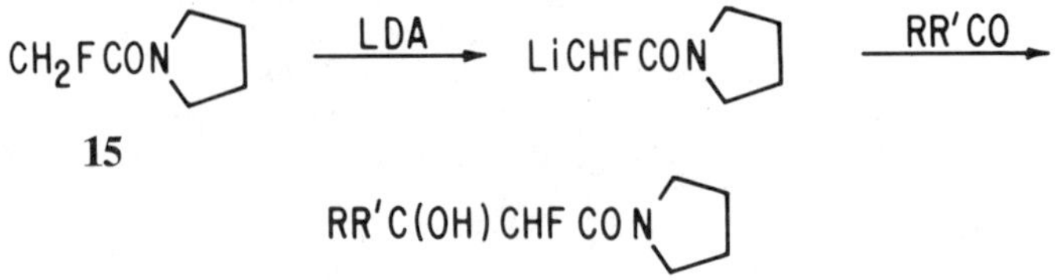

$$CH_2F\,CON \xrightarrow{\text{LDA}} LiCHF\,CON \xrightarrow{RR'CO}$$
15

$$RR'C(OH)\,CHF\,CON$$
16

(7-10)

TABLE 7-4. Products of Directed Aldol Reaction of Lithium Enolate of
N,N-Dimethylfluoroacetamide
$LiCHFCON(CH_3)_2 + RR'CO \rightarrow RR'COHCHFCON(CH_3)_2$

R	R'	Yield	Internal asymmetric induction[a]
H	Ph	90%	1 : 2.5
H	$(CH_3)_3C$	63%	1 : 1
H	CH_2CH_3	63%	1 : 1
H	$(CH_3)_2CH$	67%	1.6 : 1
CH_3	Ph	89%	1 : 1.9
Ph	Ph	98%	—
	2-adamantyl	99%	—
	3,3-dimethyl-2,4-dioxol-1-yl	86%	1 : 1

[a] As determined by ^{13}C NMR.

TABLE 7-5. Products of Directed Aldol Reaction of Lithium
Enolate of N-Fluoroacetylpyrrolidine
$LiCHFCON(CH_2)_4 + RR'CO \rightarrow RR'COHCHFCON(CH_2)_4$

R	R'	Yield	Internal asymmetric induction[a]
H	Ph	75%	1 : 1.6
H	$(CH_3)_3C$	95%	1 : 1
H	CH_2CH_3	89%	1 : 1
H	$(CH_3)_2CH$	50%	1.1 : 1
Ph	Ph	97%	—

[a] As determined by ^{13}C NMR.

TABLE 7-6. Products of Directed Aldol Reaction of Lithium
Enolate of Ethyl Fluoroacetate
$LiCHFCONCH(CH_3)_2 + RR'CO \rightarrow$
$RR'COHCHFCONCH(CH_3)_2$

R	R'	Yield	Internal asymmetric induction[a]
H	Ph	81%	1.1 : 6
H	$(CH_3)_3C$	68%	4.0 : 1
H	$CH_2CH_2CH_3$	42%	1.3 : 1
H	$(CH_3)_2CH$	86%	1.2 : 1
CH_3	Ph	74%	1.2 : 1
Ph	Ph	78%	—
	2-adamantyl	88%	—
	2-norbornyl	80%	1.7 : 1

[a] As determined by ^{13}C NMR.

$$CH_2FCON(CH(CH_3)_2)_2 \xrightarrow{\text{LDA}} LiCHFCON(CH(CH_3)_2)_2 \xrightarrow{\text{RR'CO}}$$

17

$$RR'C(OH)CHFCON(CH(CH_3)_2)_2$$

(7-11)

18

Failing to find conditions with amides or esters where good diastereoselectivity was possible suggested that fluorination was prohibiting stereoselectivity in either the deprotonation step or was disrupting the Zimmerman-Traxler transition state.

$$CH_2FCOC(CH_3)_3 \xrightarrow{\text{LHMDS}} LiCHFCOC(CH_3)_3 \xrightarrow{\text{RCHO}} RCH(OH)CHFCOC(CH_3)_3$$

19   **20**

(7-12)

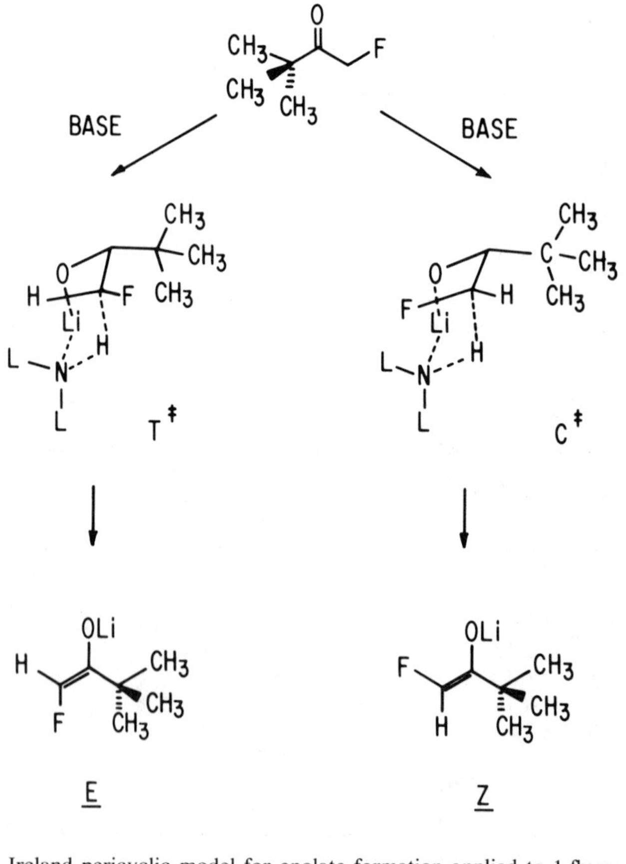

Figure 7-3. Ireland pericyclic model for enolate formation applied to 1-fluoro-3,3-dimethylbutanone.

TABLE 7-7. Products of Directed Aldol Reaction of Lithium Enolate of
1-Fluoro-3,3-Dimethyl-butanone

$$LiCHFCOC(CH_3)_3 + RR'CO \rightarrow RR'COHCHFCOC(CH_3)_3$$

R	R'	Yield	Internal asymmetric induction[a]
H	Ph	70%	7:1
H	(CH₃)₃C	62%	49:1
H	CH₂CH₃	63%	19:1
H	(CH₃)₂CH	50%	24:1
H	CH₂CH₂CH₃	74%	19:1
	3,3-dimethyl-2,4-dioxol-l-yl	90%	32:1

[a] As determined by ^{13}C NMR.

In order to determine if it was possible to form a fluorinated enolate stereose-
lectively, 1-fluoro-3,3-dimethyl-butanone, **19**, was prepared and deprotonated with LHMDS.[15] When the Ireland model for deprotonation is considered, it is clear that 1-fluoro-3,3-dimethyl-butanone would provide the most
severe steric challenge yet. From Table 7-7, it is clear that as the diastereoselectivity is excellent, the stereoselectivity of deprotonation was probably
very good as well.

B. Enol Silyl Ether Formation

Since it has been clearly established that fluorination does not prohibit diastereoselectivity in the aldol reaction, the question remained to be determined
if excellent diastereoselectivity with 1-fluoro-3,3-dimethyl-butanone resulted
from stereoselective reaction of a single enolate. If so, then the failure of the
fluoroacetates and fluoroacetamides to react diastereoselectively may have
resulted from an inability to form a single enolate or from the failure of the
enolate to react diastereoselectively. To answer these questions, the fluoroacetate and fluoroacetamide enolates were quenched with chlorotrimethylsilane.

Where the lithium enolate of ethyl fluoroacetate reacted very cleanly with
aldehydes and ketones to form aldolates in high yield, when treated with
chlorotrimethylsilane complex mixtures of products were formed. ^{1}H NMR
analysis of the product mixture contained a pair of resonances at δ 6.2 ppm
($J_{H,F}$ = 77Hz) and δ 6.32 ppm ($J_{H,F}$ = 74Hz) in a 1:1 ratio which could be
attributed to a mixture of the E and Z enol silyl ethers[9] **21** and **22**.

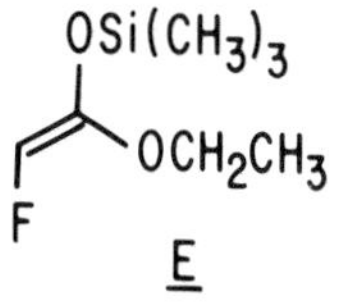

21 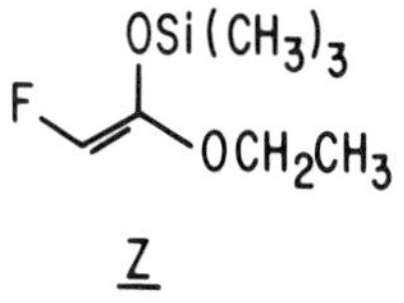   22

Trapping experiments with the lithium enolate of either benzyl or 2,6-di-*t*-butyl-4-methyl-phenyl fluoroacetates were unsuccessful.[16] No material which could be identified as an enol silyl ether **23** could be detected.

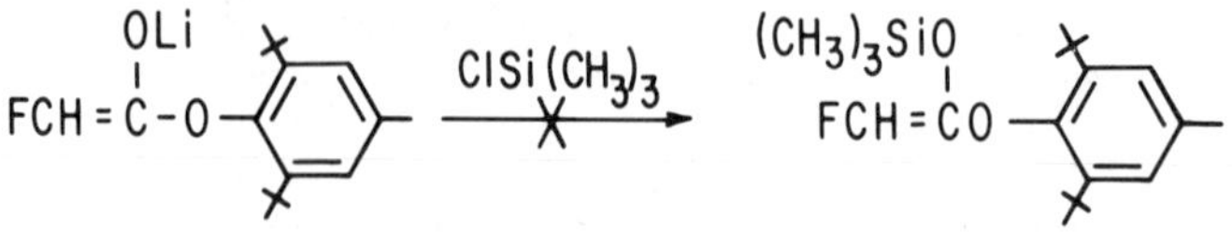

$$(7\text{-}13)$$

23

Attempted *O*-silylation of the enolate from *N,N*-dimethylfluoroacetamide with chlorotrimethylsilane, *t*-butyl-dimethylsilyl chloride or *t*-butyl di-methylsilyl triflate was not successful.[17] Presumably only *C*-silylated products such as **24** were formed on reaction of the enolate with chlorotrimethyl-silane.[17]

$$LiFCHCON(CH_3)_2 \xrightarrow{\text{ClSi(CH}_3)_3} (CH_3)_3SiCHFCON(CH_3)_2 \qquad (7\text{-}14)$$

24

In contrast to these findings the lithium enolate of 1-fluoro-3,3-dimethyl-butanone was cleanly trapped to form a single enol silyl ether **25**. Irradiation of the resonance at δ 5.97 ppm of the vinyl group.[18] Such as NOE would only be possible if the enol silyl ether formed was the Z compound.

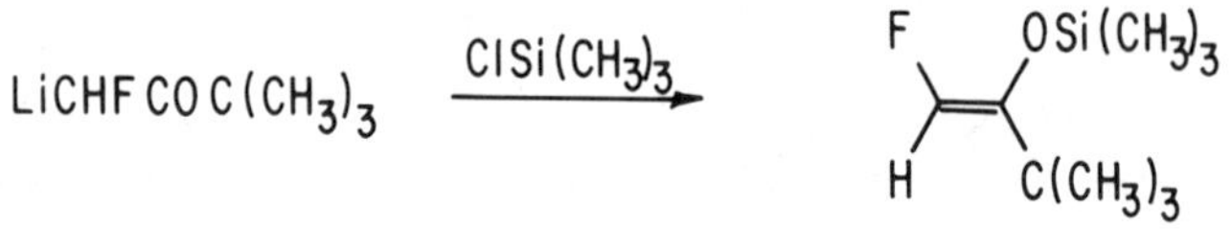

$$(7\text{-}15)$$

$\underline{Z}$

25

Although the Z isomer could be predicted on the basis of steric interactions, an examination of molecular models does not reveal steric effects in either the starting ketone or the enol silyl ether. However, it is clear from the molecular models that *C*-silylation would be severely sterically encumbered.

$$LiCHFCOC(CH_3)_3 \xrightarrow{\text{ClSi(CH}_3)_3} (CH_3)_3SiCHFCOC(CH_3)_3 \qquad (7\text{-}16)$$

26

C. ¹⁹F NMR Studies of the Enolates

Direct examination of the ¹⁹F NMR of the enolates at low temperature would avoid complications associated with isolation of reactive silyl ketene acetals and competing O vs C silylation. Unfortunately all attempts at the direct observation of the lithium enolate of ethyl fluoroacetate, benzyl fluoroacetate, or 2,6-di-*t*-butyl-4-methyl-phenyl fluoroacetate at temperatures between $-100°$ and $-75°$ were unsuccessful. No ^{19}F resonances were observable, not even those which could be attributed to undeprotonated starting material. However, ^{19}F NMR studies of the lithium enolate of *N,N*-dimethyl-fluoroacetamide at $-85°$ clearly indicated two distinct fluorine resonances in a 1:1 ratio at $\delta -213.15$ ppm (d, $J_{H,F} = 85.4$ Hz) and $\delta -214.3$ ppm (d, $J_{H,F} = 71.7$ Hz) assigned to a mixture of the E and Z enolates[17] **27** and **28**.

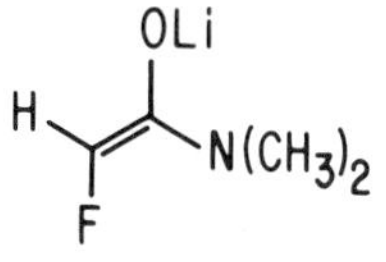

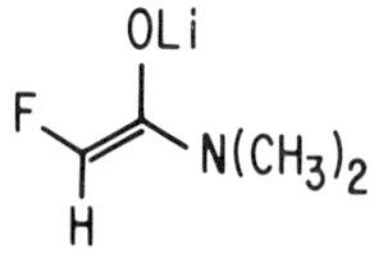

Careful preparation of the lithium enolate of 1-fluoro-3,3-dimethyl-butanone at $-70°$ made possible the observation of a single signal at $\delta -192.35$ ppm (d, $J_{H,F} = 85$ Hz). This suggested that the selectivity of the reaction with chlorotrimethylsilane to form **25** resulted from selectivity in the formation of the enolate.[18]

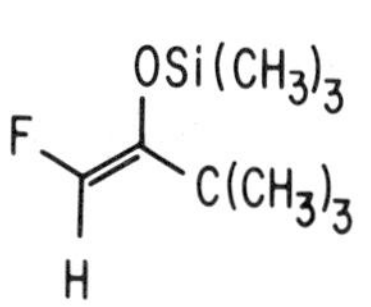

Our inability to observe the ^{19}F NMR for the ester enolates under conditions where we had previously found that these enolates would give excellent yields in the aldol reaction, suggested that a paramagnetic species was being formed. ESR spectra of the lithium enolate of ethyl fluoroacetate at $-198°$ in a THF glass indicated that this was the case.[16]

D. Fluoroacetate Ester Enolate Claisen Rearrangement

Even though we had been unable to trap the lithium enolate of ethyl fluoroacetate cleanly with chlorotrimethylsilane and have since demonstrated the intermediacy of paramagnetic species which interfered with measurement of the NMR spectrum of the enolate, trapping of lithium enolate of 2-butenyl fluoroacetate, **29**, with chlorotrimethylsilane was attempted. The resultant allyl silyl ketene acetal **30** would be expected to undergo a facile sigmatropic rearrangement.

$$FCH_2CO_2CH_2CH=CH-CH_3 \xrightarrow[ClSi(CH_3)_3]{LDA} FHC=\overset{\displaystyle OSi(CH_3)_3}{\overset{|}{C}}OCH_2CH=CHCH_3 \quad (7\text{-}17)$$

29   **30**

Previously it had been shown that isolation of the silyl ketene acetal prepared from ethyl fluoroacetate was very difficult. However, the products of the sigmatropic rearrangement of the allyl fluoroacetate ester would be stable and would retain the stereochemical information about the geometry of the enolate.

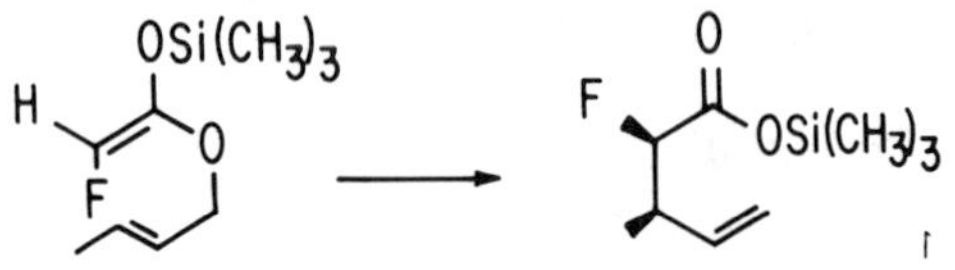

$$(7\text{-}18)$$

31

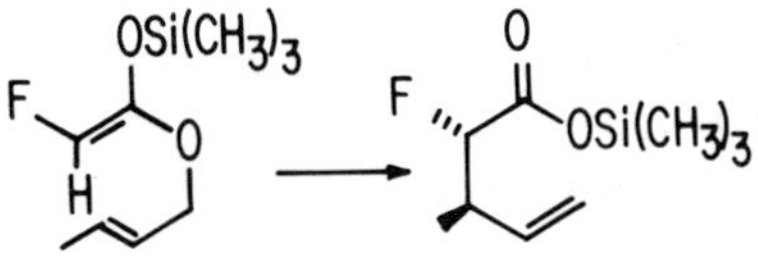

$$(7\text{-}19)$$

32

Surprisingly the rearrangement was very diastereoselective (Table 7-8).[19] Such diastereoselectivity could only arise from a single silyl ketene acetal. The formation of a single silyl ketene acetal was very difficult to rationalize. The very good yield of the rearrangements were also difficult to understand considering the propensity of the fluoroacetate ester and fluoroacetamide enolates to undergo C-silylation. A careful ^{19}F NMR study of the rearrangement helped elucidate the origin of the diastereoselectivity observed.[16]

At $-85°$ treatment of the 2-butenyl fluoroacetate with LDA followed by chlorotrimethylsilane yielded principally the C-silylated ester **33** as deter-

TABLE 7-8. Ester Enolate Claisen Rearrangement of Allyl Fluoroacetates

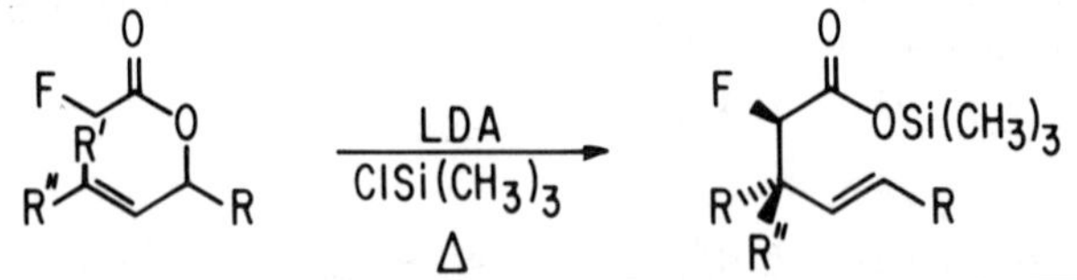

R	R'	R''	Base	Internal asymmetric induction	Yield
CH₃	H	H	3 × LDA	—	88%
H	H	H	3 × LDA	—	72%
H	CH₃	H	3 × LDA	1 : 1	81%
H	CH₃	H	1.1 × LDA	1 : 20	43%
H	H	CH₃	1.2 × LDA	4 : 1	65%
H	H	CH₃	0.9 × LDA	9 : 1	15%
H	H	CH₃O	1.1 × LDA	—	90%

mined by low temperature ^{19}F NMR. On warming to 20°, resonances assigned to the product of sigmatropic rearrangement increased while the signal attributed to the *C*-silylated material decreased. At −85° there was a signal assigned to fluoroketene silyl acetal **34**, whose relative intensity to *C*-silylated material increased, but decreased relative to the signal attributable to the rearranged products.[16] Apparently the initially formed *C*-silylated material isomerizes in a slow step to form the reactive fluoroketene acetal which undergoes sigmatropic rearrangement in a second fast step.

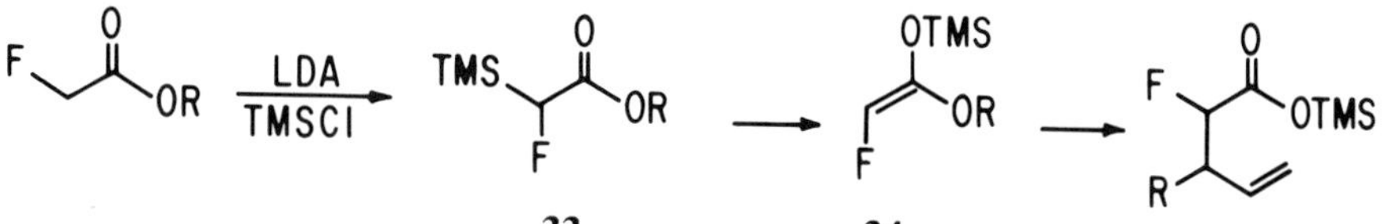

$$(7\text{-}20)$$

A necessary consequence of the isomerization reaction is formation of the E fluoroketene silyl acetal. Migration of the trimethyl silyl group would necessarily occur in a syn periplanar fashion (Equation 7-21).

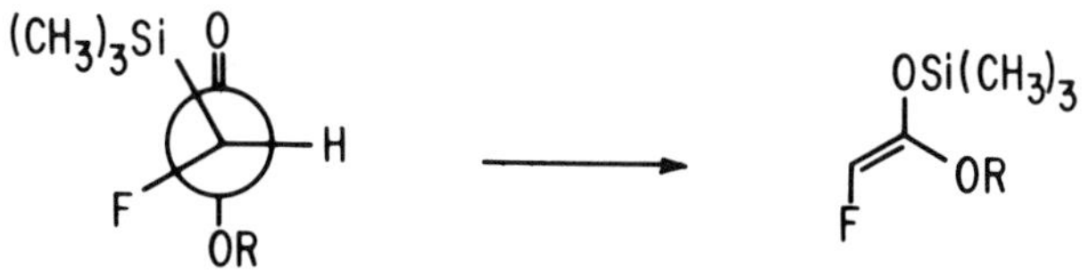

$$(7\text{-}21)$$

e. Fluorinated Azaenolates

Azaenolates, prepared by deprotonation of imines, have been widely utilized in asymmetric synthesis.[20] Control of the geometry of the azaenolate is important in stereoselective reactions just as in the reactions of enolates. When studying fluoroketone imines, regiocontrol of deprotonation is also important. The selectivity of deprotonation has been reported to be dependent upon temperature, the nitrogen alkyl substituent, and the steric bulk of the base among other factors.[21] As fluorine is not a sterically demanding substituent, fluorination is a severe test of the sensitivity of deprotonation to steric effects while at the same time fluorination will have pronounced electronic effects on the reaction.

The cyclohexylimine of fluoroacetone, **36**, simply prepared by condensation of fluoroacetone, **35**, with cyclohexylamine in the presence of anhydrous potassium carbonate, was greater than 95% a single C=N isomer. NOE studies by irradiation of the resonance at 3.35 ppm assigned to methine proton of the cyclohexyl group resulted in a 4–6% enhancement of the methyl resonance at 1.78 ppm without concomitant enhancement of the signal at 4.63 ppm assigned to the fluoromethyl group. The geometry was therefore firmly established as E CN.

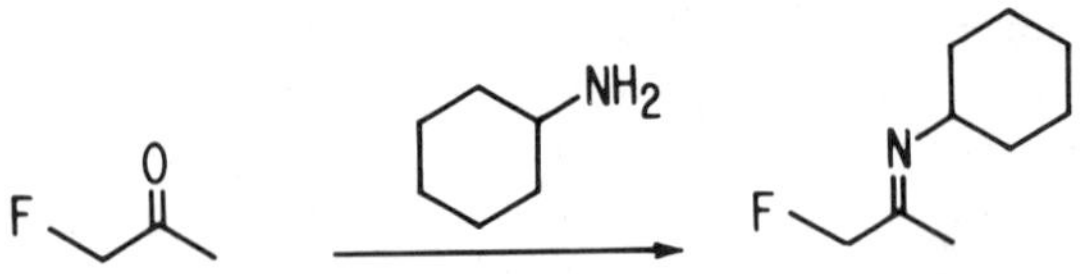

(7-22)

It was possible to effect regiospecific deprotonation of the fluoroacetone imine by careful choice of the reactant base and reaction temperature.

(7-23)

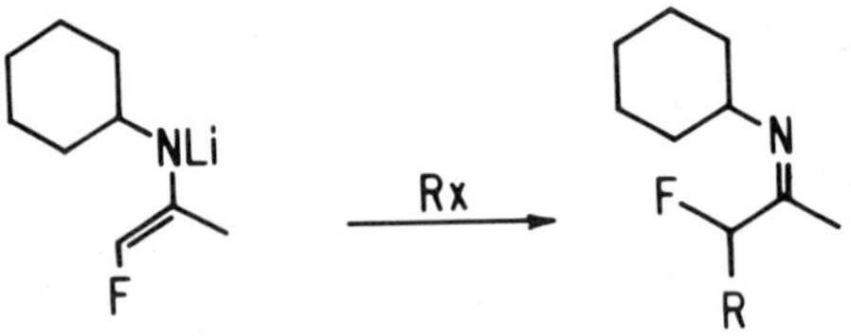

(7-24)

Deprotonation with *tert*-butyl lithium at low temperature favored exclusive deprotonation of the fluoromethyl group to form **37**, whereas treatment of the imine with LHMDS at −30° resulted in deprotonation of the methyl group (Table 7-9). The tendency of the metalated ketimine to alkylate on carbon-bearing fluorine at low temperatures may be rationalized by suggesting that the increased acidity associated with the protons near fluorine acts in concert with previously established steric interactions of the hindered base with the N-alkyl group[22] which would hinder syn approach. Regardless of the site of deprotonation isomerization to form the more stable syn configuration must be rapid.[22]

At higher temperatures, alkylation was directed away from the fluoromethyl group. It has not been possible to equilibrate the metalated ketimine, deprotonated on the fluoromethyl group **37**, to the other isomer deprotonated on methyl **39** by warming to −20° for four hours. Higher temperatures caused extensive decomposition of the anion. Formation of fluoroimine deprotonated on the nonfluorinated carbon to give the thermodynamically favored less substituted aza-allyl lithium reagent would be predicted given the enhanced stability of the syn anion with the N-alkyl substituents away from the substituents on carbon. In contrast to studies with 2-butanone imine it is not necessary to invoke isomerization of the imine to accommodate the observed results.[23]

TABLE 7-9. Temperature-dependent Deprotonations of Fluoroacetone Cyclohexylimine

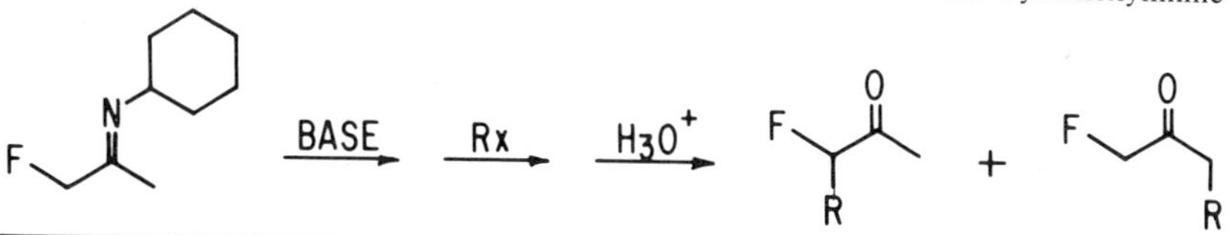

			Product composition[b]		
Alkyl halide	Base[a]	Temperature	RCHFCOCH₃	CH₂FCOCH₂R	Yield %
CH₃I	A	−30°	11	89	48
CH₃I	B	−80°	96	4	43
CH₂=CHCH₂Br	A	−30°	9	91	62
CH₂=CHCH₂Br	B	−80°	97	3	69
CH₂=CHCH₂I	A	−30°	3	97	60
CH₂=CHCH₂I	B	−80°	97	3	81
C₆H₅CH₂Br	A	−30°	18	82	71
C₆H₅CH₂Br	B	−80°	93	7	81
C₆H₅CH=CHCH₂Br	A	−30°	8	92	88
C₆H₅CH=CHCH₂Br	B	−80°	28	72	79

[a] A = Lithium hexamethyldisilazide; B = Lithium diisopropylamide
[b] As determined by ¹⁹F NMR.

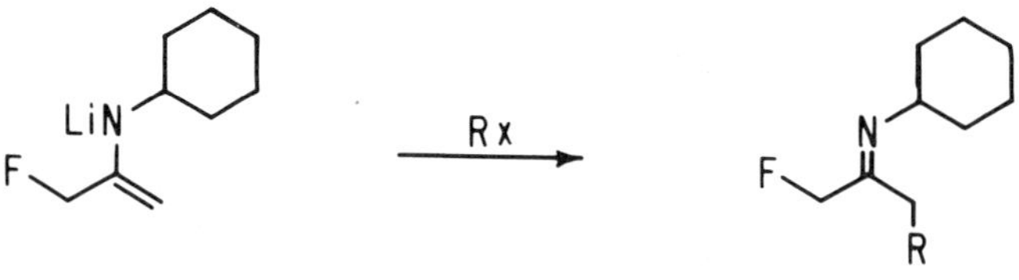

With the development of methods for the regioselective deprotonation of fluoroacetone cyclohexylimine, it has become possible to effect the regioselective deprotonation of chiral fluoroacetone imines and to determine the stereochemistry of the alkylated products. Such experiments are not only useful as a synthetic procedure for the preparation of selectively fluorinated compounds, but also in the study of azaenolate geometry. Good enantioselectivity is not possible without control of the azenolate stereochemistry. Chiral imines of fluoroacetone were prepared using the methyl ether of *L*-phenylaninol and *L*-valinol, **41a** and **41b**, chiral auxiliaries which have been shown by Meyers to be excellent and efficient agents for the induction of asymmetry[24]. Slow addition of a carbon tetrachloride solution of fluoroacetone to a solution of the methyl ether of L-phenylalaninol or L-valinol in carbon tetrachloride at 0° in the presence of activated 4A molecular sieves resulted in a smooth condensation to form the imine 42 in 96% to 91% yield.

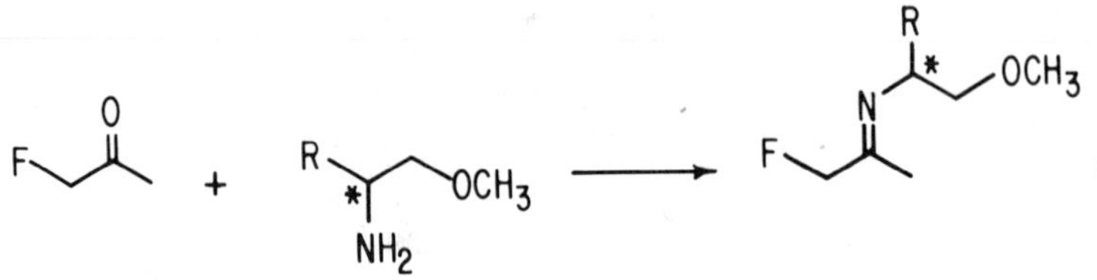

(7-26)

35

41a R = CH₂Ph
41b R = CH(CH₃)₂

42

Analysis of the product by NMR illustrated clearly that only a single carbon-nitrogen double bond isomer was formed. Comparison of the measured chemical shifts with those reported in the literature for the carbon syn to the nitrogen alkyl substituent suggest the imine formed is the E C=N isomer shown.

Regioselective deprotonation with *tert*-butyl lithium was followed by alkylation and isolation of the substituted imine 43 in 40–80% yield. It was apparent that the deprotonation of the fluoromethyl group was favored more than 20 : 1 (Table 7-10).

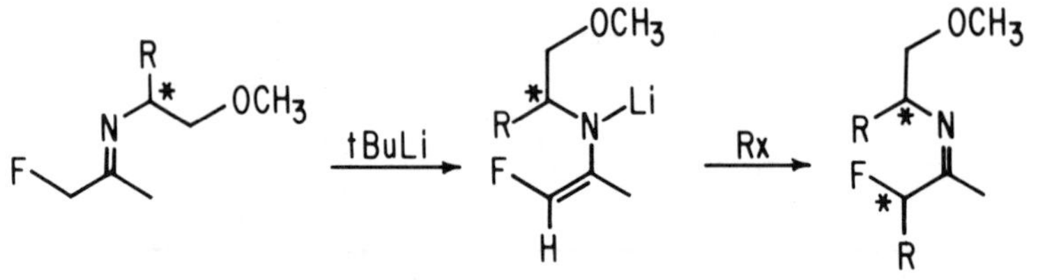

(7-27)

42   43

TABLE 7-10. Alkylation of Deprotonated Chiral Fluoroacetone Imines

Alkyl halide	Chiral imine[a]	Yield	Diastereomeric excess[b]
CH₃I	Phenylalaninol	—	26%
CH₂=CHCH₂Br	Valinol	60%	37%
CH₂=CHCH₂Br	Phenylalaninol	69%	33%
PhCH₂Br	Valinol	64%	32%
PhCH₂Br	Phenylalaninol	38%	35%
CH₃CH₂CH₂CH₂Br	Valinol	58%	33%
CH₃CH₂CH₂CH₂Br	Phenylalaninol	71%	52%
CHκCCH₂Br	Valinol	36%	46%
CHκCCH₂Br	Phenylalaninol	14%	66%

[a] The methyl ether of the amino alcohol was used to prepare the imines.
[b] Determined by ¹⁹F NMR.

The enantiomeric purity of the products was disappointing but it was in good agreement with that reported when the kinetically formed lithioenamines were alkylated.[25] When we attempted to equilibrate the deprotonated fluorinated lithioenamines by warming to −30° to form the thermodynamically more favored azaenolate isomer, only products of decomposition were isolated. It has been proposed that the kinetically formed Z lithioenamines may be poorer substrates for enantioselective alkylation because the Z substituent causes steric disruption of the crucial lithium-nitrogen-oxygen interaction.

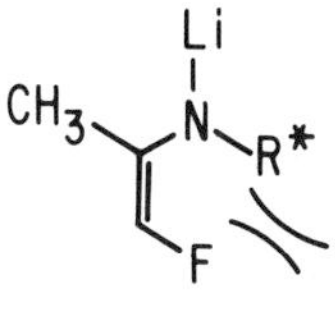

It was unlikely, however, that fluorine, with a van der Waals radii of 1.35Å and a carbon fluorine bond length of 1.39Å[26], should be capable of sufficient nonbonded interactions to disrupt the transition state. Failure to obtain kinetic control of the lithioenamine geometry, and our inability to equilibrate the lithioenamines, could also account for the low enantioselectivity observed. These problems could be overcome by deprotonation of fluorinated cyclic ketimine, where only a single azenolate geometry was possible.

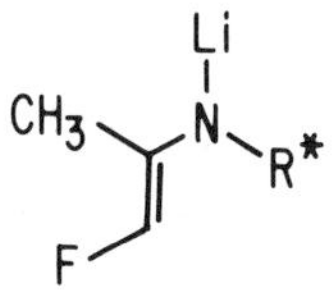

2-Fluorocyclohexanone, **44**, was successfully converted to the chiral imines **45** by the method described previously in 84% yield. ^{19}F NMR clearly showed the product to be a nearly 1 : 1 mixture of C=N isomers. Deprotonation with *tert*-butyl lithium at −85° followed by alkylation occurred without loss of regioselectivity. However, even with a fixed lithioenamine geometry, the enantioselectivity remained modest (See Table 7-11). Apparently in spite of its small fluorine is influencing the transition state in an undesirable manner. Surprisingly, when a polar additive such as DMPU is introduced to further disrupt intramolecular chelation, the enantioselectivity of the alkylation increases.

TABLE 7-11. Alkylation of Deprotonated 2-Fluorocyclohexanone O-Methyl Phenylalaninol Imine

Alkyl	Solvent	Temp.	Yield	Diastereomeric excess[a]
CH₃I	THF	−85°	41%	56%
CH₃I	THF-DMPU[b]	−80°	—	63%
PhCH₂Br	THF	−80°	48%	32%
PhCH₂Br	THF-DMPU	−80°	61%	68%
PhCH₂Br	THF	−100°	21%	22%
CH₃CH₂CH₂CH₂Br	THF-DMPU	−80°	78%	58%
CH₃CH₂CH₂CH₂Br	THF	−80°	59%	—

[a] Determined by ^{19}F NMR.
[b] N,N'-Dimethylpropyleneurea, 1,3-dimethyl-3,4,5,6-tetrahydro-2(1H)-pyrimidinone

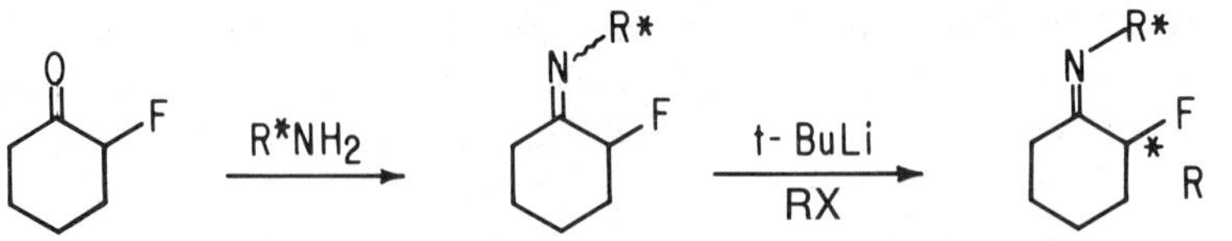

(7-28)

44   **45**   **46**

3. DISCUSSION OF RESULTS

A. Effect of Fluorination on Deprotonation

From the results presented, there is no clear evidence of a steric effect of fluorination of the geometry of deprotonation. Steric repulsions of the incoming base by fluorine or developing allylic strain in a postulated pericyclic transition state seem equally unlikely.

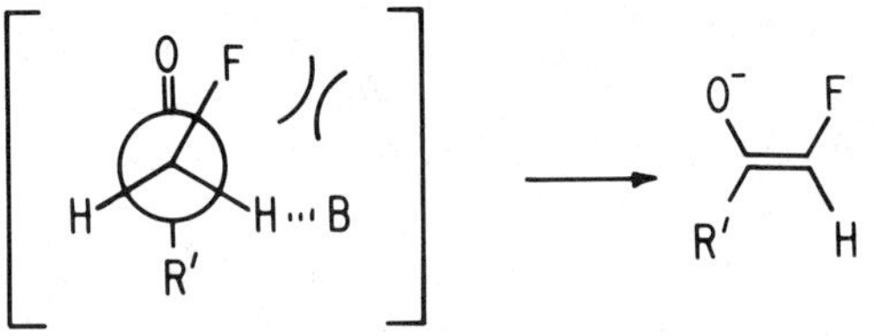

(7-29)

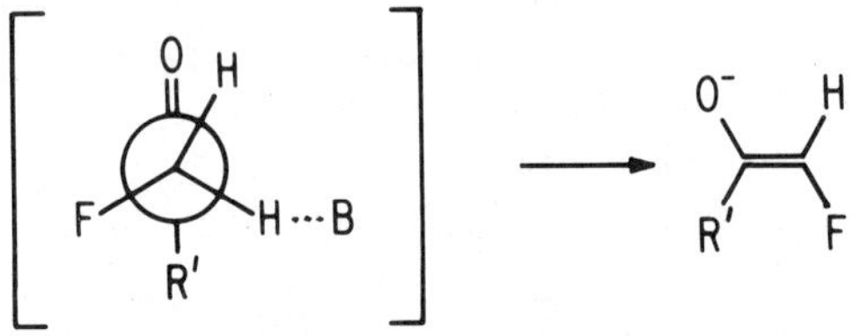

(7-30)

If the geometry of deprotonation were governed by either of these factors little difference would be expected between 1-fluoro-3,3-dimethyl-butanone and N,N-diisopropylfluoroacetamide in their stereoselectivity toward deprotonation. N,N-dialkylamides are very selectively deprotonated to form Z enolates as a consequence of ground state allylic strain considerations which strongly disfavor amide conformation B and consequently transition state T‡ (see Figure 7-1).

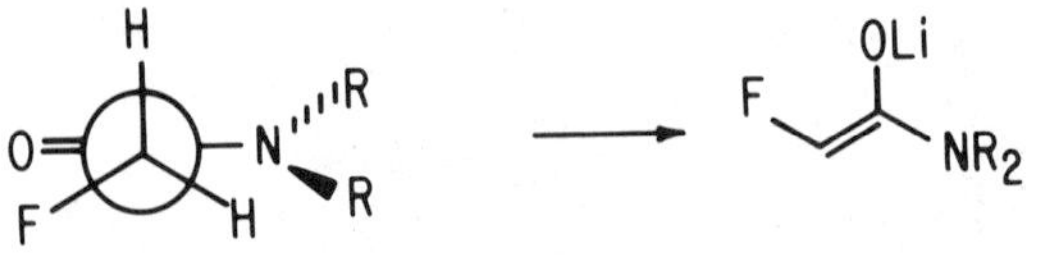

(7-31)

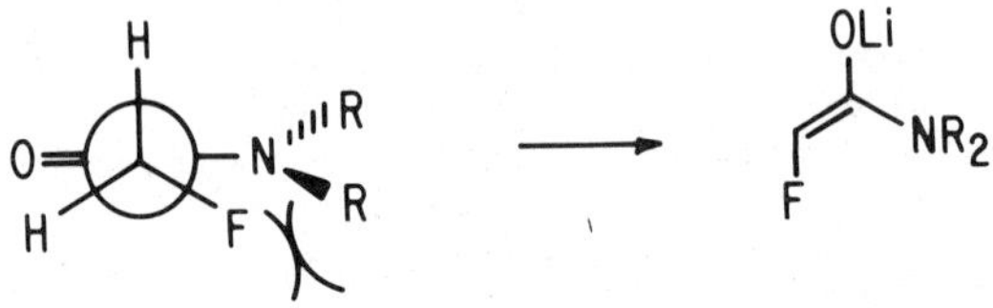

(7-32)

The same steric effects clearly would be operating with 1-fluoro-3,3-di-methyl-butanone.

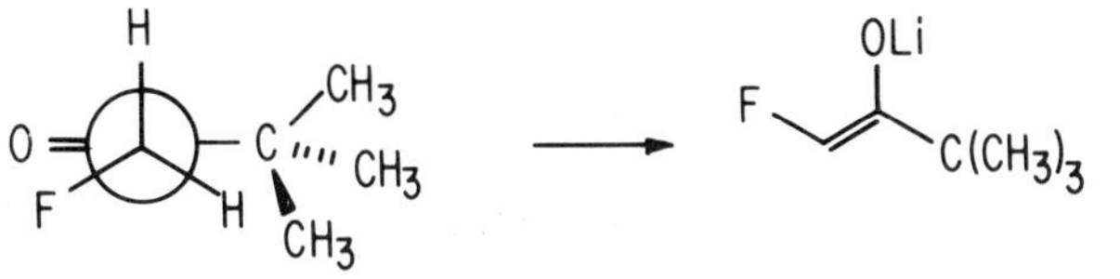

$$(7\text{-}33)$$

$$(7\text{-}34)$$

The selectivity observed with the ketone and not with amide may be attributed to electronic effects. Such electronic effects favoring a *gauche* configuration such as would be required in transition state $A^{\ddagger}$ (Equation 7-4) have been previously reported for 1,2-dihaloethanes.[27]

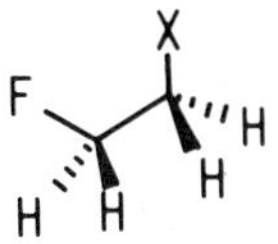

It is known, for example, when fluorine assumes a synperiplanar relationship with the carbonyl (as shown below) that this conformer is favored over one where fluorine is in an antiperiplanar relationship by 200 cal/mol.[28]

Theory has yet to provide a completely satisfactory rationale for this result. A similar tendency of ethylenes, 1,2-disubstituted with electronegative substituents with a lone pair of electrons or electrons available for π-bonding to favor the cis stereoisomer has been described as the "cis effect."[29] Again, theoretical arguments are unable to completely explain these observations.

These earlier findings nicely mesh with our observation of the differences between 1-fluoro-3,3-dimethylbutanone and *N,N*-diisopropylfluoroace-tamide. In 1-fluoro-3,3-dimethylbutanone, there is a single polar substituent, oxygen, so there is a strong preference in both the ketone and in the enolate to favor both the synperiplanar and syn relationship. However in *N,N*-diisopropylfluoroacetamide, there are two polar substituents, oxygen and nitrogen, so there is no preferred conformation.

B. Effect of Fluorination on the Stereochemistry of Imine Formation

As established there is little reason to attribute the pronounced stereoselec-tivity of C=N geometry in the preparation of fluoroketone imines to steric effects. Rather secondary orbital interactions in the intermediate animal may lead to stabilization of a conformation from which loss of water can only form the E C=N imine.

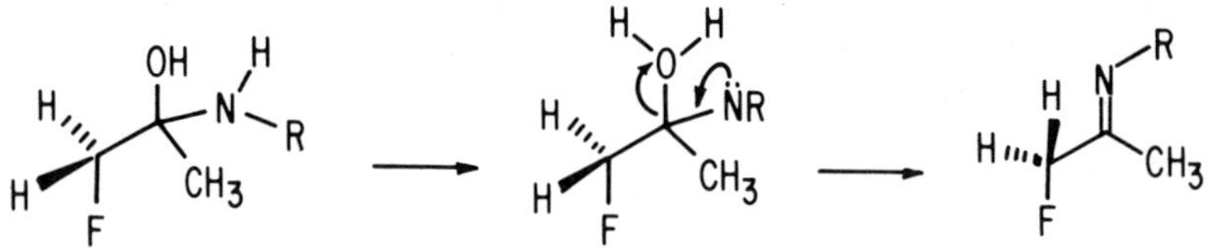

(7-35)

C. Effect of Fluorination on Regiochemistry of Imine Deprotonation

The propensity of fluoroketone imines to be deprotonated on the carbon bearing fluorine at low temperature with a reactive, hindered base such as *t*-butyl lithium strongly suggests that steric hindrance to approach of the base results in initial attack at the carbon bearing fluorine anti to the nitrogen alkyl substituent. Once deprotonated isomerization to the more stable syn confor-mation must be rapid.[22]

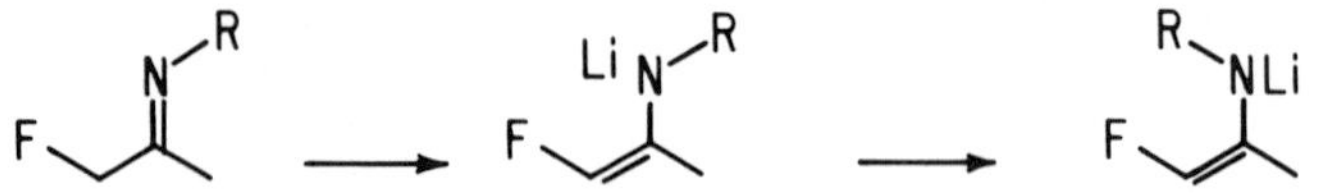

(7-36)

However, the observation of deprotonation of the nonfluorinated carbon at higher temperature with somewhat less reactive but equally hindered LHMDS indicates that fluorination may decrease the pKa of the nearby hydrogen as well. As mentioned earlier there is no evidence for equilibration of the anions.

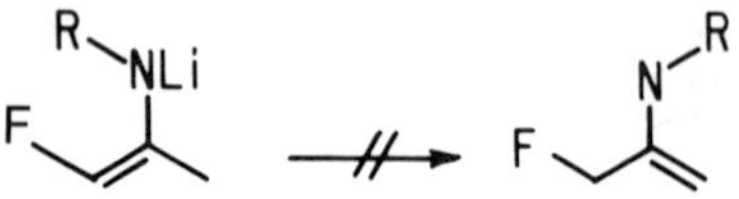

(7-37)

D. Effect of Fluorination on C vs O Silylation

The propensity of fluoroacetate and fluoroacetamide enolates to silylate on carbon suggests that I-π destabilization by fluorine of the enolate contributor decreases the significance of the enolate resonance contributor in fluoroacetates and fluoroacetamides.

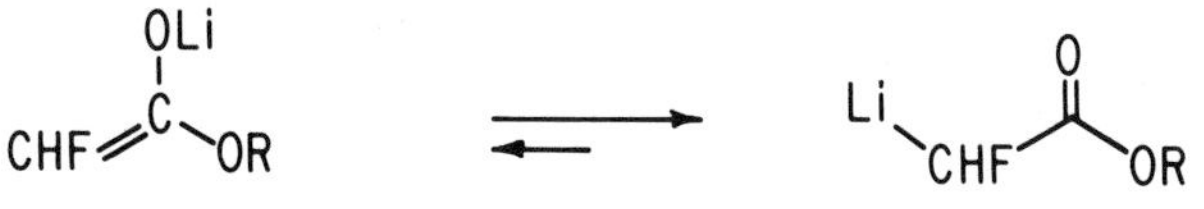

(7-38)

Iπ greater — Iπ decrease

This tendency toward I-π repulsion results in greater charge density residing on the fluoromethyl group. In complimentary manner, such an anion might tend to be pyramidal and hence stabile relative to the planar anion.[30]

Iσ stabilizing — Iπ destabilizing

As a practical result of the decreased ability of fluorinated enolates to distribute electron density to oxygen, silylation occurs predominantly on carbon.

4. SUMMARY

To investigate the well-known fact that fluorine can have pronounced electronic effects on molecules into which it has been substituted, in this work we have systematically studied the influence of fluorination on a variety of enolates and enolate equivalents. In those cases where selectivity was observed, it cannot be easily rationalized on the basis of steric effects. Stereoelectronic effects clearly are responsible for the selectivity determined. Fluorination has also increased the rate of the C—O silicon transfer rearrangement which has not been previously reported to occur with the facility described here. The combination of these findings illustrates that fluorination of reactive intermediates such as enolate or their equivalents can be a productive technique for probing the origin of stereoselectivity in their reactions.

REFERENCES

1. (a) Heathcock, C. H. In "Asymmetric Synthesis"; Morrison, J. D., Ed.; Academic: Orlando, 1984, Vol. 3, Chapter 2. (b) Heathcock, C. H. In "Comprehensive Carbanion Chemistry"; Buncel, E.; and Durst, T., Eds.; Elsevier: Amsterdam, 1984; Part B, Chapter 4. (c) Heathcock, C. H. In "Current Trends in Organic Synthesis"; Nozaki, H., Ed.; Pergamon: New York, 1983, pp. 27–43. (d) Evans, D. A.; Nelson, J. V.; Taber, T. R. *Top. Stereochem.* **1982,** *13,* 1–115. (e) Mukaiyama, T. *Org. React.* (NY) **1982,** *28,* 203–331.

2. Evans, D. A. In "*Asymmetric Synthesis*"; Morrison, J. D., Ed.; Academic: Orlando, 1984, Vol. 3, Chapter 1.

3. Fataftah, Z. A.; Kopka, I. E.; Rathke, M. W. *J. Am. Chem. Soc.* **1980,** *102,* 3959–3960.

4. Zimmerman, H. E.; Traxler, M. D. *J. Am. Chem. Soc.* **1957,** *79,* 1920–1923.

5. Kobayashi, Y.; Taguchi, T. In "Biomedical Aspects of Fluorine Chemistry"; Filler, R.; and Kobayashi, Y., Eds.; Kodansha: Tokyo, 1982, pp. 33–53.

6. (a) Blank, I.; Mager, J.; Bergmann, E. D. *J. Chem. Soc.* **1955,** 2190–2193. (b) Bergmann, E. D.; Szinai, S. *J. Chem. Soc.* **1956,** 1521–1524. (c) Bergmann, E. D.; Schwarcz, J. *J. Chem. Soc.* **1956,** 1524–1527. (d) Bergmann, E. D.; Cohen, S.; Shahak, I. *J. Chem. Soc.* **1959,** 3278–3285. (e) Bergmann, E. D.; Cohen, S.; Shahak, I. *J. Chem. Soc.* **1959,** 3286–3289. (f) Bergmann, E. D.; Cohen, S. *J. Chem. Soc.* **1961,** 3537–3538. (g) Bergman, E. D.; Chun-Hsu, L. *Synthesis* **1973,** 44–47.

7. (a) Poulter, C. D.; Mash, E. A.; Argyle, J. C.; Muscio, O. J.; Rilling, H. C. *J. Am. Chem. Soc.* **1979,** *101,* 6761–6763. (b) Brandange, S.; Dahlman, O.; March, L. *J. Am. Chem. Soc.* **1981,** *103,* 4452–4458.

8. (a) Molines, H.; Massoudi, M. H.; Cantacuzene, D.; Wakselman, C. *Synthesis* **1983,** 322–324.

8b. Nakai, H.; Hamanaka, N.; Miyake, H; Hayashi, M. *Chem. Lett.* **1979,** 1499–1502.

9. Welch, J. T.; Seper, K.; Eswarakrishnan, S.; Samartino, J. *J. Org. Chem.* **1984,** *49,* 4720–4721.

10. Samartino, J. S. The State University of New York at Albany, unpublished results, 1984.

11. Samartino, J. S. The State University of New York at Albany, unpublished results, 1985.

12. Heathcock, C. H.; Pirrung, M. C.; Sohn, J. E. *J. Org. Chem.* **1979,** *44,* 4294–4299.

13. Tamaru, Y.; Harada, T.; Nishi, S.; Mizutani, M.; Hioki, T.; Yoshida, Z. *J. Am. Chem. Soc.* **1980,** *102,* 7806–7808.

14. Welch, J. T.; Eswarakrishnan, S. *J. Org. Chem.* **1985,** 5403–5405.

15. Welch, J. T.; Seper, K. W. *Tetrahedron Lett.* **1984,** *25,* 5247–5250.

16. Samartino, J. S. The State University of New York at Albany, unpublished results, 1986.

17. Eswarakrishnan, S. Ph.D. Dissertation, The State University of New York at Albany, Albany, NY, 1986.

18. Seper, K. W. The State University of New York at Albany, unpublished results, 1984.

19. Welch, J. T.; Samartino, J. S. *J. Org. Chem.,* **1985,** *50,* 3663–3665.

20. (a) Bergbreiter, D. E.; Newcomb, M. In "Asymmetric Synthesis"; Morrison, J. D., Ed.; Academic: Orlando, 1983, Vol. II, Chapter 9. (b) Fraser, R. R. In "Comprehensive Carbanion Chemistry"; Buncel, E.; and Durst, T., Ed.; Elsevier: Amsterdam, 1984, Part B, Chapter 2.

21. Smith, J. K.; Newcomb, M.; Bergbreiter, D. E.; Williams, D. R.; Meyers, A. I. *Tetrahedron Lett.* **1983,** *24,* 3559–3562.

22. Houk, K. N.; Stozier, R. W.; Rondan, N. G.; Fraser, R. R.; Chaqui-Offermanns, N. *J. Am. Chem. Soc.* **1980,** *102,* 1426–1429.

23. Smith, J. K.; Bergbreiter, D. E.; Newcomb, M. *J. Am. Chem. Soc.* **1983,** *105,* 4396–4400.

24. Meyers, A. I.; Williams, D. R.; Erickson, G. W.; White, S.; Druelinger, M. *J. Am. Chem. Soc.* **1981,** *103,* 3081–3087.

25. Meyers, A. I.; Williams, D. R.; Erickson, G. W.; White, S.; Druelinger, M. *J. Am. Chem. Soc.* **1981,** *103,* 3088–3093.

26. Pauling, L. "The Nature of the Chemical Bond", 3rd ed.; Cornell University Press: Ithaca, 1960.
27. (a) Fernholt, L.; Kveseth, K. *Acta Chem. Scand. A*, **1980,** *34*, 163–170. (b) Wolfe, S.; *Acc. Chem. Res.* **1972,** *5*, 102–111.
28. Brown, T. L. *Spectrochimica Acta* **1962,** *18*, 1615–1623.
29. (a) Epiotis, N. D.; Cherry, W. R.; Shaik, S.; Yates, R.; Bernardi, F. *Topics in Current Chemistry*, **1977,** *70*, 1–242. (b) Liberles, A.; Greenberg, A.; Eilers, J. E. *J. Chem. Ed.* **1973,** *50*, 676–678. (c) Bingham, R. C. *J. Am. Chem. Soc.* **1976,** *98*, 535–540. (d) Pitzer, K. S.; Hollenberg, J. L. *J. Am. Chem. Soc.* **1954,** *76*, 1493–1496. (e) Kollman, P. *J. Am. Chem. Soc.* **1974,** *96*, 4363–4369.
30. Chambers, R. D. "Fluorine in Organic Chemistry"; Wiley: New York, **1973,** p. 86.

CHAPTER 8

Stereospecific Preparation, Reactivity, and Utility of Polyfluorinated Alkenyl Organometallics*

Donald J. Burton

University of Iowa, Department of Chemistry, Iowa City, Iowa

CONTENTS

* Presented in part at the Symposium on Reactivity and Mechanism in Fluorine Chemistry, 191st ACS Meeting, New York, Spring 1986, Abst. FLUO 18.

1. INTRODUCTION

Although the preparation of polyfluoroalkenyl lithium and Grignard reagents have been described in the literature,[1] the synthetic utility of these reagents has been impeded by the restricted thermal stability of these organometallic compounds. Generally, the preparations must be carried out at low temperature to prevent elimination of metal fluoride, and consequently serious difficulties often arise when scale-up processes are attempted. In addition, only reactive substrates can be employed with these reagents, since it is not possible to force sluggish substrates via the use of higher temperatures. Our previous experience with vinylic polyfluorinated lithium reagents[2] prompted us to focus on the preparation of alternative organometallic reagents that might exhibit superior thermal behavior and which might also manifest disparate chemical behavior, such as coupling processes or addition to carbon–carbon multiple bonds.

Our initial attention was focused on a general route to stable R_F-vinyl (all atoms attached to carbon are fluorine) organometallic compounds that would satisfy the following criteria: (1) the reagents must possess thermal stability at room temperature or above, so that the reactions could be readily scaled-up and pregeneration of the organometallic compound could be accomplished with minimal experimental difficulty in standard glassware apparatus; (2) stereochemical integrity with E/Z-precursors should be retained in formation of the organometallic reagent, so that stereochemical control could potentially be accomplished in further elaboration of these intermediates; (3) the reagents must possess functionalization (or exchange) capability, so they could be utilized in preparative synthetic transformations to yield polyfunctionalized materials; and (4) formation of these reagents would be achievable in a one-step (or one-pot) procedure from accessible precursors.

Our initial target compounds were:

$$F_2C{=}CFCu; \qquad \underset{F}{\overset{CF_3}{\diagdown}}C{=}C\underset{Cu}{\overset{F}{\diagup}} \quad ; \quad \underset{F}{\overset{CF_3}{\diagdown}}C{=}C\underset{F}{\overset{Cu}{\diagup}}$$

These initial targets were selected to test the criteria cited above and because we anticipated that such vinylic copper reagents could potentially participate in coupling processes or addition reactions to multiple bonds. The utility of nonfluorinated alkenyl copper and/or alkenyl cuprate reagents for regio- and stereochemical control in organic synthesis is well documented, and the addition of these organometallic compounds to the organic chemist's arsenal of reagents has permitted synthetic transformations which were difficult or impossible to accomplish effectively with any other reagent.[3] A notable omission, however, from the list of known copper moieties was the absence of fluorinated vinyl copper reagents, although such moieties could be invaluable in the construction of fluorine-containing bioactive drugs, polymers, and agricultural chemicals.[4]

The only previous report of a fluorinated vinylic copper reagent was the preparation of E-CF_3CF=$C(CF_3)Cu$ from E-CF_3CF=$C(CF_3)Ag$ by Miller and co-workers.[5]

$$CF_3C\equiv CCF_3 + AgF \rightarrow E\text{-}CF_3CF=C(CF_3)Ag \xrightarrow{Cu}$$
$$E\text{-}CF_3CF=C(CF_3)Cu \quad (8\text{-}1)$$

This reported vinyl copper reagent exhibited some thermal stability, thus providing confidence that our initial target compounds might also exhibit thermal stability. However, the Miller approach presented several problems for a general approach: (1) it demanded an F-alkyne as a precursor; (2) with unsymmetrical F-alkynes a problem of regiospecificity was introduced; and (3) only E-isomers were obtained by this methodology.

Consequently, we decided to pursue a strategy that utilized vinylic halides as precursors with the expectation that formation of the vinyl organometallic reagent would proceed with stereospecificity. Therefore, by proper selection of an E- or Z-precursor, regio- and stereochemical control could be achieved.

2. INITIAL STUDIES

Our initial approach involved attempts at the direct preparation of trifluorovinyl copper via reaction of iodotrifluoroethene with various types of activated copper. However, all these attempts proved unsuccessful; only

$$2\ CF_2=CFI + Cu \rightarrow F_2C=CF-CF=CF_2 + 2\ CuI \quad (8\text{-}2)$$

varying yields of F-1,3-butadiene was observed depending on the type of copper employed. It readily became apparent that the slow step of the reaction was the reaction of the iodoethene with copper to presumably form trifluorovinyl copper; then a fast reaction of trifluorovinyl copper with more iodotrifluoroethene produced the butadiene product (Scheme I).

$$F_2C=CFI + 2\ Cu \xrightarrow{slow} F_2C=CFCu + CuI$$
$$fast \Big| CF_2=CFI$$
$$F_2C=CF-CF=CF_2 \quad (8\text{-}3)$$

Scheme I

Consequently, in order for the initially formed vinyl copper reagent to survive, it must be produced in the absence of the vinylic halide precursor, thereby precluding the coupling process to form the symmetrical diene.[6] We were able to successfully accomplish such a strategy via initial formation of the vinylic lithium reagent followed by exchange of the lithium reagent with copper (I) trifluoroacetate.[7] However, this approach was temporarily abandoned because it did not avert the problem of the low temperature preparation of the thermally labile lithium reagents.[8]

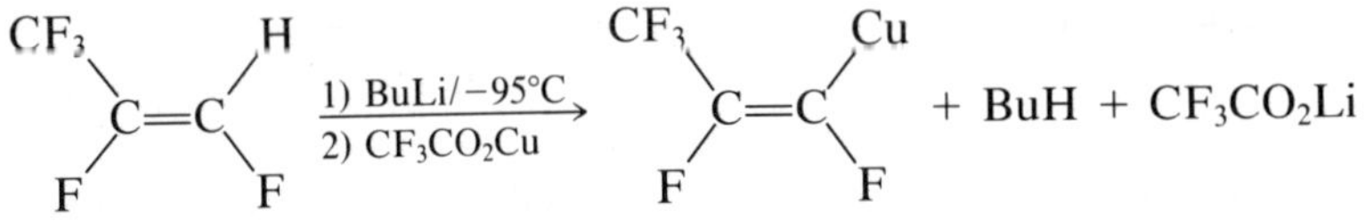

$$\text{(8-4)}$$

3. VINYL CADMIUM REAGENTS

Our initial observation that the vinyl copper reagent rapidly coupled with additional vinyl halide to give symmetrical dienes indicated that an additional prerequisite in our approach was the selection of an organometallic compound that not only met the previously noted criteria, but one that would not readily couple with the vinyl halide precursor. In addition, this proxy organometallic compound must be easily converted to our targeted vinyl copper compounds.

Thus, we selected organocadmium reagents for our initial model studies. We anticipated that the vinyl cadmium reagents would exhibit stability and would not easily couple with vinyl halides. Additionally, cadmium has two NMR active isotopes, ^{111}Cd and ^{113}Cd, of high natural abundance ($\sim 12.5\%$ for each isotope). These NMR active isotopes would assist us to monitor the spectroscopic detection of the vinylic cadmium reagent and also assist us in the mono or bis structural assignments. Also, we expected that a metathetical process would permit a facile conversion to the copper reagent.

Our expectations were successfully met and we found that the alkenyl cadmium reagents can be easily prepared via direct reaction of F-vinyl iodides or bromides with cadmium metal in dimethylformamide (DMF). The vinyl iodides react readily (short induction period) at room temperature, and the vinyl bromides require mild heating (60°C). The cadmium reagents are formed as a mono/bis mixture (as determined by ^{19}F and ^{113}Cd NMR) and this ratio varied with structure of the vinylic halide. However, no significant difference in mono/bis rates of reaction was detected in functionalization of these reagents. Table 8-1 summarizes the preparation of some typical cadmium reagents.

$$R_FCF\!\!=\!\!CFX + Cd \xrightarrow[\text{RT} - 60°C]{\text{DMF}} \underset{\text{mono}}{R_FCF\!\!=\!\!CFCdX} + \underset{\text{bis}}{(R_FCF\!\!=\!\!CF)_2Cd}$$

$$X = I, Br \qquad\qquad\qquad\qquad\qquad + CdX_2 \quad \text{(8-5)}$$

TABLE 8-1. Preparation of Alkenyl Cadmium Reagents from F-Vinyl Halides and Cadmium

$$R_FCF{=}CFX + Cd \xrightarrow[\text{RT-60°C}]{\text{DMF}} R_FCF{=}CFCdX + (R_FCF{=}CF)_2Cd + CdX_2$$

Olefin	Cadmium Reagent[a]	Yield[b]
$CF_2{=}CFI$	$CF_2{=}CFCdX$	99
$Z{-}CF_3CF{=}CFI$	$Z{-}CF_3CF{=}CFCdX$	96
$E{-}CF_3CF{=}CFI$	$E{-}CF_3CF{=}CFCdX$	92
$Z{-}CF_3(CF_2)_4CF{=}CFI$	$Z{-}CF_3(CF_2)_4CF{=}CFCdX$	95
$CF_3CF{=}CICF_3{}^c$	$CF_3CF{=}C(CF_3)CdX^d$	91
$CF_3CF{=}C(Ph)CF{=}CFBr^e$	$CF_3CF{=}C(Ph)CF{=}CFCdX^f$	61
$E{-}CF_3C(Ph){=}CFI$	$E{-}CF_3C(Ph){=}CFCdX$	77

[a] Mixture of mono and bis reagent; X = halogen or another F-alkenyl group.
[b] ^{19}F NMR yields vs $PhCF_3$
[c] E/Z mixture; E/Z = 39/61
[d] E/Z = 36/63
[e] E,Z : Z,Z = 90/10
[f] E,Z : Z,Z = 90/10

The preparation of the cadmium reagents proceeds with total retention of configuration, as determined from ^{19}F NMR and/or ^{113}Cd NMR analysis of the cadmium compound or via ^{19}F NMR analysis of the resultant hydrolysis product, $R_FCF{=}CFH$. All the cadmium reagents prepared in our study were readily hydrolyzed by water; consequently, dry solvents are imperative in their preparation.

The thermal stability of these vinyl cadmium reagents is remarkable. At room temperature (in absence of moisture) these reagents are stable indefinitely.[9] Even at temperatures greater than 100°C these reagents only slowly lose activity. Indeed, a short path distillation of reaction mixtures followed by recrystallization from CH_2Cl_2/pentane (1 : 5) permits isolation of the bis reagents as solvates. Thus, the following solvates were isolated via distillation/recrystallization to illustrate the remarkable thermal stability of these reagents;

$(ZCF_3CF{=}CF)_2Cd \cdot DMF$ $(ZCF_3CF{=}CF)_2Cd \cdot TG$ $(F_2C{=}CF)_2Cd \cdot TG$

mp: 77–80°C   mp: 64–65°C   mp: RT

bp: 137°C/0.09 mm Hg   bp: 130°C/0.08 mm Hg   bp: 121°C/0.10 mm Hg

The vinyl cadmium reagents are readily functionalized with allyl bromide, $(EtO)_2PCl$ and Ph_2PCl. However, attempted acylation with acyl halides failed to give the α,β-unsaturated ketone; but instead gave $(CH_3)_2NCH(R_F)_2$ via a solvent participation reaction.

4. VINYL ZINC REAGENTS

Our success with the preparation of vinyl cadmium reagents prompted us to investigate similar reactions with zinc metal.[10] Fluorinated vinyl iodides reacted smoothly with zinc metal at room temperature in a variety of solvents, including DMF, DMAC, THF, glymes, and acetonitrile. *F*-vinyl bromides also readily reacted with zinc metal in DMF. The induction periods for these reactions vary from a few seconds to several minutes. The zinc reagents are formed as a mono/bis mixture, and this ratio varied with the structure of the vinylic halide and the solvent (Table 8-2).

$$R_FCF{=}CFX + Zn \xrightarrow[RT]{solvent} R_FCF{=}CFZnX$$
$$\text{mono}$$

$$X = I, Br \qquad\qquad + (R_FCF{=}CF)_2Zn + ZnX_2 \quad (8\text{-}6)$$
$$\text{bis}$$

$$(R_FCF{=}CF)_2Zn + ZnX_2 \rightleftharpoons 2\, R_FCF{=}CFZnX \qquad\qquad (8\text{-}7)$$

The mono/bis zinc reagents were distinguished by ^{19}F NMR by enhancement of the signal for the mono reagent on addition of the appropriate zinc halide at the expense of the signal for the bis reagents as expected for the above equilibrium. Unfortunately, zinc has no useful NMR active isotopes like cadmium to provide unequivocal identification of the organometallic species.

The vinyl zinc reagents exhibit exceptional thermal stability. A sample of $ZCF_3CF{=}CFZnX$ in triglyme (TG) showed no loss of activity after 3 days at room temperature, and a loss of only 10% activity after 36 days at room temperature. Similarly, $ZCF_3CF{=}CFZnX$ showed only a 5% loss of activity after 3 days at 65°C, and 25% loss of activity after 36 days at 65°C. A sample of $CF_3C(Ph){=}CFZnX$ in DMF lost only 1% activity after 16 hours at 70°C. This excellent stability permits these reagents to be formed on a large scale and utilized over an extended period in a variety of synthetic reactions without any significant change in activity of the stock reagent. For example, stock reagents of $[CF_2{=}CFZnX]$ can be routinely prepared on a 0.5 M scale, and the resulting solution can be stored at room temperature or in a refrigerator under nitrogen for several days to several weeks.

As noted in Table 8-2, the stereochemical integrity of the *E*, *Z*-vinyl halide is preserved at all times. The stereochemistry is conveniently monitored by ^{19}F NMR. Cis and trans vinyl fluorines in the zinc reagent are readily distinguished by the large coupling constant between trans vinyl fluorines (typically 100–113 Hz vs 0–32 Hz for cis vinyl fluorines).

The zinc reagents are also moisture sensitive and are readily hydrolyzed to $R_FCF{=}CFH$. The zinc reagents also react rapidly with I_2 to yield the corresponding iodides. This reaction allows the more accessible *F*-vinyl bromides[11] to be conveniently converted to *F*-vinyl iodides.

TABLE 8-2. Preparation of Vinyl Zinc Reagents from F-Vinyl Halides and Zinc Metal

$$R_FCF{=}CFX + Zn \xrightarrow{\text{solvent/RT}} R_FCF{=}CFZnX + (R_FCF{=}CF)_2Zn + ZnX_2$$

Vinyl halide	Solvent[a]	Zinc reagent[b]	Mono:Bis	Yield[c]
$CF_2{=}CFI$	DMF	$CF_2{=}CFZnX$	80:20	79%
	DMAc		84:16	97%
	MG			60%
	TG		68:32	95%
	TetG		67:33	85%
$CF_2{=}CFBr$	DMF	$CF_2{=}CFZnX$	95:5	72%
$CF_2{=}CBr_2$	DMF	$CF_2{=}CBrZnX$	68:32	97%
$Z\text{-}CF_3CF{=}CFI$	THF	$Z\text{-}CF_3CF{=}CFZnX$	68:32	98%
	TG		65:36	96%
	DMF		74:26	100%
	CH_3CN		81:19	90%
$E\text{-}CF_3CF{=}CFI$	TG	$E\text{-}CF_3CF{=}CFZnX$	59:41	100%
$Z\text{-}CF_3CF_2CF{=}CFI$	TG	$Z\text{-}CF_3CF_2CF{=}CFZnX$		90%
$Z\text{-}CF_3(CF_2)_4CF{=}CFBr$	DMF	$Z\text{-}CF_3(CF_2)_4CF{=}CFZnX$		77%
$Z\text{-}CF_3(CF_2)_4CF{=}CFI$	TG	$Z\text{-}CF_3(CF_2)_4CF{=}CFZnX$		74%
$CF_3C(Ph){=}CFBr$[d]	DMF	$CF_3C(Ph){=}CFZnX$[d]		94%
$CF_3C(Ph){=}CBr_2$	DMF	$CF_3C(Ph){=}CBrZnX$[e]		95%
$E\text{-}CF_3C(Ph){=}CFI$	THF	$E\text{-}CF_3C(Ph){=}CFZnX$	67:33	78%
$Z\text{-}CF_3(C_6F_5)C{=}CFI$	THF	$Z\text{-}CF_3(C_6F_5)C{=}CFZnX$	57:43	86%
$CF_3CF{=}C(Ph)CF{=}CFBr$[f]	DMF	$CF_3CF{=}C(Ph)CF{=}CFZnX$[f]		71%
$E\text{-}CF_3CH{=}CFI$	TG	$E\text{-}CF_3CH{=}CFZnX$	67:33	89%
$E\text{-}CF_3CF{=}C(CF_3)I$	TG	$E\text{-}CF_3CF{=}C(CF_3)ZnX$	80:20	75%

[a] MG = monoglyme, TG = triglyme, TetG = tetraglyme.
[b] Mixture of mono and bis reagent; X = halogen or another F-vinyl group.
[c] ^{19}F NMR yield vs PhCF$_3$
[d] E/Z mixture; E/Z = 59/41
[e] E/Z = 67/33
[f] $E,Z:Z,Z$ = 90:10

As noted earlier, vinyl cadmium reagents are formed via an analogous reaction and also meet our stated criteria. However, the zinc reagents do have some superior qualities. The zinc reagents may be prepared at room temperature from the often more accessible F-vinyl bromides as well as the F-vinyl iodides. There is also less concern over toxicity and cost with zinc. Finally, the most important advantage is the possibility for functionalization in solvents other than DMF. This factor may be of concern when the functionalized product is extremely reactive and may react with DMF (cf. preparation of $F_2C{=}CFC(O)R$ described later).

5. VINYL COPPER REAGENTS

Copper (I) halide metathesis of the vinyl cadmium and/or zinc reagents provides a high yield preparation of stable polyfluorinated vinyl copper re-

TABLE 8-3. Preparation of Alkenyl Copper Reagents

Olefin	M	$R_FCF{=}CFCu$	% Yield[a]
$CF_2{=}CFI$	Cd	$CF_2{=}CFCu$	99%
$CF_2{=}CFBr$	Zn	$CF_2{=}CFCu$	72%
$Z\text{-}CF_3CF{=}CFI$	Cd	$Z\text{-}CF_3CF{=}CFCu$	92%
$Z\text{-}CF_3CF{=}CFI$	Zn	$Z\text{-}CF_3CF{=}CFCu$	76%
$E\text{-}CF_3CF{=}CFI$	Cd	$E\text{-}CF_3CF{=}CFCu$	83%
$Z\text{-}CF_3(CF_2)_4CF{=}CFI$	Cd	$Z\text{-}CF_3(CF_2)_4CF{=}CFCu$	87%
$Z\text{-}CF_3CCl{=}CFI$	Cd	$Z\text{-}CF_3CCl{=}CFCu$	78%
$CF_3C(Ph){=}CFBr$	Zn	$CF_3C(Ph){=}CFCu$	84%
E/Z = 59/41		E/Z = 59/41	
$CF_3CF{=}C(Ph)CF{=}CFBr$	Zn	$CF_3CF{=}C(Ph)CF{=}CFCu$	63%
E/Z = 90/10		E/Z = 90/10	

[a] Overall ^{19}F NMR yield based on starting olefin.

agents (Table 8-3). The preparation of the vinyl cadmium or zinc precursors proceeds stereospecifically and the metathesis with Cu(I)X occurs with retention of configuration.[12] Consequently, stereochemical integrity is retained throughout the synthetic sequence and one can confidently prepare either isomeric copper reagent with total stereochemical control by proper choice of the appropriate vinyl halide.

$$R_FCF{=}CFX + M \qquad\qquad X = Br, I$$
$$\Big\downarrow {\scriptstyle DMF \atop RT\text{-}60°C} \qquad\qquad M = Cd, Zn$$
$$R_FCF{=}CFMX + (R_FCF{=}CF)_2M + MX_2$$
$$\Big\downarrow {\scriptstyle Cu(I)X \atop RT}$$
$$R_FCF{=}CFCu$$
$$68\text{–}99\%$$

$$(8\text{-}8)$$

The vinyl copper reagents exhibit excellent stability at room temperature in the absence of oxygen and/or moisture. At higher temperatures (more than 50°C) they undergo rapid decomposition. And, since they are formed in the absence of vinyl halide, no coupling to form symmetrical diene is observed.

The vinyl copper reagents participate in a variety of alkylation, coupling, and acylation reactions as illustrated in Scheme II.

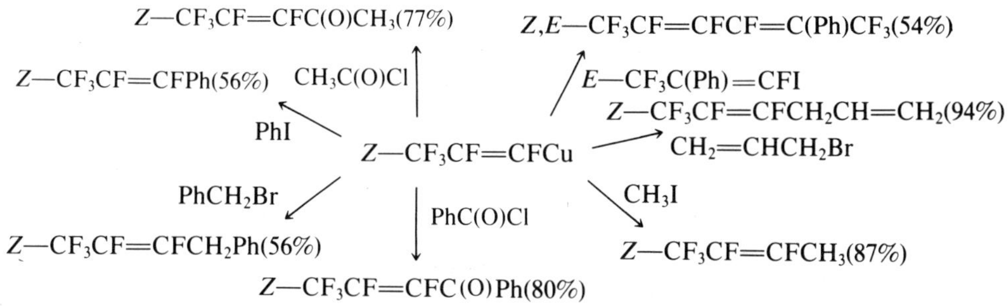

Z—CF$_3$CF=CFC(O)CH$_3$(77%) Z,E—CF$_3$CF=CFCF=C(Ph)CF$_3$(54%)

Z—CF$_3$CF=CFPh(56%)   CH$_3$C(O)Cl E—CF$_3$C(Ph)=CFI
                                                   Z—CF$_3$CF=CFCH$_2$CH=CH$_2$(94%)
              PhI   CH$_2$=CHCH$_2$Br
                       Z—CF$_3$CF=CFCu

      PhCH$_2$Br   CH$_3$I
Z—CF$_3$CF=CFCH$_2$Ph(56%)   PhC(O)Cl Z—CF$_3$CF=CFCH$_3$(87%)

              Z—CF$_3$CF=CFC(O)Ph(80%)

Scheme II

6. SYNTHETIC APPLICATIONS OF THESE STABLE *F*-ORGANOMETALLIC REAGENTS

These organometallics have found widespread utility in a number of useful applications. For example, the E- and Z-perfluoroprop-l-enylcopper reagents were utilized to provide model precursors that were employed to study the contrasteric, electrocyclic ring opening of fluorinated cyclobutenes,[13] in which it was demonstrated that the rules employed to predict the ring-opened product of hydrocarbon analogs were not applicable to the fluorinated system.

In the following paragraphs, several illustrative examples of additional applications are briefly described.

A. Palladium Catalyzed Coupling of *F*-Vinyl Zinc Reagents

1. With Aryl Iodides

Previous reports on the preparation of α,β,β-trifluorostyrenes have either given low yields, produced significant amounts of side products, or have utilized multi-step procedures.[14] We have now found that the *F*-vinyl zinc reagents, produced in situ from iodo- or bromotrifluoroethene, react readily with aryl iodides in the presence of 1–2 mol percent (Ph$_3$P)$_4$Pd to give excellent yields of the α,β,β-trifluorostyrenes (Table 8-4), under mild conditions.[15,16]

$$CF_2=CFX + Zn \xrightarrow[\text{THF}]{\text{DMF}} CF_2=CFZnX + (CF_2=CF)_2Zn$$

X = Br, I   or
                TG

$$\left| \begin{array}{l} \text{ArI, (Ph}_3\text{P)}_4\text{Pd} \\ 60\text{–}80°C \\ 1\text{–}8 \text{ hr} \end{array} \right.$$

$$CF_2=CFAr \qquad\qquad (8\text{-}9)$$

TABLE 8-4. Isolated Yields of α,β,β-Trifluorostyrenes

$$CF_2{=}CFZnX + ArI \xrightarrow[\substack{1-2 \text{ mol \%} \\ 60°C\ 1-3\ hr}]{(Ph_3P)_4Pd} CF_2{=}CFAr$$

α,β,β-Trifluorostyrene	Solvent	Isolated yield
$C_6H_5CF{=}CF_2$	DMF	74%
$o\text{-}NO_2C_6H_4CF{=}CF_2$	DMF	73%
$p\text{-}MeOC_6H_4CF{=}CF_2$	THF	61%
$o\text{-}(CH_3)_2CHC_6H_4CF{=}CF_2$	THF	70%
$2,5\text{-}Cl_2C_6H_3CF{=}CF_2$	DMF	75%
$o\text{-}CF_3C_6H_4CF{=}CF_2$	TG	73%
$m\text{-}NO_2C_6H_4CF{=}CF_2$	TG	81%
$p\text{-}ClC_6H_4CF{=}CF_2$	DMF	77%
$p\text{-}CF_2{=}CFC_6H_4CF{=}CF_2$	DMF	56%

A wide variety of substituted aryl iodides readily participate in the coupling process (Table 8-4). Even bulky ortho substituents such as $(CH_3)_2CH\text{-}$, $CF_3\text{-}$, and $NO_2\text{-}$ gave excellent yields of styrene.

Extension of this methodology to the E- and Z-1-iodo-F-propenes stereospecifically produces the Z- and E-1-phenyl-F-propenes, respectively.[15,16] Earlier routes to these olefins gave mixtures of olefinic products.[17] The Pd° catalyzed coupling process works well for both electron-withdrawing and electron-releasing groups (Table 8-5).

TABLE 8-5. Isolated Yields of
E-1-Arylperfluoropropenes and
Z-1-Arylperfluoropropenes[a]

$$\underset{E \text{ or } Z}{CF_3CF{=}CFZnX} + ArI \xrightarrow[1-2 \text{ mol \%}]{(Ph_3P)_4Pd} \underset{Z \text{ or } E}{CF_3CF{=}CFAr}$$

1-Arylperfluoropropene	Isolated yield
$E\text{-}C_6H_5CF{=}CFCF_3$	80%
$E\text{-}p\text{-}CH_3C_6H_4CF{=}CFCF_3$	65%
$E\text{-}p\text{-}ClC_6H_4CF{=}CFCF_3$	61%
$E\text{-}m\text{-}NO_2C_6H_4CF{=}CFCF_3$	80%
$Z\text{-}C_6H_5CF{=}CFCF_3$[b]	82%
$Z\text{-}p\text{-}CH_3C_6H_4CF{=}CFCF_3$[c]	70%
$Z\text{-}p\text{-}ClC_6H_4CF{=}CFCF_3$[b]	74%

[a] All preparations were conducted in triglyme with
yellow catalyst prepared in our laboratory.
[b] 97/3 Z/E
[c] 92/8 Z/E

$$E\text{-}CF_3CF{=}CFZnX + ArI \xrightarrow[\substack{TG \\ 80°C \\ 8\ hr}]{(Ph_3P)_4Pd} Z\text{-}CF_3CF{=}CFAr \qquad (8\text{-}10)$$

$$Z\text{-}CF_3CF{=}CFZnX + ArI \xrightarrow[\substack{TG \\ 80°C \\ 8\ hr}]{(Ph_3P)_4Pd} E\text{-}CF_3CF{=}CFAr \qquad (8\text{-}11)$$

2. With F-Vinyl Iodides

The F-vinyl zinc reagents also couple smoothly with iodotrifluoroethene in the presence of $(Ph_3)_4Pd$ and provide a stereospecific route to substituted F-1,3-butadienes.[16] These dienes have also been utilized as model precursors in the electrocyclic interconversion of F-3-methylcyclobutene with Z- and E, F-1,3-pentadiene.[18]

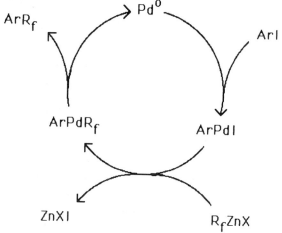

These palladium catalyzed reactions presumably proceed through a catalytic cycle as depicted in Scheme III.

$$ArR_f \qquad Pd^0 \qquad ArI$$

$$ArPdR_f \qquad ArPdI$$

$$ZnXI \qquad R_fZnX$$

Scheme III.   Palladium catalytic cycle

B. Preparation of Trifluorovinyl Ketones

One of the most important and useful reactions of α,β-unsaturated carbonyl systems is nucleophilic attack at the β-carbon, and incorporation of electronegative substituents on the vinylic portion of such functionality generally enhances attack towards nucleophiles.[19] Thus, one would anticipate that trifluorovinyl ketones would exhibit high reactivity and provide useful precursors as building blocks for fluorinated materials. However, the paucity of general synthetic methodology for the preparation of trifluorovinyl carbonyl compounds has hampered detailed investigations of these highly reactive and interesting compounds.

Initial attempts at the preparation of such systems were reported by Russian workers,[19] who carried out the aluminum chloride catalyzed acylation of trifluoroethene with acetyl chloride. Dehydrochlorination of the resultant product with triethylamine in dimethyl ether gave only impure material, tentatively identified as the hydrate of 3,4,4-trifluoro-3-buten-2-one. Ishikawa later improved this procedure and isolated the methyl trifluorovinyl ketone via dehydrochlorination of 4-chloro-3,4,4-trifluoro-2-butanone with triethyl amine in diglyme.[20]

$$CH_3C(O)Cl + FHC{=}CF_2 \xrightarrow[RT]{AlCl_3} CH_3C(O)CHFCF_2Cl \qquad (8\text{-}14)$$
$$35\%$$

$$CH_3C(O)CHFCF_2Cl + Et_3N \xrightarrow{diglyme} CH_3C(O)CF{=}CF_2 \qquad (8\text{-}15)$$
$$85\%$$

We recently reported[21] a general route to this class of ketones via the copper (I) mediated acylation of trifluorovinyl zinc reagents in glyme solvents.

$$CF_2{=}CFI + Zn \xrightarrow[TetG]{TG \text{ or}} CF_2{=}CFZnI + (CF_2{=}CF)_2Zn$$
$$85\text{--}95\%$$
$$\Big\downarrow \begin{array}{l} RC(O)Cl \\ CuBr, RT \end{array}$$
$$CF_2{=}CFC(O)R \qquad (8\text{-}16)$$

Since the zinc reagents can be produced in situ from iodotrifluoroethene, this route provides a convenient entry to these valuable reactive compounds.[22] Table 8-6 illustrates typical examples prepared via this approach.

In the absence of the Cu(I) halides, no reaction is observed. Thus, the presumed intermediate is the trifluorovinyl copper reagent, although this

TABLE 8-6. Preparation of Trifluorovinyl Ketones ($CF_2=CFC(O)R$)

R	Yield[a]	Solvent	bp. (°C)
CH_3	76% (71%)	triglyme	62–66 (510 mm)
CH_3CH_2	83%	triglyme	55–57 (190 mm)
$CH_3CH_2CH_2$	73% (79%)	triglyme	57–59 (100 mm)
$(CH_3)_2CH$	87% (94%)	triglyme	45–47 (90 mm)
$(CH_3)_3C$	81% (83%)	triglyme	54–58 (106 mm)
$CH_3CH_2CH_2CH_2CH_2$	67% (84%)	triglyme	63–67 (10 mm)
$C(O)OCH_2CH_3$	50% (98%)	triglyme	57–59 (15 mm)
CH_2Cl	27% (91%)	triglyme	79–81 (60 mm)
CH_2Br	— (35%)	triglyme	
$CH_2CH_2C(O)OCH_3$	44% (76%)	tetraglyme	40–45 (0.1 mm)
CF_3	dec.	triglyme	
$CF_2CF_2CF_3$	dec.	triglyme	

[a] Yields in parentheses were determined by ^{19}F NMR integration of the product signals in the reaction mixture relative to internal benzotrifluoride standard.

species is not observed spectroscopically. It is necessary to utilize the glyme system in the preparation of these reactive ketones. If $CF_2=CFCu$ in DMF is employed, acylation results; however, the trifluorovinyl ketone products are immediately attacked by DMF and destroyed. Thus, the ability to prepare the zinc reagents in alternative solvent systems (alluded to earlier) contributes to the success in this particular application.

C. Addition to Multiple Bonds

As noted in the introduction, one of the reactions of interest for our target compounds was the potential capability to add to carbon–carbon multiple bonds. To avoid any regiochemical problems, we chose F-2-butyne as a model alkyne to study this type of reaction. If reaction occurs, three possible products might be detected: (a) stereospecific anti addition product; (b) stereospecific syn addition product; (c) a mixture of anti and syn addition products.

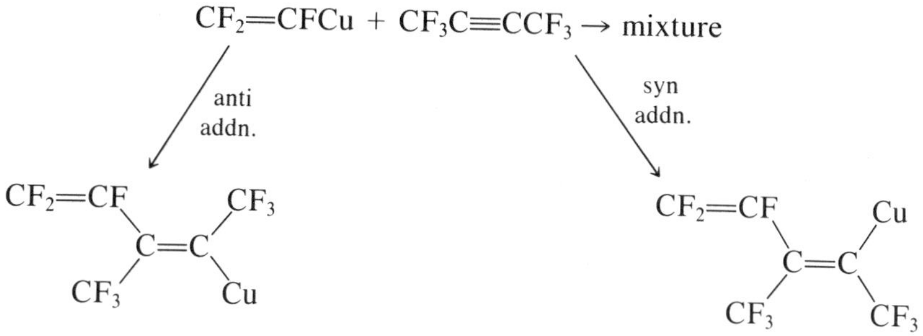

$$CF_2=CFCu + CF_3C{\equiv}CCF_3 \rightarrow \text{mixture}$$

(8-17)

The selection of F-2-butyne as the model alkyne also permits one to readily determine the mode of addition via J_{CF_3,CF_3} in the addition product:

$$\text{trans} \qquad\qquad \text{cis}$$
$$(J_{CF_3,CF_3}\ 10\text{–}12\ \text{Hz};\ J_{CF_3,CF_3}\ 0\text{–}1\ \text{Hz}).$$

We found that upon addition of F-2-butyne to a solution of $CF_2{=}CFCu$ in DMF it was readily absorbed and the addition adduct was easily detected by ^{19}F NMR.[23] Quenching of the addition adduct with acid or iodine followed by ^{19}F NMR analysis of the isolated products demonstrated that with this alkyne, a stereospecific syn addition occurred. No anti addition product was detected.

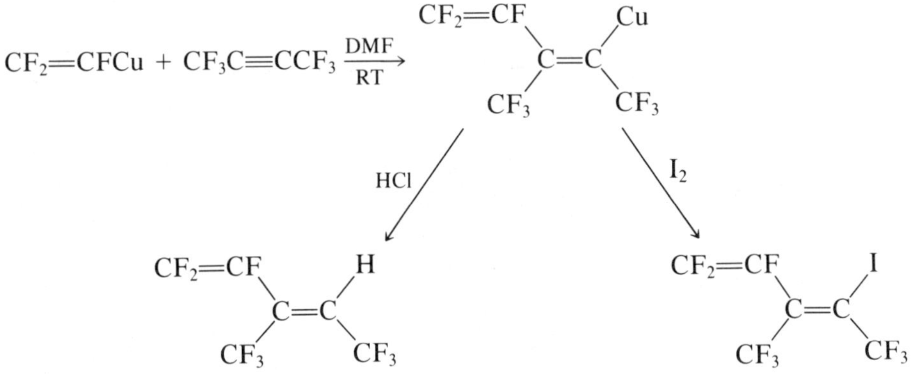

$$CF_2{=}CFCu + CF_3C{\equiv}CCF_3 \xrightarrow[RT]{DMF}$$

(8-18)

Thus, this type of addition process provides a new methodology to build stable fluorinated organometallic reagents in which the metal is part of a conjugated diene system. More importantly, this new type of reagent can be prepared stereospecifically.

7. SUMMARY

We have demonstrated that F-vinyl cadmium and zinc reagents can be readily prepared via direct reaction of F-vinyl iodides and/or bromides with cadmium and zinc metal. In contrast to the F-vinyl lithium and magnesium reagents, the cadmium and zinc analogs exhibit remarkable thermal stability. Metathesis of the F-vinyl cadmium and zinc reagents with copper (I) salts provides the first equivocal pregenerative route to the F-vinyl copper reagents. In all cases, stereochemical integrity is retained throughout all synthetic sequences.

These stable F-vinyl cadmium, zinc, and copper reagents are readily functionalized and serve as useful synthetic intermediates for the preparation of

F-dienes, α,β,β-trifluorostyrenes, trifluorovinyl ketones, and as building blocks to new organometallics that contain a conjugated diene system. Future applications will no doubt extend and elaborate the utility of these reagents.

The initial criteria that we established for our target compounds has been more than adequately met, and the methodology established in this work will undoubtedly prove successful in the development of other new F-alkenyl organometallic reagents that will find successful application in organofluorine chemistry.

ACKNOWLEDGMENT

I thank the National Science Foundation, the Air Force Office of Scientific Research, and the Gas Research Institute for financial support of organofluorine research at the University of Iowa. I also am deeply indebted to the talented research of S. W. Hansen, P. L. Heinze, and T. D. Spawn on whose work this paper was predicated.

REFERENCES

1. Chambers, R. D. "*Fluorine In Organic Chemistry*". Wiley-Interscience: New York, 1973, Chapter 10.
2. Hahnfeld, J. L.; Burton, D. J. *Tetrahedron Letters* **1975,** 773.
3. Posner, G. H., Ed. "An Introduction to Synthesis Using Organocopper Reagents". Wiley-Interscience: New York 1980.
4. Filler, R., Ed. *ACS Symp. Ser. 1976, No. 28*; Filler, R.; and Kobayashi, Y., Eds. "Biomedicinal Aspects of Fluorine Chemistry". Kodasha/Elsevier: New York, 1982; Banks, R. E., Ed. "Organofluorine Chemicals and Their Industrial Applications". Ellis Harwood Ltd.: Chichester, 1979.
5. Miller, W. T. "Abstracts of the 9th International Symposium on Fluorine Chemistry". Avignon, France, 1979, p. 027.
6. Symmetrical dienes can be prepared via reaction of F-vinyl iodides and copper metal without prior pregeneration of the F-vinyl copper reagents. Several examples of this approach have been previously reported in the literature.
7. Bailey, A. R. Unpublished work, University of Iowa.
8. In some cases, when suitable precursors, other than F-vinyl halides, are available, this method may provide a useful alternative and we are pursuing this approach with model substrates.
9. Burton, D. J.; Hansen, S. W. *J. Fluorine Chemistry* **1986,** *31,* 461.
10. Hansen, S. W.; Spawn, T. D.; Burton, D. J. *J. Fluorine Chemistry,* **1987,** *35,* 415.
11. Howells, R. D.; Gilman, H. *J. Fluorine Chemistry* **1974,** *4,* 247; **1975,** *5,* 99; Dua, S. S.; Howells, R. D.; Gilman, H. *J. Fluorine Chemistry* **1974,** *4,* 409; Smith, C. F.; Soloski, E. J.; Tamborski, C. *J. Fluorine Chemistry* **1974,** *4,* 35; Thoai, N. *J. Fluorine Chemistry* **1975,** *5,* 115; Moreau, P.; Albandri, R.; Commeyras, A. *Nouveau J. Chimie* **1977,** *1,* 497.
12. Hansen, S. W.; Burton, D. J. *J. Am. Chem. Soc.* **1986,** *108,* 4229.

13. Dolbier, W. R., Jr.; Koroniak, H.; Burton, D. J.; Bailey, A. R.; Shaw, G. S.; Hansen, S. W. *J. Am. Chem. Soc.* **1984,** *106,* 1871.
14. Cohen, S. G.; Wolosinski, H. T.; Scheuer, P. J. *J. Am. Chem. Soc.* **1949,** *71,* 3439; **1950,** *72,* 3953; Prober, M. *J. Am. Chem. Soc.* **1953,** *75,* 968; Dixon, S. *J. Org. Chem.* **1956,** *21,* 400; Antonucci, J. M.; Wall, L. A. *SPE Trans* **1963,** *3,* 225; Kazenikova, G. V.; Talalaeva, T. V.; Zimin, A. V.; Simonov, A. P.; Kocheshkov, K. A. *Izv. Akad. Nauk SSSR* **1961,** 1063; Hogdow, R. B.; MacDonald, D. I. *J. Polymer Science (Pt. A)* **1968,** *6,* 711.
15. Heinze, P. L.; Burton, D. J. *J. Fluorine Chemistry* **1986,** *31,* 115.
16. Heinze, P. L. Ph.D. thesis, University of Iowa, 1986.
17. Dmowski, W. *J. Fluorine Chemistry* **1981,** *18,* 25; **1985,** *29,* 273 and references therein.
18. Dolbier, W. R., Jr.; Koroniak, H.; Burton, D. J.; Heinze, P. *Tetrahedron Letters* **1986,** *27,* 4387.
19. Knunyants, I. L.; Sterlin, R. N.; Pinkina, L. N.; Kyatkin, B. L. *Izv. Akad. Nauk SSSR Ser. Khim* **1958,** 296.
20. Nakayama, Y.; Kitazume, T.; Ishikawa, N. *J. Fluorine Chemistry* **1985,** *29,* 445.
21. Spawn, T. D.; Burton, D. J., presented in part at the Centenary of the Discovery of Fluorine Symposium, Paris, France, August 1986, Abstract 0–15.
22. Spawn, T. D.; Burton, D. J. *Bull. Soc. Chim. France,* **1987,** 876.
23. Hansen, S. W. Ph.D. thesis, University of Iowa, **1984.**

CHAPTER 9

The Inorganic Chemistry of Hydrogen Fluoride

Albert W. Jache

Department of Chemistry, Marquette University, Milwaukee, Wisconsin

CONTENTS

1. INTRODUCTION: HYDROGEN FLUORIDE

Hydrogen fluoride or aqueous solutions of it have been known for some time. Simons[1] states that it was first obtained in 1768 but first definitely characterized in 1771 by Scheele. Davy showed that it contained no oxygen and showed that after removing the water by electrolysis his sample did not conduct electricity.[2] Thus, he had prepared anhydrous hydrogen fluoride. Later Fremy prepared anhydrous hydrogen fluoride by thermal decomposition of potassium hydrogen fluoride ($KF \cdot HF$).[3] This remained the method of choice until rather good hydrogen fluoride became an article of commerce and effective methods for further purification were developed. Anhydrous hydrogen fluoride whose conductivity was enhanced by dissolved potassium fluoride was used by Moissan in the first preparation of fluorine.[4]

This chapter deals primarily with the anhydrous substance. The term "hydrogen fluoride" will be reserved for the anhydrous substance or solutions which the investigators considered to be anhydrous. In those few cases where aqueous solutions are discussed, the term "hydrofluoric acid" will be used. It is acknowledged that some authors have chosen to specify anhydrous hydrogen fluoride (AHF) when they were dealing with anhydrous material. Some have allowed the term hydrofluoric acid to be ambiguous.

There have been several reviews of the chemistry of the substance. Since it is not intended that this be a complete review, reference is made to several of them. The classic works of the Fredenhagens, Cadenbach, and Klatt were reviewed by H. Fredenhagen in 1939.[5] Simons devoted two chapters[6] of the first volume of his series on "Fluorine Chemistry" to hydrogen fluoride and updated these in a chapter in the fifth volume.[7] The 1965 treatment[8] by Hyman and Katz was a good summary of the situation at that time. It was followed two years later by a treatment[9] by Kilpatrick and Jones. Jander and Lafrenz prepared a fairly short treatment[10] in which they compared hydrogen fluoride and the higher hydrogen halides. Dove and Clifford have published an extensive review, "Inorganic Chemistry in Liquid Hydrogen Fluoride."[11] It appears that the promised parallel chapter dealing with organic chemistry never did appear. Waddington's chapter[12] on "The Halogen Halides" was published the following year. O'Donnell nicely devoted several pages to "The Hydrogen Fluoride Solvent System"[13] in his chapter entitled "Fluorine." Olah and his co-workers have made extensive use of hydrogen fluoride and solutions of it, particularly in organic chemistry, and have recently published a book entitled "Super Acids."[14] In addition, they have published a large number of papers, many of which deal with organic reactions. Since the "Gmelin Handbook of Inorganic Chemistry," the reputable critical compendium of inorganic chemistry, uses a system of supplements to bring matters up to date, it is fortunate that two Gmelin supplements to Fluorine, System No. 5, have come out early in this decade. The second,[15] "Compounds with Hydrogen" is claimed to have a "closing date" of mid 1980. The first[16] deals peripherally with hydrogen fluoride in a section on "Reactions with Hydrogen."

2. PREPARATION

Industrially, most hydrogen fluoride is made from the reaction of fluorspar (calcium fluoride) with sulfuric acid. Much of the drive for this comes from the volatility of the product. There is a large potential source from silicon tetrafluoride or fluorosilicates which arise as by-products in the production of wet process phosphoric acid or the production of "defluorinated phosphate rock" from apatite, which commonly has about half of its hydroxide groups substituted by fluoride. A variety of processes have been described. Many are discussed in earlier reviews. Gmelin[15] has the most recent and most nearly complete review. While the laboratory method of preparation by the thermal decomposition of potassium hydrogen fluoride still gives a useful product, most workers choose to purify and use the commercial product. Indeed, if they go through the acid fluoride it is usually purchased or prepared from the commercial product. Purification methods that have been used include fractionation, electrolysis, treatment with $SOCl_2$, and treatment with fluorine, chlorine trifluoride, cobalt trifluoride or other oxidizing fluorides. The emphasis has been on the elimination of water since it is quite basic in the medium and has a significant effect on its behavior. Other impurities that may be found at low levels include sulfur dioxide and silicon tetrafluoride. Those in the semiconductor industry look for trace levels of compounds of certain elements, such as arsenic.

Specific conductance has been the principle measure of purity since it is sensitive to water or other electrolyte content and is relatively easily measured. Reported minimum specific conductances dropped by a factor of approximately one hundred over about two decades and have been nearly constant at 1×10^{-6} ohm^{-1} cm^{-1} for nearly that long. The decrease reflects significant improvement in purification schemes and in materials of construction. Again, Gmelin[15] is the most recent source on purification of the substance for laboratory use.

3. PHYSICAL PROPERTIES

Many of the properties of hydrogen fluoride reflect its strong tendency toward association. It is a common and often useful classroom activity to list the melting and boiling points and the dielectric constants of pnicogen, chalcogen, and halogen hydrides while pointing to the apparently anomalous properties of the first members, attributing their properties to association resulting from a strong inclination toward hydrogen bonding.

Some of the properties of hydrogen fluoride are listed in Table 9-1.

TABLE 9-1.   Physical Properties of Hydrogen Fluoride

Freezing Point	−83.55°C
Cryoscopic Constant	1.55°C/mol/kg
Boiling Point	19.51°C
Vapor Pressure	364 mm at 0°C; 34.8 mm at −50°C
Ebullioscopic Constant	1.9°C/mol/kg
Heat of Vaporization	7.485 kJ/mol at 19.45°C
Heat of Fusion at fp	3.929 kJ/mol
Entropy of Vaporization	25.59 J/mol · K
Entropy of Fusion	20.68 J/mol · K
Critical Temperature*	188°C
Critical Pressure	6.48 MPa
Critical Density	0.29 g/ml
Critical Volume	69.0 ml/mol
Vapor Pressure	7891 mm at 100°C; 363.8 mm at 0°C
Density	0.796 g/ml at 100°C; 1.015 g/ml at 0°C
Dielectric Constant	83.6 at 0°C; 175 at −73°C
Specific Conductance	1×10^{-6} ohm^{-1}cm^{-1} at 0°C
Viscosity	0.26 cP at 0°C; 0.23 cP at 10°C
Ion Product	2×10^{-12} mol^2 L^{-2}
Surface Tension	8.6 dyn/cm at boiling point

* Also reported as 230°C

4. CONSTITUTION OF THE VARIOUS STATES

While one often chooses to begin a discussion with the gaseous state, as it is often the simplest, this discussion will begin with the solid state since it has fewer complications than do the others.

A. Solid State

In 1954 Atoji and Lipscomb[17] reported that their X-ray diffraction work at 148 K showed that the lattice of F atoms in solid HF was orthorhombic with unit cell dimensions a = 3.42, b = 4.32, c = 5.42 Å. Z = 4. They calculated the density to be 1.663 g/cm^3. Johnson and co-workers[18] studied DF with neutron diffraction techniques. They obtained the following unit cell dimensions: a = 3.31, b = 4.26, c = 5.22 Å at 4.2 K, and a = 3.33, b = 4.27, c = 5.27 Å at 85 K. The calculated densities are 1.89 g/cm^3 at the lower and 1.86 g/cm^3 at the higher temperature. DF, at least with respect to the F atom lattice, is, as expected, isomorphous with HF. Hydrogen bonded F · · · F run through the lattice parallel to the (100) plane. The F · · · F distance is 2.49 Å and the F · · · F · · · F angle is 121.1° in HF at 148 K. Distances in adjacent chains between an F and its nearest and next-nearest neighbors are 3.12 and 3.20 Å, respectively. In DF at 4.2 K the F · · · F distance is 2.50 Å while the angle is 116°. At 85 K the distance is slightly larger, 2.51 Å. [19]F

NMR studies[19,20] showed the F—H bonded distance to be 0.95 Å and the F—H direction not more than 20° off line and showed the F—D bond length to be 0.97 Å at 4.2 K and 0.95 Å at 85 K while the hydrogen-bonded D—F lengths were determined to be 1.53 and 1.56 Å, respectively. The angle between the F · · · F direction and the chain axis is, within experimental error, constant at 31.6° at the two temperatures, while the orientation of the D—F axes with respect to the chain axis is 29.6° at 4.2 K and 34.4° at 85 K.

There remains some uncertainty about the positions of the protons in spite of the considerable work which has gone into this problem. It is conceivable that the FH molecules in adjacent chains may be exclusively parallel, exclusively antiparallel or randomly parallel or antiparallel. The neutron diffraction study of DF[18] showed the parallel arrangement for DF molecules. The vibration spectra[21,22] of HF and DF are closely correlated and their lattice constants are essentially the same. Thus, they are considered to be isomorphous. There have been other studies arguing for disorder or an antiparallel arrangement. Arguments against these have been generally accepted. These are discussed further in Gmelin.[23] A consequence of the parallel arrangement is that HF should be pyroelectric.

B. Gaseous State

In 1924 Simons and Hildebrand published papers entitled "Preparation, freezing point and vapor pressure of hydrogen fluoride"[24] and "Density and molecular complexity of gaseous hydrogen fluoride."[25] Association was much in evidence. They were able to fit the much earlier data of Thorpe and Hambly,[26] as well as their own, to the hypothesis that there is a single kind of polymerized species, $(HF)_6$, in equilibrium with the monomer over the range from 234 K to 361 K and from 56 mm (Torr) to 750 mm. Since then, there has been a variety of studies attempting to adjust parameters to fit various polymerization schemes to observed physical properties. Gmelin[27] gives a good account of these up to the time the summary was written. Essentially, a continuous-polymerization model and a few species model have been those models taken seriously. Strohmeier and Briegleb[28] first proposed the continuous-polymerization model with its stepwise association, $HF + (HF)_n \rightleftarrows (HF)_{n+1}$. The few species model involves the dimer and hexamer primarily. The continuous-polymerization model failed to correctly predict heat capacity C_p[29] and saturated vapor densities[30] near room temperature. IR and Raman studies[31,32] and NMR studies, as well as an electron diffraction study, all suggest that there is at least one cyclic polymer, probably the hexamer. Selective mass spectral work[35] and variation of microwave dielectric constant with pressure and temperature lead to the conclusion that all polymers higher than the dimer are not polar, hence, cyclic. While there is considerable more work in the literature than presented here, it appears that, at least below 300 K, the best model is a monomer-dimer-hexamer model. Indeed, in this region, the continuous-polymerization model reduces to this, along with minor constituents of noncyclic chains. It seems that the early thoughts of Simons and Hildebrand were close to the mark.

C. Liquid State

There is a rather little change in entropy upon melting. The entropy of boiling is rather low (ca 26 J/mol/K) (1 cal ≡ 4.184J) for an associated liquid. These suggest that there is not a remarkable change in structure on melting or a sizable degree of depolymerization on boiling. It is not necessarily true that there is no change in structure throughout the liquid range. The strong bonding within the chains could well persist while melting, while the weaker attractive forces between chains could be overcome giving way to a large decrease in density upon melting and a liquid of low viscosity. IR, Raman, ^{1}H NMR spectroscopy, and neutron scattering all show the expected hydrogen bonds. Raman studies[36] suggest the existence of several polymeric species. That the degree of polymerization is decreased by dilution with liquid SO_2 is reflected in IR[37] and ^{1}H NMR[38] spectra. Assuming that the only polymer in dilute solutions is the dimer, equilibrium constants, hence enthalpy of dimerization (28.9 kJ/mol[38]) can be calculated. Fredenhagen and Dahmlos[39] determined that the dielectric constants at 273 K and 200 K are 84 and 175, respectively. These values require polymerization for an explanation. Cole[40] analyzed these in terms of the Kirkwood correlation factor and argued for extended linear chains and stated that ". . . unless the early dielectric constant data or the interpretations are greatly in error it is difficult to escape the conclusion that association in liquid HF is much more like a disordered form of the solid than the vapor." He surmised that if the closed rings are more stable in the vapor because of the extra bond, this extra stability is compensated in the liquid by effects of packing and configurational entropy. Desbat and Huong[41] studied liquid HF, DF, and mixtures of these by means of infrared and Raman spectroscopy. They interpreted their data to mean that at 273 K the most probable structure is zigzag chains six or seven molecules long with an angle of 42° with the average axis while, just before solidification (190 K), the chain length is eight with an angle of about 35°. The opening of a puckered ring and the addition of one unit is not a drastic step as one goes from the gaseous to liquid state nor is the lining up of chains upon solidification.

D. The Monomer, Dimer, and Higher Oligomers

Many papers have been written on the electron configuration and spectra of the monomer. A discussion of them is beyond the scope of this chapter.

Gmelin[42] deals with this literature up to mid 1980. The dipole moment has been of interest to many. Muenter,[43] using the radio-frequency spectrum measured by a molecule beam electric resonance approach found the permanent dipole moment of HF in its lowest vibrational, first rotational state to be 1.826178 D. He and Klemperer[44] found the quadrupole coupling constant of the deuteron to be 354.238 kHz. This is consistent with an earlier measurement and in the range of values reported from quantum mechanical calculations. Again, the literature abounds with work resulting from interest in magnetic constants and nuclear magnetic resonance of this interesting molecule. Hindermann and Cornwall[45] have measured both the ^{19}F and ^1H NMR spectra of gaseous HF and DF at 34°C over a wide range of pressures. The monomer chemical shifts, obtained by extrapolation to zero pressure, are for ^{19}F, 46.85 ppm in HF, and 49.35 ppm for DF, upfield from gaseous SiF_4. These authors attributed the 2.5 ppm difference to differences in vibrational energies. For ^1H, the shift is -2.10 ppm downfield from gaseous CH_4. The shifts for a cylindrical liquid sample at 34°C are ^{19}F, 25.53 ppm and ^1H, -8.67 ppm with respect to gaseous SiF_4 and CH_4. (These authors determined that the ^{19}F shift in $CFCl_3$, relative to the gaseous SiF_4 standard, is 174.5 ppm. This brings the liquid HF shift to the 190 ppm relative to $CFCl_3$ shown in Hudlicky's paper.[48] The proton resonance in CH_4 is shifted 4.60 ppm relative to $(CH_3)_4Si$.) Adding 0.85 ppm will convert these to shifts for a spherical sample.[44] Solomon and Bloembergen[46] determined $J_{HF} = 615$ Hz from a theoretical treatment of nuclear magnetic resonance studies. Their limits of error were 50 Hz. MacLean and Mackor,[47] using HF solutions of BF_3 and $NaBF_4$, were able to reduce chemical exchange sufficiently to allow the observation of the doublet splittings of the ^{19}F and ^1H resonances of $J_{HF} = 521$ Hz at -70°C. More recently, Hudlicky has published a paper[48] summarizing and adding to the reported ^{19}F NMR frequencies observed in HF and other fluoride-containing solutions.

There has been generally good agreement among the fairly recent determinations of the dissociation energy of the molecule. The spectroscopically determined values, $D_o = 5.8684$, $D_o = 6.1221$ eV[49] are in line with those earlier determined by Johns and Barrow,[50] $D_o = 5.895$, $D_o = 6.11$ eV as well as those reported by Berkowitz and co-workers for HF. These include: $D_o = 5.84$,[51] not significantly less than 5.86,[52] 5.845,[53] and 5.87 eV.[54]

Legon and Millen[55] have recently published a paper entitled "Gas-Phase Spectroscopy and the Properties of Hydrogen Bonded Dimers: HCN . . . HF as the Spectroscopic Prototype," in which they discuss work published on hydrogen-bonded "heterodimers" and "homodimers." The first are made of two molecules of different types, while the second is of two of the same. The $(HF)_2$ dimer is a homodimer. HF furnishes the proton for hydrogen bonding in the majority of the cases studied. There is a strong tendency for it to form hydrogen bonds, even to itself.

Dyke and co-workers[35] have prepared molecular beams containing $(HF)_2$,

(DF)$_2$, or HFDF (not DFHF) by effusion of the appropriate compound or mixture through a supersonic nozzle. They investigated these by means of radiofrequency and microwave spectroscopy. In order to interpret their data they used a rigid rotor model, assuming that the inner hydrogen was on the F—F axis and that the monomer HF units had the same internuclear distance as they had as the free monomer ($r_o = 0.917$ Å). The calculated F—F distance, 2.79 Å, was good to 0.05 Å. This is considerably longer than the X-ray diffraction value of 2.49 Å for the solid and the electron diffraction value of 2.55 Å for the gas phase hexamer. Their value agrees nicely with an ab initio value of 2.78 Å calculated for the HF dimer.[56] They noted the monotonic decrease in the bond length with degree of polymerization. They reported that the calculated FFH angle of 108° is sensitive to the motion of the inner and the outer hydrogen. They considered that reasonable agreement with the X-ray diffraction determination of 120° for the solid. It did not agree well with the ab initio value of 160°. The potential here was flat with respect to changes in the angle. The observed dipole moment, 1.6 D, is consistent with the angle. They also carried out some not completely conclusive mass spectroscopy and checked for polarity of species in their beam by electric deflection and concluded that any polymers greater than the dimer were not polar, hence cyclic. They liked the continuous polymerization model rather than the monomer-dimer-hexamer model. This is consistent with the calculations of Del Bene and Pople[57] who reported that for polymers larger than the dimer, cyclic structures should be more stable than the corresponding chains and that the high stability of cyclic structures suggests that a mixture of cyclic polymers in addition to monomer and dimer exists. Gaw, Yamaguchi, Vincent, and Schaefer,[58] working on four theoretical levels, reported respectable agreement with previously reported F—F distance in the dimer having calculated 2.73, 2.76, 2.80, and 2.72 Å and an FFH bond angle between 130° and 138°. They found good agreement with vibrational frequency shifts for the dimer, relative to the monomer, with the experimental determination of Pine and Lafferty.[59] Andrews and co-workers[60] have observed infrared spectra of dimers, trimers in solid argon matrices. Some observations concerning a tetramer are also reported.

The microwave spectrum, between 9 and 24 GHz, of the gas at 203 K and 0.01 torr was used for structural investigations.[61] Planar (HF)$_6$ chains were found to have an F—H distance of 0.9997 Å, an F · · · F distance of 2.449 Å and an F · · · F · · · F angle of 104.00°, while the planar (HF)$_7$ chain's corresponding parameters are 0.9640 Å, 2.5745 Å, and 103.73°. In the electron diffraction work discussed earlier,[30] corresponding values were 0.973 Å, 2.525 Å, and ca 104° for the cyclic hexamer at 254 K. The F . . . F distance was reported to be 2.533 Å at 295 K. The angle in the cyclic hexamer is between that expected for a boat and that for a chair conformation. Apparently these cyclic hexamer molecules randomly and rapidly move through these conformations.

5. DISCUSSION OF PROPERTIES

Huber and Herzberg[62] derived the equilibrium internuclear distances, r_o for HF, DF, and TF from previously published vibrational-rotational spectra and found the r_o to be 0.91681, 0.91694, and 0.9176 Å, respectively. De Lucia[63] and his co-workers found the distance in DF to be 0.916914 Å by means of microwave spectroscopy.

Gmelin[64] points out that spectroscopic values for the dissociation energy of the molecule, D(HF), and those derived from photoionization-mass spectrometry and by photoelectric spectroscopy are in excellent agreement. Various values and the methods used are referenced therein. The values for D_o generally accepted range from 5.84 to 5.87 eV. Huber and Herzberg[62] report a $D_o°$ of 5.869, 5.938, and 5.968 eV for HF, DF, and TF, respectively. A value of about 5.86 eV for HF seems likely to be close to correct.

The situation with respect to proton affinity is not quite as nice. Values from ab initio SCF-MO calculations differ widely. Experimental values also differ but not so widely, ranging from 4.09 to 5.30 eV. Probably uncertainty over the nature of the products produced upon dissociation is responsible for the experimental discrepancies. Again, Gmelin[65] should be consulted for further discussion and references.

The average polarizability values have been calculated from Perkins[66] molar refraction data, R_D, of liquid HF at 25°C and from Batsanov and Vesnin's[67] data at 10°C as 0.84 and 0.87 Å^3, respectively.

There has been considerable activity involved in permanent dipole moment measurements of the molecule. Radiofrequency spectroscopy gives a value of 1.826178 D for HF in its lowest vibrational, first rotational state and 1.826178 D for DF in its lowest vibrational second rotational state.[68,69,70] The value, 1.7965 D, has been derived[71] for the vibrationless molecule. IR dispersion measurements give a value of 1.84 D. Further discussion may be found in Gmelin.[72]

The melting point curve: p = 307 (T − 190.1), p in bar, T in K, was derived from observed melting points.[73] The cryoscopically determined triple point is 189.58 K.[74] The cryoscopic constant calculated from this and the heat of fusion is 1.52 K/mol/kg. This agrees quite well with the experimentally observed value of 1.55 K/mol/kg.[72]

Vanderzee and Rodenburg[75] have recommended that the boiling point be accepted as 292.69 K after considering fairly recently and earlier reported values. Sheft, Perkins, and Hyman[76] and Jarry and Davis[77] determined vapor pressure curves, hence boiling points, of 292.90 and 292.67 K. The critical temperature and critical densities of 461 K and 0.29 g/cm^3 are derived from the liquid-vapor coexistence curve. The critical pressure then is 66.2 kg/cm.[278] The molal boiling point elevation constant has been determined to be 1.90 K/mol/kg.[79,80]

Frank and Spalthoff[78] carried out calorimetric determinations of the heat of vaporization as did Hu[81] and co-workers earlier. Jarry and Davis[75] obtained values from the vapor pressure curve from 273 to 378 K while Sheft and co-workers[76] did this at the boiling point. Vanderzee and Rodenburg[75] showed that it is likely that the vapor density used by both is incorrect. Correcting the work of Jarry and Davis leads to the conclusion that the heat of vaporization is $7490 - 43.9$ ($T_K - 292.69$) J/mol (a mole is considered to be based on the monomer). This equation nicely fits the calorimetric value of Hu and co-workers and, when extrapolated, gives values in agreement with Franck and Spalthoff up to 390 K.

Jarry and Davis offered two equations describing the vapor pressure-temperature relationship. They seem not to have a definite preference for one over the other. There remains some uncertainty about values as the liquid nears the freezing point. Their equations are:

(1) $\log_{10}p(\text{mm}) = 8.38036 - 1952.55/(335.52 + T)$ ($T = °C$)

(2) $\log_{10}p(\text{mm}) = -1.91173 - 918.24/T + 3.21542 \log_{10}T$ ($T = K$)

Some pairs of values of T (in °C) and p (in mm) are: -50, 34.82; 0, 363.8; 25, 921.4; 100, 7891. Gillespie and Humphreys[74] reported an approximate value of 24 mm for the molar depression of vapor pressure. This is to be compared with the value calculated from Raoult's law, 7.13 mm. This suggests that at 273 K the solvent under measurement conditions has an association factor of 3.3. Since the presence of solute would effect the association, one ought not to be distressed that this differs from the 4.72 value from the Jarry and Davis[77] work. They also reported average association factors of 3.58 and 2.45 at 298 and 373 K, respectively.

Considerable attention has been given to measurements of temperature-pressure-vapor density relationships. Early on, they indicated considerable association, especially at lower temperatures and modest or higher pressures. These relationships have been among the tools used in trying to sort out the degree of oligomerization and distribution of oligomers. As early as 1889, Thorpe and Hambly[26] investigated the relationship. Since then a number of investigators[25,77,82,83,84,85] have worked on this problem. Four decades ago Hildebrand and co-workers[84] showed that the relationship, except at low pressures, could be nicely accounted for by the monomer–hexamer relationship which had been proposed by the group two decades earlier. Indeed, curves following $\log K = 8910/T - 43.65$ for $(HF)_6$ and $\log K = 8970/T - 43.65$ for $(DF)_6$ fit experimental points quite nicely except at higher temperatures where some deviation, probably arising from lower oligomers, occurred. They calculated heats of polymerization for HF as -171 J and -172 J for DF. Figure 9-1 plots association factors as a function of HF pressure at 299, 305, and 311 K. Curve 1 is from Reference 25, curves 2 from Reference 82, curves 3 from Reference 84, and curves 4 from Reference 77.

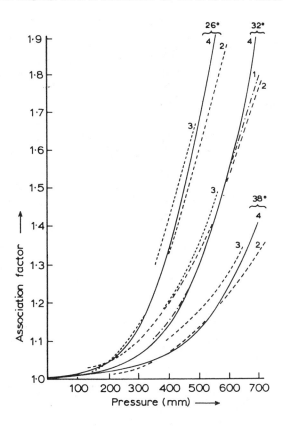

Figure 9-1.   Association factors of gaseous HF as a function of total HF pressure, at 26°, 32°, and 38°C.

Reprinted with permission from M. F. A. Dove and A. F. Clifford, "Inorganic Chemistry In Liquid Hydrogen Fluoride", Pergamon Press, 1971.

Yarbroff and co-workers[86] have critically reviewed the thermodynamic properties of HF from 233 to 513 K and 2 to 1000 psi. The data are nicely presented in chart form.

Electrical properties have had important influences on the use of the liquid as a solvent or reaction medium. To the casual observer, the specific conductance may have seemed to be time dependent. As methods of preparing, purifying, and containing the substance have improved, so has the reported specific conductivity dropped. The lowest value reported,[87] to date, seems to be 1×10^{-6} ohm^{-1}cm^{-1} at 298 K. The value may indeed turn out to be somewhat lower as even better methods for handling and purification are developed. There have been reports that on standing, lower unsaturated polymers of CF_2CFCl, the monomer from which it is made, may be leached from Kel-F, a frequently used material of construction, become protonated, and add to the conductivity. Traces of water (quite basic in this medium) markedly enhance conductivity. Hence, conductivity has been an important criterion for purity. Hyman and Katz[88] have noted that the Fredenhagen and Cadenbach's[89] acid, which was considered quite good, had a conductivity of 1.4×10^{-5} ohm^{-1} cm^{-1}, which corresponds to a water content of less than 0.0002%.

While some of the conductivity may arise from the presence of impurities, it is unlikely that the major part of it does. It must arise from autoionization leading to extensively solvated protons and fluoride ions. The limiting ionic mobilities for the solvated proton[90] and for the solvated fluoride ion[91] are 350 ohm^{-1} cm^2 and 273 ohm^{-1} cm^2, respectively. There is considerable reason to believe that as each ion exists as part of a hydrogen-bonded network, conductance involves a chain mechanism. If one applies the molar conductance, 623 ohm^{-1} cm^2, to the specific conductivity, 1×10^{-6} ohm^{-1} cm^{-1}, one arrives at an ion product of 2.6×10^{-12} mol^{2}l^{-2} or hydrogen ion and fluoride ion concentrations of 1.6×10^{-6} mol l^{-1}. To the extent that the conductance is caused by impurities, the calculated ion product and concentrations are too high. More recently, an autoprotolysis constant at 273 K of $10^{-13.7}$ has been reported.[92] Conductivities and ionic mobilities of a variety of species have been measured in this solvent. Dave and Clifford[93] nicely summarize the situation at the time of their writing. Conductivity has been a most useful tool when investigating behavior in the solvent.

The dielectric constant is another property that contributes to the usefulness of this remarkable solvent. Fredenhagen and Dahmlos,[39] as early as 1929, reported the dielectric constant at 273 K to be 80. This figure is frequently quoted when comparing HF with water. While this temperature is close to that where many measurements have been made, it should be pointed out that the dielectric constant is much higher as temperatures approach the freezing point, where some work has been done and the condition under which most aqueous work is carried out. More recently, Jones[94] reported that the dielectric constant follows $69,000/T - 167.5$. The dielectric constant reflects the general head to tail chain polymeric structure of the chains in the liquid.

Those who have had occasion to work with the liquid contained in a vessel with reasonably transparent walls are likely to have been impressed with its low viscosity. At 20°C it is about one tenth that of water. The viscosity reflects the two-dimensional nature of the polymer as contrasted with the three-dimensional nature of water. Jones[94] found:

$$\log_{10} \eta = 370/T - 1.95 \text{ (T in K, } \eta \text{ in } 10^{-2} \text{gcm}^{-1}\text{s}^{-1}$$
(centipoise))

Perkins[95] measured the refractive index at 298 K throughout the visible region and found that it follows: $1.15436 + 0.001025/\lambda^2$, where λ is the wavelength expressed in angstroms.

6. LASER ACTION

While the first chemical laser was based on HCl, it was not long before laser emission was reported from HF and DF.[96,97] One of the important reactions involved in a H_2, F_2 laser is: (generation of fluorine atoms) $F_2 + M \rightarrow 2F + M$. A large number of compounds have been examined as potential fluorine sources, since elemental fluorine, while a simple source, is not considered to be as easily handled as desired. The desirability of storable, relatively easily handled sources has furnished some of the impetus for synthesis and characterization of several new fluorine-containing compounds. Considerable work involving metathesis, in liquid HF, of NF_4^+ containing compounds, has been carried out.[98] Since the NF_4^+ species used in these methatical reactions is usually made by reacting NF_3, F_2, and a strong Lewis acid in HF under pressure, one might expect that compounds containing this anion may be decomposed to give fluorine.

$$NF_3 + 1/2\ F_2 + SbF_5\ (or\ SbF_3) \rightarrow NF_4^+SbF_6^-$$

Pumping reactions, which populate upper vibrational-rotational levels of HF, are needed: $F + H_2 \rightarrow HF + H$ (HF vibrational levels 3 or less); $H + F_2 \rightarrow HF + F$ (HF vibrational levels 9 or lower). Fluorine atoms are able to abstract hydrogen atoms from a variety of compounds, forming HF in higher vibrational-rotational states (the H—F bond is about 565 kJ/mol and stronger than other bonds involving hydrogen). Therefore a number of hydrogen-containing compounds have been investigated as potentially easily handled hydrogen sources. Vibrational-vibrational and vibrational-rotational and translational energy transfer processes are involved.

$$HF(v) + HF(v') \rightarrow HF(v - 1) + HF(v' + 1);$$

$$HF(v) + M \rightarrow HF(v - 1) + M$$

Stimulated emission then results in the output of the laser.

$$HF(v) \rightarrow HF(v - 1) + h\nu.$$

A variety of other reactions may be profitably involved. They include elimination reactions, photoelimination reactions, radical combination reactions, addition reactions, insertion reactions, and mercury-sensitized elimination reactions.

Pulsed and continuous wave HF and DF lasers have become interesting from both a practical and theoretical viewpoint. Readers who want to learn about these would do well to consult the many pages of Gmelin[99] devoted to the subject, as well as the many references therein and the considerable literature that has been written since this part of Gmelin was put together.

7. THE SOLVENT

Practically every review dealing with HF has had something to say about the HF solvent system. For many, this has been the emphasis.

Ionization has already been discussed as properties were under consideration. It is convenient to write the self-ionization equation as if protons and fluoride ions are produced, although we know that the ions are extensively solvated. Some have chosen to treat the solvated fluoride as FHF^- to emphasize the strong hydrogen-bonding tendency and since the existence of the ion, as well as the $H_2F_3^-$, $H_3F_4^-$ and the $H_4F_5^-$ ions in the solid state have been well established.[100] In general, these ions are seen as arising through hydrogen bonding of HF to F^- where the maximum coordination number of F^- is four. The stability of these species drops off in the order $H_4F_5^-$, $H_3F_4^-$, $H_2F_3^-$, HF_2^-. When the cation is H_3O^+ or NH_4^+, $H_4F_5^-$ and $H_2F_3^-$ species are not seen. The absence of the former can be explained by the taking on of the coordinating position by a hydrogen bond from the cation, while the absence of the latter is not so glibly explained. There have been several calculations of the enthalpy of the gas phase reaction $HF_2^- \rightarrow HF + F^-$. Probably the best value is that calculated by a Born-Haber treatment by Jenkins and co-workers.[101] They reported a value of 176 kJ/mol. O'Donnell[102] has been able to account for the change in conductivity and free F^- concentration with change in water (or HF) concentration using a model based on equilibrium constants for these species. However, it may well be that much of the original chain structure remains in the nearly anhydrous solvent initially.

According to the waterlike analogy, fluorides which ionize to give fluoride ion should be basic. Those that accept a fluoride are acidic. Indeed, the alkali and alkaline earth metal fluorides are bases in this solution. Water is also a base, since it is protonated by this strongly protonating very strong acid, leaving behind a solvated fluoride. Kongpricha and Clifford[103] report a minimum ionization constant for water $(F^-)(H_3O^+)/(H_2O)$, of 0.55 at 273 K. Many other compounds, which we ordinarily think of as acidic, are basic since they become protonated.

$$H_2O \rightleftarrows H^+ + OH^-$$

$$HF \rightleftarrows H^+ + F^-$$

There are few protonic acids in HF solution. However, several of the common Lewis acids remain Lewis acids in HF since they take up a fluoride and liberate a proton. Antimony pentafluoride is among the strongest.

$$HF + SbF_5 \rightarrow H^+ + SbF_6^-$$

The order, in decreasing acidity, of commonly discussed acids is: SbF_5, AsF_5, TaF_5, BF_3, NbF_5, PF_5. While the order of basicities has been less discussed, there is evidence[104,105] that, generally, members of Group I are more basic than are those of Group II and that basicity increases going down the group. Those of us who like the simplicity of the waterlike analogy find that rather pretty.

O'Donnell[102] states that all Hammett function, H_o, determinations for the solvent prior to 1976 are not sufficiently negative since Kel-F or similar polymers leach out a basic impurity. This kind of polymer has been commonly used in constructing apparatus to contain HF. Liang,[106] working with Gillespie, circumvented this problem and found an H_o value of -15.05. A comparison with that of H_2SO_4 (-11.93),[107] HSO_3F (-15.07),[108] and CF_3SO_3F (-14.00)[109] shows that it is indeed a quite acidic substance!

The values compared here were obtained using comparable approaches. Antimony pentafluoride-HF solutions are strongly acidic. A 0.6 mole % SbF_5 solution has an H_o of -21.13[106] and there are reasons to believe that more concentrated solutions are even more acidic.[102] O'Donnell points out that there is a wide range of acidities (or basicities) in which to work in the solvent and that the acidity or basicity should be controlled. Not only does a solution become less acidic (or more basic) by the adventitious occurrence of water by reactions with oxide surfaces or other means, but it also may when elemental F_2 is used as an oxidizing agent. As it becomes reduced, it becomes the basic F^-.

O'Donnell and his co-workers[110] have looked at the spectra of a variety of species in HF. Since the fluorides of many of the cations involved are essentially insoluble, they found it necessary and possible to put them into solution upon the addition of strong Lewis acids. They were also able to prepare solutions of some metals in low oxidation state by reacting the metals with HF solutions of these acids. O'Donnell[102] has argued that the solvent is an especially good one in which to work or bring about unusually high or low oxidation states since the range for oxidation or reduction of the solvent is 4.5 V. This is to be compared with water with a range of 2 V. Overvoltages can be significant. Practically, one does not evolve fluorine until about 7 V is reacted.

There have been many qualitative and quantitative determinations of solubilities in HF. Dove and Clifford[11] have systematically presented those available to them in their review and Gmelin[15] is essentially up to date. Jache and Cady[104] published an extensive list of solubilities of fluorides at three different temperatures, pointed out trends as a function of position in the periodic table and compared these solubilities with those of the corresponding hydroxides or oxides in water. Clifford and his co-workers, as they worked with apparently higher purity HF, redetermined some of the solubilities. Differences, where they exist, were ascribed to traces of water. Discussion of these can be found in the Dove and Clifford[11] review. Sheft, Hyman, and Katz[111] determined solubilities of metal fluorides in BrF_3 and compared these with the corresponding solubilities in HF. Seeley and Jache[112] determined the solubilities at 298.15 K of Group I and Group II fluorosulfates in HSO_3F and compared them with the earlier determined solubilities of corresponding fluorides in HF.

Table 9-2 abstracts some of the data from this study. The solubilities of the Group I and Group II fluorides in HF, at temperatures close to 285 K, are given, as are the solubilities of the corresponding fluorosulfates at 298.15 K. The solubilities are expressed in g mol/100 g of solvent. For HF and for HSO_3F, 100 g is a rather convenient unit since 100 g is very nearly 5 mol (4.998 mol) of HF and essentially 1 mol (0.9993 mol) of HSO_3F. While 100 g of water is not an integral number of moles the number is close (5.556) to that of HF, hence convenient to use when establishing ratios of solubilities. In the table, Ratio I is the ratio of the solubility of the fluoride in HF at the temperatures indicated to that of the corresponding fluorosulfate in fluorosulfuric acid at 298.15 K. Similarly, Ratio II is the solubility of the fluoride in HF divided by that of the corresponding hydroxide in water at 298 K (in most cases).

Since one would expect solvation energies to follow the same trend as hydration energies, one would also expect, considering only solvation energies, that within a group, solubilities would decrease with increasing atomic number. For Group I and II fluorides, lattice energies alone would argue an opposite trend. The nature of the solid phase in equilibrium with the solution will also have a significant influence. The increase in solubility for Group I

TABLE 9-2. Solubilities of Compounds in Their Parent Acids

Metal	Fluoride* mol/100g	Fluorosulfate** mol/100g	Ratio I ***	Ratio II ****
Li	0.397	0.319	1.24	0.74
Na[a]	0.716	0.658	1.09	0.25
K[b]	0.628	0.461	1.36	0.30
Rb[c]	0.667	0.486	1.37	0.5
Cs[d]	1.31	0.992	1.32	0.64
Mg	4.8×10^{-3}	5.4×10^{-4}	8.9	8.7
Ca	10.5×10^{-3}	6.86×10^{-2}	0.153	6.8
Sr	118×10^{-3}	5.10×10^{-2}	2.31	16.5
Ba	31.9×10^{-3}	1.39×10^{-2}	2.29	1.15

* T = 12.2°C except where noted
** T = 25°C
*** $C_{fluoride\ in\ HF}/C_{fluorosulfate\ in\ fluorosulfuric\ acid}$
**** $C_{fluoride\ in\ HF}/C_{hydroxide\ in\ water}$:
[a] T = 11°C.
[b] T = 8°C Ref. 113.
[c] T = 20°C Ref. 114.
[d] T = 10°C Ref. 115.

and Group II fluorides in HF and for Group I fluorosulfates in fluorosulfuric acid seems to argue that lattice energy effects nearly predominate here. There may be some crossover of predominance of effects at about sodium for the fluorosulfate and strontium for the fluoride. The solubilities of the Group II fluorosulfates decrease as atomic number increase while with the fluorides, except strontium, the opposite trend occurs. The effect for these fluorosulfates is consistent with the expected trend in solvation energy and may also reflect a tendency for more favorable lattice energies with bigger cation than with very small ones (anion–anion repulsion). It is not surprising that the magnesium salts do not fit into the scheme nicely, since the high charge/size ratio often makes the behavior of magnesium compounds like those of aluminum. While there is nothing compellingly obvious to tell us that the ratio of solubilities of salts in their parent acids should be constant, it is gratifying, perhaps even surprising, to note the small spread in Ratio I (fluoride/fluorosulfate comparison) for Group I compounds. While the ratios have a greater spread as we deal with Group II solubilities, they do not widely deviate within the group. Since the trend for the fluorides is essentially opposite to that of the fluorosulfate for this group it is surprising that the spread is as little as it is. It should be noted that Ratio I for calcium would be somewhat higher, diminishing the spread somewhat were Kongpricha and Clifford's more recent value[103] for the solubility of CaF_2 used. It is probably a better value but was not used here in order to keep as many values as possible from one source.

Jache and Cady[104] also reported on the solubilities of a variety of other fluorides in HF and compared their solubilities with those of the corresponding hydroxides or oxides in water. The researchers found that, in general, there is a parallel. As long as one stays in the part of the periodic table where fluorides behave as bases, solubilities tend to drop off as one goes from left to right. Solubilities are greater in lower than in higher oxidation states, sometimes dramatically. This is consistent with trends in basicity or size/charge ratio of cation. On the other hand, in the region where fluorides are acids in the base, HF, the opposite occurs. For example, SbF_5 is infinitely soluble while SbF_3 is soluble only to the extent of 0.536 g/100 g of HF at 11.9°C. The pentafluoride is exceedingly acidic ($SbF_5 + F^- \rightleftarrows SbF_6^-$) while the trifluoride is not. There have been many more reports on solubilities in HF. They are well covered to preparation date in Gmelin[15] and in the Dove and Clifford review.[11]

Cady and Hildebrand[116] and later, Pawlenko,[117] studied the solid-liquid phase diagram of the H_2O—HF system. That there is strong interaction is attested by the finding of the following solid phases: H_2O, $H_2O \cdot HF$, $H_2O \cdot 2HF$, $H_2O \cdot 4HF$, and HF. Since Ferriso and Hornig[118] have found the infrared spectrum of the 1:1 phase at 78 K consistent with $H_3O^+ F^-$, it is suggested that the phases containing water be considered to be $H_3O^+ H_nF_{n+1}^-$. Fluoride is solvated by HF while the proton is solvated by one water.

The boiling points and vapor compositions for the H_2O—HF system have been determined. Vieweg[119] critically reviewed the thermodynamic properties of the system and reviewed and integrated the available data on vapor pressures over mixtures.[120] Vdovenko and co-workers[121,122] contributed data on compositions over aqueous hydrofluoric acid. The system is a fairly simple one with an azeotrope, composition by weight 37.37% HF, boiling at 384.55 K at atmospheric pressure. As the pressure is lowered, the azeotrope becomes richer in HF and the boiling point drops. For example, at 5 mm pressure the composition is 39.75% HF and the boiling point 283.2 K.

A sizable number of binary systems involving fluorides and HF have been investigated. Similarly, there has been considerable activity with ternary systems involving a fluoride, hydrogen fluoride, and water. Many of these involve Group I or Group II fluorides. That a strong tendency toward hydrogen bonding and persistence of chain fragments involving the fluoride and HF exists is evidenced by the large number of HFates reported. Boinon and co-workers[123] have studied liquid-solid equilibria in the LiF—HF and the NaF—HF systems up to 673 K by means of thermal analysis. They observed LiF · HF and NaF · HF, NaF · 2HF, NaF · 3HF, and NaF · 4HF. There have been earlier studies involving these systems which are referenced in the Boinon study. Tananaev[124] and Morrison and Jache[125] and Vouillon[126] have studied the ternary system NaF—HF—H_2O. Morrison and Jache found the one, two, three, and four HFates at 273 and 258 K as did Vouillon at 263 K. Tananaev found only the mono HFate at 313 K. It is interesting to note that at the lowest of these temperatures NaF · HF is formed at quite low HF concentrations. Indeed, it was found at the lowest concentration studied, 0.365% HF in the liquid phase. Sodium fluoride pellets are frequently used to remove HF from gas streams and, occasionally, from aqueous or other solutions.

The crystal structure of NaF · HF has been investigated by X-ray and neutron diffraction studies.[127] The HF_2^- ion is linear and has a F(H)F distance of 2.265 Å. That[128] for the lithium analog has been reported as 2.27 Å. The thermodynamic properties of $NaHF_2$ have been discussed by Higgins and Westrum.[129] They report the molal heat of solution in HF of $NaHF_2$ to be −20.3 kJ and that of NaH_2F_3 as −9.00 kJ.

Cady's[130] 1934 investigation of the phase diagram of the KF—HF system from 0.46–1.00 mole fraction HF turned out to be an especially important one since it defined the composition and temperature ranges over which chemists have found it feasible to produce fluorine electrochemically from KF—HF melts. Westrum and Pitzer[131] have also dealt with the system. The HFates found were: KF · HF, KF · 2HF, KF · 3HF, and KF · 4HF. There is a higher and a lower temperature form of the mono HFate. Tananaev[132] found KF · HF, KF · 2HF, KF · 2.5HF, KF · 3HF, KF · 4HF, and KF · $2H_2O$ when he studied the KF—HF—H_2O system at 273, 293, and 313 K. All of these solvates were formed at each of the temperatures. Westrum and his co-workers[129,133] have discussed the thermodynamic properties of the KHF_2—KF—HF system. The standard heats of formation are listed in the

Dove and Clifford review.[134] The molal heat of solution of KF · HF in HF is −39.9 kJ.[128] The lower temperature form of the mono HFate is tetragonal[135] while the higher temperature form is cubic.[136] Much attention[137] has been given to the HF_2^- ion. The general conclusion is that it is symmetrical and that the F(H)F distance is 2.27^7 Å. Single crystal X-ray diffraction work[138] on the di HFate suggests that the $H_2F_3^-$ ion is bent. There are two nonequivalent ions in the unit cell with F—F—F angles of 130° and 139°. The average F(H)F distance is 2.33 Å. An NMR study[139] of the solid gives results consistent with the above. The authors of the NMR work point out that it is not consistent with the notion of a symmetrical hydrogen bond.

Euler and Westrum[140] have studied the NH_4F—HF system between 40 and 100 mole % HF and confirmed Ruff and Staub's[141] earlier report of the existence of NH_4F · HF, NH_4F · 3HF, and NH_4F · 5HF. There are some discrepancies between the two sets of data. The newer study seems to be the better one. Buettner and Jache[142] investigated the ternary system NH_4F—HF—H_2O at 273 and 253 K. The only solvates which they observed were the mono and the tri HFates. These were observed at both temperatures. The molar heat of solution in HF[127] is, for NH_4HF_2, −24.1 kJ and −13.97 kJ for $NH_4H_3F_4$. The most recent determination[143] of the structure (orthorhombic) of NH_4HF_2 shows nonequivalent HF_2^- ions with F(H)F distances of 2.26^9 and 2.27^5 Å. These are shorter than have been reported earlier. The N(H)F distances, 2.79^7 and 2.82^2 Å, are longer than the 2.63 Å[144] corresponding distance in NH_4F. Presumably this reflects a weaker hydrogen bond between the nitrogen and the F of the bifluoride.

There have been several investigations[145] involving the binary RbF–HF system, the most recent and complete of which is that of Boinon and coworkers. The following HFates were observed: RbF · HF, RbF · 3/2HF, RbF · 2HF, RbF · 3HF, RbF · 4HF, and RbF · 7HF. In addition, the authors report an unstable 2RbF · 5HF. Opalovskii and Fedotova[146] have studied the system RbF—HF—H_2O at 273 K while Boinon[147] investigated it at a ten degree higher temperature. At the higher temperature Boinon found RbF · HF, RbF · 2HF, RbF · 3HF, and RbF · 4HF. The hydrate, 2RbF · 3H_2O, was found as well. Opalovskii claimed RbF · HF, RbF · 2HF, RbF · 3HF, and 2RbF · 7HF and a monohydrate. It has been reported, but not confirmed elsewhere, that the solid phase in equilibrium with HF at 293 K is $Rb_2H_9F_{11}$.[148] The mono HFate is dimorphic as is that of potassium and of cesium. The lower temperature form is tetragonal while the higher is cubic.[136]

The phase diagram CsF–HF was first investigated by Mathers and Stroup[149] and then by Winsor and Cady.[115] Winsor and Cady found the following solvates: CsF · HF, CsF · 2HF, CsF · 3HF, and CsF · 6HF. The mono HFate again seems to exist in two forms. The solid in equilibrium with the solution at 293 K is the tri HFate. Opalovskii and Fedetova[146] have investigated the CsF—HF—H_2O system at 273 K. They reported that they found CF · HF, CsF · 2HF, and CsF · 3HF.

Since the chemistry of Tl^+ is often compared to that of Ag^+, it might be informative to look briefly at what is known about behavior of the fluorides in the solvent. Boinon[150] has investigated the binary TlF—HF system, as well as the ternary TlF—HF—H_2O system at 253.7 and 244.6 K. Earlier, the ternary system at 263 K was looked at in the same laboratory.[149] In the binary system the following solvates were observed: TlF · HF, 2TlF · 3HF, TlF · 2HF, TlF · 3HF, TlF · 5HF, 2TlF · 13HF, and TlF · 7HF. In addition to ice and TlF, Boinon found the solid phases TlF · HF, 2TlF · 3HF, TlF · 2HF, and TlF · 3HF in the ternary system at both temperatures at which he worked. Four decades earlier Hassel and Kringstad[151] reported that they had identified the compound 2TlF · 4HF · H_2O. Thomas and Jache[152] and Buettner and Jache[153] investigated the ternary system AgF—HF—H_2O at 273 K and at 258 K. Considerably earlier, Guntz and Guntz[154] investigated parts of the system, using only aqueous hydrofluoric acid. In addition to unsolvated AgF, at 273 K, the solid phases are: AgF · $2H_2O$, AgF · $4H_2O$, 3AgF · 2HF, AgF · 2HF, AgF · 3HF, AgF · 5HF, and the mixed solvate, 6AgF · 7HF · $2H_2O$, while at fifteen degrees lower, the solvates are AgF · $2H_2O$, AgF · $4H_2O$, AgF · HF, AgF · 3HF, and AgF · 5HF. A variety of studies which give some insight into the basicity of AgF and of TlF have been carried out.[155,156,157] While at lower concentrations of AgF it appears that it is a strong electrolyte, there is evidence that it is weaker at higher concentrations, although probably stronger than is water. Thallium (I) fluoride is also not always completely ionized in HF. At 273 K the ionization constant for AgF, activity of Ag^+ × activity of F^-/activity of AgF, is 0.087 while that of TlF is 1.04.[157]

A number of other systems have been investigated. Of particular interest is the NiF_2—HF—H_2O system which was completed by Clifford and Tulumello.[158] The region of lower concentrations of HF had been studied by Kurtenacker, Finger, and Hey[159] three decades earlier. They found that at 273 K when the water concentration is over about 0.2 M, the nickel-containing species in solution is predominately $Ni(H_2O)_6^{2+}$. The solid phase never is more hydrated than NiF_2 · $4H_2O$ at this temperature. Clifford and his coworkers showed that at water concentrations between 0.2 and 0.3 M the equilibrium constant for the dissociation of the hexaaquo ion to Ni^{2+} and water is $4.6 × 10^{-10}$. Several transition metal fluorides show an increase in solubility with increase of concentration of HF in aqueous solutions rather than the decrease experienced upon addition of KF. This argues for the formation of aquo complexes. Clifford and Sargent[160] demonstrated the existence of Ni^{2+}—CH_3CN complexes in the NiF_2—HF—CH_3CN system.

Gmelin[16] updates the Dove and Clifford discussion[161] of complex ions which have been shown to exist in HF solution. Considerable work has been done with ligands as arenes, cyanides, etc. The chemistry of mercury and silver cyanides, as well as hexacyano iron complexes and HCN, has received considerable attention.[162] Dove and Clifford offer a useful table of complexing agents and metal ions and indicate which metal ions react with each ligand.

8. SOME INTERESTING SOLUTION CHEMISTRY

There is considerable interesting solution chemistry described in the literature.

A. Metathesis Reactions

While NF_3 had been known for some time, there existed a general understanding that NF_5 could not be made. Indeed many of us used its nonexistence and the existence of the pentafluorides of the other pnicogens to illustrate a point while discussing orbitals and trends in the periodic table. Speculation regarding the possibility of the existence of NF_4^+ containing salts and whether $NF_4^+F^-$, which would get around the lack-of-d orbital problem, could be considered NF_5 was common until 1966. Price and colleagues,[163] on the basis of electron affinities and size, concluded that NF_4^+ could not form crystalline salts. Wilson[164] estimated its heat of formation and opined that the BF_4^- salt might exist at low temperatures. Encouraged by this, Tolberg and colleagues[165] prepared the SbF_6^- salt by reacting NF_3, F_2, and SbF_5 in HF at moderately high temperatures and pressures. Christe and co-workers[166] confirmed the existence of the NF_4^+ ion by publishing an account of their preparation, via glow discharge, of the AsF_6^- salt. The initial accounts of the preparation of this remarkable-at-the-time species appeared in the same issue of the journal. Since then, Christe and co-workers have prepared a number of NF_4^+ containing salts from HF solution in an esthetically pleasing series of metathetical reactions. In general, they have used $NF_4^+SbF_6^-$ in HF solution and have precipitated the desired salt by adding the appropriate reagent. When they have chosen to eliminate the SbF_6^- ion from the reaction zone they have been able to precipitate it as the Group I metal salt and filter it off, leaving effectively, $NF_4^+F^-$ in solution and available for further reaction. NF_4CrF_6 has been quite recently made this way.[98] References to other preparations may be found in Christe paper and in those references found in it.

Similarly Christe and Wilson[167] prepared $ClF_6^+BF_4^-$ from $ClF_6^+SbF_6^-$ and $CsBF_4$ in HF solution.

Malm and Carnall[168] showed that they could precipitate NpF_5 from HF solutions of metal NpF_6^- salts by displacement with the acid, BF_3.

$$M^+ + NpF_6^- + BF_3 \rightarrow M^+ + BF_4^- + NpF_5$$

Alternately, this might be looked upon as acid-base chemistry.

B. Disproportionation Reactions

Jache and his co-workers have shown interest in reactions of some elements in HF solutions. There was a particular interest in seeing if disproportionation would occur. Kongpricha and Jache[105] investigated the reaction of phosphorus in HF and in HF solutions made basic with Group I or Group II fluorides. Using the waterlike solvent approach, they anticipated that if phosphorus behaved in HF in a manner analogous to that in water, it might behave as a metal:

$$P_4 + 16H_2O(g) \rightarrow 10H_2 + P(OH)_3O \ (H_3PO_4) \ (\text{above } 553 \ K)$$

$$P_4 + 12HF \rightarrow 4PF_3 + 6H_2 \ \text{or}$$

$$P_4 + 20HF \rightarrow 4PF_5 + 10 \ H_2$$

or it might disproportionate:

$$P_4 + 3OH^- + 3H_2O \rightarrow PH_3 + 3H_2PO_2^- \ (\text{boiling})$$

$$P_4 + 6HF \rightarrow 2PF_3 + 2PH_3 \ \text{or}$$

$$2P_4 + 15HF \rightarrow 3PF_5 + 5PH_3$$

Since the heat of formation of PH_3 is slightly positive, one might guess that formation of hydrogen is favored. Since a PF bond in PF_3 is a bit stronger than one in PF_5, the favoring of the former might be anticipated. Of course, since all of the considered reactions are thermodynamically favored, the outcome could be kinetically controlled. It turns out that in the absence of added (the reaction vessel walls were of Monel) metal fluoride, the products were hydrogen and phosphorus trifluoride. Small amounts of added meal fluoride increased the percent conversion. Stoichiometric amounts increased the amount of the trifluoride found in shorter runs but might actually decrease the amount in longer ones since the metal fluoride became converted to hexafluorophosphate. Apparently disproportionation of the trifluoride to the pentafluoride and elemental phosphorus occurred in significant amounts when the pentafluoride was taken up by the metal fluoride to form metal hexafluorophosphate.

$$P_4 + 12HF \rightarrow 6H_2 + 4PF_3$$

$$10PF_3 + 6MF \rightarrow P_4 + 6MPF_6$$

$$10PF_3 + 3MF_2 \rightarrow P_4 + 3M(PF_6)_2$$

where M = Li, Na, K or M = Ca, Ba. The experimental work was carried out at temperatures ranging from 473 to 493 K. Note that while a critical temperature of 461 K seems to be the one favored today, Bond and Williams[169] reported a value of 503 K.

The conversion to hexafluorophosphate salts increases as the size of the cations increase from Li^+ to Na^+ to K^+ in Group I and from Ca^{++} to Ba^{++} in Group II. (Other fluorides in these groups were not tried.) The overall conversion of phosphorus to fluorine-containing compounds can also be correlated with the size/charge ratio of the cations. When those of low ratio are involved more PF_3, presumably, is formed, but it is likely to be converted to PF_6^- salts if enough time is involved. The correlation of yield of the PF_6^- salts with position in the periodic table is probably due to lattice energy. Metal fluorides of cations of high charge/size density have high lattice energy and will likely have a greater tendency to polarize hexafluorophosphates making them relatively less stable. The ionic radii[170] of F^- and of PF_6^- are 1.36 and 2.69 Å, respectively. Indeed, yields of the hexafluorophosphates correlated reasonably well with other size/charge related properties as solubility or basicity in HF. Table 9-3 illustrates the correlations.

This work was followed by a look at the behavior of halogens in HF media. That none of the other halogens are good enough oxidizing agents to liberate fluorine from HF, as chlorine is able to slowly liberate oxygen from water, seemed fairly obvious. However, at the time it was not so obvious that disproportionation could or could not occur. In water the following may occur and further disproportionation is encouraged as the system is made more basic:

$$H_2O + X_2 \rightarrow HOX + HX$$

The parallel reaction in HF is:

$$HF + X_2 \rightarrow FX(XF) + HX$$

TABLE 9-3.   Correlation of the Conversion of PF_6 Salts With Lattice Energy and Solubility

Cations	Ionic radii (Å)(a)	Lattice energy of fluoride salts kcal mole^{-1} (b)	Solubility of fluoride salts in HF, in g/100 g HF	(c)	% Conversion to PF$_6$ salts
Li^+	0.60	240.1	10.3	(12.2°)	19.2
Na^+	0.95	215.0	30.1	(11.0°)	37.4
K^+	1.33	190.4	36.6	(8.0°)	90.0
Ca^{++}	0.99	617.2	0.817	(12.2°)	2.0
Ba^{++}	1.35	547.1	5.60	(12.2°)	32.2

[a] L. Pauling, *The Nature of the Chemical Bond*, Cornell University Press, 3rd Ed., Ithaca, New York, 1960, p. 514.

[b] J. Sherman, *Chem. Rev., 11* (1932) 93.

[c] A. W. Jache and G. H. Cady, *J. Phys. Chem., 56* (1952) 1106.

From *J. Fluorine Chemistry*, 1971, *1*, 79 with permission Elsevier Sequoia S.A.

Further disproportionation is conceivable.

When X is chlorine, the heat of reaction is about +28 kJ/mol. Hence, it is not surprising to observe that the reverse reaction goes nicely even in a solution made basic with alkali metal fluoride. The addition of the Lewis base, AgF, which ties up the reduction product, Cl^-, as AgCl and furnishes F^- causes disproportionation, yielding ClF.[171,172,173] Le Chatelier's principle applies as the lattice energy of AgCl contributes significantly to the overall thermodynamic driving force. Were not AgCl one of the exceptions to the generalization that metal chlorides react with HF to yield metal fluoride and HCl, one could regenerate the AgF and recycle it leading to the overall reaction:

$$HF + Cl_2 \rightarrow ClF + HCl$$

Of course, had the silver chloride reacted with HF it would most likely not have formed in the first place. Incidentally, Cl_2 will react with AgF and some other silver salts (which probably decompose to fluoride) even in the absence of HF at elevated temperatures.

Were it possible to separate the reaction of a metal fluoride and chlorine to form ClF and MCl from one in which the metal fluoride is regenerated by reaction with HF, we would have an overall scheme to form ClF from Cl_2 and HF.

$$Cl_2 + MF \rightarrow ClF + MCl$$

$$HF + MCl \rightarrow MF + HCl$$

$$HF + Cl_2 \rightarrow ClF + HCl \text{ overall}$$

A quick look at heats and free energies of formation will quickly lead to the conclusion that the reaction of chlorine with a solid Group I or Group II fluoride will not go spontaneously. The formation of the metal fluorides from the reaction of the chlorides with HF is well known. The difference between the free energy of formation of the fluoride and the corresponding chloride is greater than the free energy of formation of ClF. However, this difference becomes smaller as one goes down the periodic table in either of the two groups. Subtracting the lattice energies from the free energy results in the residual energies of the fluorides and the corresponding chlorides becoming nearly the same. In each case the residual energy difference is less than the free energy of formation of ClF. This is illustrated in Table 9–4. Essentially, this says that were both the fluoride and corresponding chloride of an alkaline metal liquids in the standard state, the reaction would be mildly thermodynamically favored. Chlorine passed through a molten eutectic mixture of LiF, NaF, and KF at temperatures far from the standard state did give essentially quantitative conversion to metal chloride. Recoveries of ClF were significant but lowered by corrosion of the apparatus.[171,173]

TABLE 9-4.   Comparison of 25°C Heat of Formation, Free Energy of Formation, and Lattice Energy Differences for Fluorides and Corresponding Chlorides

M	$\Delta H_f(MF) - \Delta H_f(MCl)$	$\Delta G_f(MF) - \Delta G_f(MCl)$	$U(MF) - U(MCl)$
Li	-208.7 kJ/mol	-204.63 kJ/mol	196 kJ/mol
Na	-164.3	-161.0	145
K	-131.9	-130.2	109
Cs	-111.8	-111.0	100

They also found that when Cl_2 was added to an HF solution containing SbF_5 at dry ice temperature a precipitate formed. At about 253 K the precipitate changed color, melted, and ClF began to evolve from the reaction vessel. At about 293 K, HCl was evolved while at some intermediate temperatures Cl_2 was observed. Presumably this arose from the reaction of ClF with HCl. The pertinent reactions were:

$$Cl_2 + 2HF + 3SbF_5 \rightarrow Cl_2F^+SbF_6^- + 2HSbF_5Cl$$

$$Cl_2F^+SbF_6^- \rightarrow 2ClF + SbF_5$$

$$HSbF_5Cl \rightarrow HCl + SbF_5$$

Again Le Chatelier's principle was at work through the energetics of tying up both products which could then be separately decomposed, yielding ClF and HCl and freeing the SbF_5 for future use.[172,173]

Thermodynamic considerations suggest that disproportionation of Br_2 and I_2 to halogen fluoride and halide in HF solutions of AgF would go as well as or better than does Cl_2. Since BrF is unstable with respect to Br_2 and BrF_3 and IF with respect to IF_3 and I_2 while IF_3 ultimately goes to IF_5 and iodide, it was not a great surprise to find that the products recovered were BrF_3 and AgBr and IF_5 and AgI, respectively.[174]

Since ClF is known to be in equilibrium with Cl_2 and ClF_3, ClF_3 with ClF and ClF_5, and ClF_5 with ClF_3 and F_2, speculation over the possibility of producing fluorine from HF via a series of disproportionations and Le Chatelier's principle is intriguing.

Speculation over the potential of HF—Cl_2 mixtures to accomplish the same or similar ends as did ClF followed the determination[175] that ClF would react with some chalcogens tungsten and molybdenum and compounds thereof to form useful fluorides and chlorine.[176] The rationale was that if one could use Le Chatelier's principle to make ClF from Cl_2 and HF there must be a small amount of ClF in equilibrium with the Cl_2, HF, and Cl^-. If one added a substance with which the ClF could react in situ the ClF would be consumed as it was formed as the system worked toward establishing equilibrium conditions. Could one accomplish the same overall chemistry with mixtures of HF, Cl_2, and reactive substrate as would occur by preparing

ClF, then reacting it with the same substrate? The heats of reaction and free energies of reaction of some elements with ClF and with Cl_2 and HF at 298 K are given in Table 9-5. The Cl_2—HF—element reaction to form hexafluorides is, save for SeF_6, favorable. The experimental work which was guided by this kind of reasoning was carried out far from standard state conditions and where solvent effects may be significant.

Tungsten hexafluoride, which is readily formed from ClF, is found in good yield at 473 K. While MoF_6 is also readily produced when ClF is used, MoF_5, a mixture of mixed molybdenum chloride fluorides and a trace of spectroscopically detected MoF_6 were found with the Cl_2—HF. Indeed, the temperature had to be raised to 573 K, the temperature at which Cl_2 is known[177] to react with Mo, before significant reaction occurred. Since $MoCl_5$, but not $MoCl_6$, is known, one is tempted to conclude that, at least in this case, the original rationale does not operate, but rather, chlorination, followed by halogen exchange is involved. It is interesting to note that McCaulay and associates[178] did not produce MF_5 when they reacted the corresponding chloride with HF while Ruff and Eisner claimed[179] to produce MoF_6 from the same reaction.

At 473 K metallic uranium is quantitatively converted to the tetrafluoride while at temperatures between 673–873 K, where some hydrogen resulting from attack on the vessel was also present, some U_2F_9 was identified via X-ray powder patterns and a spectroscopic trace of UF_6 appeared. While formation of UF_6 appears to be thermodynamically favorable, formation of the lower fluorides is more so. The tetrafluoride will react with ClF to form UF_6.[175] The metal will also react with neat HF at 523 K to form UF_4.[180]

While sulfur reacts with ClF to give SF_4 at room temperature and SF_6 at higher temperatures,[175] neither was isolated from HF—Cl_2—S mixtures. A trace of SF_6 was observed spectroscopically at high temperatures. The products of every run included SOF_2, a product of reaction of SF_4 with water or

TABLE 9-5.   Calculated Standard Heats and Free Energies of Reaction

| Product | M + n ClF → MF$_n$ + n/2Cl$_2$ | | M + n/2 Cl$_2$ + nHF → MF$_n$ + nHCl | |
	ΔH (kJ)	ΔG (kJ)	ΔH (kJ)	ΔG (kJ)
SF_6	−870	−757	−163	−50
SeF_4	−460	−393	+38	+20
		minimum value		minimum value
WF_6	−1427	−1314	−720	−607
MoF_6	−1287	−1163	−515	−456
UF_6	−1824	−1683	−1125	−979
SeF_6	−690	−590	+17	+30
UF_4	−1630	−1530	−1130	−1060
U_2F_4	−3400	−3180	−2320	−2120
UF_5	−1780	−1630	−1140	−1040
MoF_5	−1100	—	−506	—

oxygen-containing substances, in spite of efforts to minimize the presence of water or oxides. Neat ClF reacts[181] with Se to give SeF_4[175] and with $SeCl_4$ at room temperature and SCl_4 at dry ice temperature to yield the tetrafluorides and Cl_2. Small amounts of the mixed $SeClF_5$ were also formed. Mixtures of Se, Cl_2, and HF gave only predominately $SeCl_4$ along with some $SeCl_2$. The latter might have arisen by decomposition of the higher chlorides while separations were being made. Halogen exchange occurs between SeF_4 and HCl but not the other way.

C. Amphoteric Behavior

When Clifford and co-workers[182] reported that the solubility of AlF_3 and of CrF_3 was markedly enhanced by the presence of the acid, BF_3, or the basic alkali metal fluorides, they offered the first report of amphoteric behavior in HF. These were particularly nice examples since the behavior of aluminum hydroxide in aqueous solution is often used to illustrate the concept of amphoterism.

In the course of the determination of the nature of the adduct of IO_2F with AsF_5 ($IO_2^+AsF_6^-$), Pitts, Kongpricha, and Jache[183] showed that they were able to isolate $AgIO_2F_2$ and IO_2AsF_6 from HF solutions effectively containing IO_2F. Iodyl fluoride, then, in this solvent is amphoteric. The fluoride ion concentration determines the concentrations of the species in the solution.

$$IO_2F \rightarrow IO_2^+ + F^-$$

$$IO_2F + F^- \rightarrow IO_2F_2^-$$

Hence one can go from IO_2F to $IO_2F_2^-$ by the addition of a F^- and to IO_2^+ by the loss of a F^-.

These observations ought not to be terribly surprising since most of the halogen fluorides will, under appropriate conditions, yield a fluoride to Lewis acid fluorides or accept a fluoride from strongly basic fluorides such as cesium fluoride.

D. Other Reactions

A large number of binary metal fluorides and complex fluorides can be made by the action of hydrogen fluoride on the chlorides or oxides. In the case of ternary compounds mixtures of the starting chlorides or oxides may be employed. Chlorides are usually the reagent of choice since the evolved hydrogen chloride is easily removed while water may be troublesome, especially in the case of moisture-sensitive products. Successive washings with HF is frequently helpful. There are cases where the approach will not work as there are chlorides which are refractory toward HF. Tin (II) hexafluorozirconate, $(SnZrF_6)$,[184] and Na_3FeF_6 are examples of compounds that can be made from the binary chlorides.

A recent example of a related approach is that of Carre and colleagues.[185] They prepared $NaTeF_5$ by introducing excess HF to a mixture of NaF and TeO_2.

As the work on this chapter was nearly complete, attention was drawn to a report to the effect that Christe[186] had chemically synthesized fluorine from compounds not made from fluorine. Effectively the fluorine came from HF. He displaced the weaker Lewis acid, MnF_4, with the stronger acid SbF_5 in the salt K_2MnF_6, thus liberating MnF_4. Since the tetrafluoride is unstable with respect to the trifluoride and fluorine under the experimental conditions, fluorine is liberated. The SbF_5 is prepared from HF and $SbCl_5$[187] while K_2MnF_6 can be prepared in aqueous HF solution.[188]

$$2KMnO_4 + 2KF + 10HF + 3\ H_2O_2 \rightarrow 2K_2MnF_6 + 8H_2O + 3O_2$$

9. ELECTROCHEMISTRY

A variety of electrochemically related researches have been carried out in HF. They include measurements of electrode potentials, conductivity, polarography, and synthetic approaches. Much of this is discussed in the previous reviews. The most recent, Gmelin,[189] touches on it briefly. Electrochemical fluorination has been used for the preparation of inorganic compounds as well as for organic compounds, several of which are commercially important. The June 1986 volume of the *Journal of Fluorine Chemistry*[190] was devoted to the memory of J. H. Simons whose name has long been associated with electrochemical fluorination. Readers interested in the development of this approach, particularly at 3 M, will find the paper of Pearlson useful.[191]

From early on, there has been speculation concerning the mechanism of electrochemical fluorination of organic substances in HF.[191] Simons originally favored the concept of the generation of fluorine atoms which substituted for hydrogen after homolysis of the C—H bond. The concept of fluorination by high oxidation state nickel [Ni (III) or Ni (IV)] formed on the anode has also been proposed. A variation of this involves the adsorption of complexes of the organic substrate on metal fluorides at the anode. The effects of conditioning of the anode have been noted. More recently an EC_bEC_n mechanism has become favored. This involves: E, the organic molecule adsorbed on the electrode becomes electrochemically oxidized to a radical cation; C_b, a saturated organic compound generally eliminates a proton; E, the radical is then oxidized to a cation; C_n, the cation then reacts with a fluoride ion. Gamberretto and co-workers[192] found their work consistent with the last mechanism although the possibility of another was not completely eliminated in a few cases.

10. OTHER TOPICS

There are several other topics that might have been profitably covered here. Most are not covered, however, since space and time are limited. A few are briefly mentioned following.

A. Intermolecular Complexes

Gmelin[193] devotes considerable space to complexes between HF and other molecules. Most of these involve bonding to molecules which might be expected to form hydrogen bonds with the HF but there are cases where bonds to ordinarily inert species are formed under special conditions. Legon and Millen, in a review mentioned earlier in this chapter,[55] discuss "Gas Phase Spectroscopy and the Properties of Hydrogen-Bonded Dimers: HCN · · · HF as the Spectroscopic Prototype" and are a source of many references to related studies. In a recent paper, Patten and Andrews[194] reported on HF complexes with acetic acid and with methyl acetate in solid argon. In these complexes the HF ligates to the carbonyl oxygen in the skeletal plane, presumably on the same side as the ester oxygen.

B. Organic Chemistry

A decent discussion of the involvement of HF with organic chemistry is beyond the scope of this chapter. HF is important as a catalyst, reaction medium, and reactant in a variety of reactions, some of them of industrial importance. Electrochemical fluorination has already been mentioned. Many organic compounds are bases in HF. This enhances their solubility. A brief exposure to solubility, thermodynamic considerations, etc, can be found in Gmelin.[195]

11. A FINAL WORD

Hydrogen fluoride is an intriguing compound. There have been many theoretical studies involving it and its reactions, as it is a relatively simple compound with many interesting properties. Experimental work is exciting, demanding, challenging, and rewarding. However, be sure to wear suitable gloves before you try your hand at it!

REFERENCES

1. Simons, J. H. In "Fluorine Chemistry"; Simons, J. H., Ed.; Academic Press: New York, 1950, Vol. 1, Chapter 6.
2. Davy, H. *Phil. Trans.* **1831**, *103,* 263.
3. Fremy, E. *Ann. chim. phys.* **1856**, *47,* 5.
4. Moissan, H. "Le Fluor et ses composes"; G. Steinheil: Paris, 1900.
5. Fredenhagen, H. *Z. anorg. u. allgem. Chem.* **1939**, *242,* 23.
6. Simons, J. H. In "Fluorine Chemistry"; Simons, J. H., Ed.; Academic Press: New York, 1950, Vol. 1, Chapters 6 & 7, p. 225.
7. Simons, J. H. In "Fluorine Chemistry"; Simons, J. H., Ed.; Academic Press: New York, 1964, Vol. 5, Chapter 1, p. 2.
8. Hyman, H. H.; Katz, J. J. In "Non-aqueous Solvent Systems"; Waddington, T. C., Ed.; Academic Press: New York, 1965, Chapter 2, p. 47.
9. Kilpatrick, M.; Jones, J. G. In "The Chemistry of Non-aqueous Solvents"; Lagowski, J. J., Ed.; Academic Press: New York, 1967, Vol. 2, Chapter 2, p. 43.
10. Jander, J.; Lafrenz, C. In "Ionizing Solvents: Special Topics for Students"; Foerst, W.; and Gruenewald, H., Eds.; John Wiley & Sons/Verlag Chemie, GmbH.: Weinheim/Bergstr., 1970, Chapter 3, p. 53 (English translation).
11. Dove, M. F. A.; Clifford, A. F. In "Chemistry in Anhydrous Prototropic Solvents"; Spandau, H.; and Addison, C. C., Eds.; Pergamon Press: New York, 1971, Vol. II, Pt. 1, Chapter 2, p. 119.
12. Waddington, T. C. In "Inorganic Chemistry, Series One"; Gutman, V., Vol Ed.; MTP International Review of Science, Buttersworths: London, Vol. 3, Chapter 3, p. 85.
13. O'Donnell, T. A. In "Comprehensive Inorganic Chemistry"; Bailar, J. C.; Emeleus, H. J.; Nyholm, R.; Trotman Dickenson, A. F., Eds.; Pergamon Press: 1973, Vol. 2, Chapter 25, p. 1038.
14. Olah, G. A.; Prakash, G. K. S.; Sommer, J. "Super Acids". Wiley-Interscience: New York, 1985.
15. Gmelin "Handbook of Inorganic Chemistry"; Koschel, D., Chief Ed.; Fluorine Supplement Vol. 3, Springer-Verlag: Berlin, 1982.
16. Gmelin "Handbook of Inorganic Chemistry"; Koschel, D., Chief Ed.; Fluorine Supplement Vol. 2, Springer-Verlag: Berlin, 1980.
17. Atoji, M.; Lipscomb, W. N. *Acta Cryst.* **1954**, *7,* 173.
18. Johnson, M. W.; Sandor, E.; Arzi, E. *Acta Cryst.* B **1975**, *31,* 1998.
19. Habuda, S. P.; Gagarinski, Yu V. *Acta Cryst.* B **1971**, *27,* 1677.
20. Gabuda, S. P.; Gagarinsky, Yu V. *Soviet Phys.-Dokl.* **1972**, *17,* 80.
21. Kittelberger, J. S.; Hornig, D. F. *J. Chem. Phys.* **1967**, *46,* 3099.
22. Anderson, A. *Chem. Phys. Letters* **1980**, *70,* 300–304.
23. Ref. 15, p. 89.
24. Simons, J.; Hildebrand, J. H. *J. Am. Chem. Soc.* **1924**, *46,* 2979.
25. Simons, J.; Hildebrand, J. H. *J. Am. Chem. Soc.* **1924**, *46,* 2183.
26. Thorpe, T. E., and Hambly, F. J. *J. Chem. Soc.* **1889**, *55,* 163.
27. Ref. 15, p. 81.
28. Strohmeier, W.; Briegleb, G. *Z. Electrochem.* **1953**, *57,* 662.
29. Frank, E. U.; Meyer, F. *Z. Electrochem.* **1959**, *63,* 571.
30. Janzen, J.; Bartell, L. S. *Ames Lab. Rep. Iowa State Univ.* **1968 IS-1940**, 1–47. *J. Chem. Phys.* **1969**, *50,* 3611.
31. Huong, P. V.; Couzi, M. *J. Chem. Phys.* **1969**, *66,* 1309.
32. Le Duff, Y.; Holzer, W. *J. Chem. Phys.* **1974**, *60,* 2175.
33. Mackor, C.; MacLean, C.; Hilbers, C. W. *Rec. Trav. Chim.* **1968**, *87,* 655.
34. Hinderman, D. K.; Cornwall, C. D. *J. Chem. Phys.* **1968**, *48,* 2017.
35. Dyke, T. R.; Howard, B. J.; Klemperer, W. *J. Chem. Phys.* **1972**, *56,* 2442.

36. Sheft, I.; Perkins, A. J. *J. Inorg. Nucl. Chem.* **1976**, *38*, 665.
37. Maybury, R. H.; Gordon, S.; Katz, J. J. *J. Chem. Phys.* **1955**, *23*, 1277.
38. Shamir, J.; Netzer, A. *Can. J. Chem.* **1973**, 2676.
39. Fredenhagen, K.; Dahmlos, J. *Z. Anorg. Allg. Chem.* **1929**, *178*, 272.
40. Cole, R. H. *J. Chem. Phys.* **1973**, *59*, 1545.
41. Desbat, B.; Huong, P. V. *J. Chem. Phys.* **1983**, *78*, 6377.
42. Ref. 15, p. 24.
43. Muenter, J. S. *J. Chem. Phys.* **1972**, *56*, 12.
44. Muenter, J. S.; Klemperer, W. *J. Chem. Phys.* **1970**, *52*, 6033.
45. Hindermann, D. K.; Cornwell, C. D. *J. Chem. Phys.* **1968**, *48*, 2017.
46. Solomon I.; Bloembergen, N. *J. Chem. Phys.* **1956**, *25*, 261. Solomon, I. *Phys. Rev.* **1955**, *99*, 559.
47. MacLean, C.; Mackor, E. L. *J. Chem. Phys.* **1961**, *34*, 2207.
48. Hudlicky, M. *J. Fluorine Chem.* **1985**, *28*, 461.
49. Di Lonardo, G.; Douglas, A. E. *J. Chem. Phys.* **1972**, *56*, 5185. *Can. J. Phys.* **1973**, *51*, 434.
50. Johns, J. W. C.; Barrow, R. F. *Proc. Roy. Soc. (London) A* **1959**, *251*, 504.
51. Brundle, C. R. *Chem. Phys. Letters* **1970**, *7*, 317.
52. Berkowitz, J. *Chem. Phys. Letters.* **1971**, *11*, 21.
53. Berowitz, J.; Chupka, W. A.; Guyon, P. M.; Holloway, J. M. *J. Chem. Phys.* **1971**, *54*, 5165.
54. Chupka, W. A.; Berkowitz, J. *J. Chem. Phys.* **1971**, *54*, 5126.
55. Legon, A. C.; Millen, D. J. *Chemical Rev.* **1986**, *86*, 635.
56. Kollmann, P. A.; Allen, L. C. *J. Chem. Phys.* **1970**, *52*, 5085.
57. Del Bene, J. E.; Pople, J. A. *J. Chem. Phys.* **1971**, *55*, 2296.
58. Gaw, J. F.; Yamaguchi, Y.; Vincent, M. A.; Schaefer, H. F., III, *J. Am. Chem. Soc.* **1984**, *106*, 3133.
59. Pine, A. S.; Lafferty, W. J. *J. Chem. Phys.* **1983**, *78*, 2154. See also Lisy, J. M.; Tramer, A.; Vernon, M. F.; Lee, Y. T. *J. Chem. Phys.* **1981**, *75*, 4733.
60. Andrews, L.; Johnson, G. L. *Chem. Phys. Letters* **1983**, *96*, 133–8. Andrews, L.; Arlinghaus, R. T.; Johnson, G. L. *J. Chem. Phys.* **1983**, *78*, 6347.
61. Reichert, U. V.; Hartman, H. *Z. Naturforsch.* **1972**, *27a*, 983.
62. Huber, K.; Herzberg, G. In "Molecular Spectra and Molecular Constants of Diatomic Molecules"; Herzberg, G., Ed., Van Nostrand-Reinhold: New York, 1979, Vol. 4, p. 304.
63. Delucia, F. C.; Helminger, P.; Gordy, W. *Phys. Rev.* **1971**, *A 3*, 1849. Helminger, P. *Phys. Rev. Letters* **1970**, *11*, 1397.
64. Ref. 15, p. 46.
65. Ref. 15, p. 30.
66. Perkins, A. J. *J. Chem. Phys.* **1964**, *68*, 654.
67. Batsanov, S. S.; Vesen, Yu. I. *Opt. Spektroskopiya* **1961**, *10*, 808–9. *Chem. Abstr.* **1963**, *58*, 4006. *Opt. Spectrosc. (USSR)* **1961**, *10*, 429.
68. Muenter, J. S. *J. Chem. Phys.* **1972**, *56*, 5409.
69. Muenter, J. S.; Klemperer, W. *J. Chem. Phys.* **1970**, *52*, 6033.
70. Weiss, R. *Phys. Rev.* **1963**, *131*, 659.
71. Wickliffe, M.; Rollefson, R. *J. Chem. Phys.* **1979**, *70*, 1372.
72. Ref. 15, p. 31.
73. Babb, S. E. *J. Chem. Phys.* **1969**, *50*, 5271.
74. Gillespie, R. J.; Humphreys, D. A. *J. Chem. Soc. A* **1970**, 2311.
75. Vanderzee, C. E.; Rodenburg, W. W. *J. Chem. Thermodyn.* **1970**, *2*, 461.
76. Sheft, I.; Perkins, A. J.; Hyman, H. H. *J. Inorg. Nucl. Chem.* **1973**, *35*, 3677.
77. Jarry, R.; Davis, W. *J. Phys. Chem.* **1953**, *57*, 600.
78. Frank, E. U.; Spalthoff, W. *Z. Electrochem.* **1957**, *61*, 348. *Z. Physik. Chem. (Frankfort)* **1956**, *8*, 255.
79. Fredenhagen, K. *Z. anorg. Chemie* **1933**, *210*, 210.

80. Klatt, W. *Z. anorg. Chem.* **1937,** *233,* 307.
81. Hu, J. H.; White, D.; Johnston, H. L. *J. Am. Chem. Soc.* **1953,** *75,* 1232.
82. Fredenhagen, K.; Klatt, W.; Kunz, H.; Butzke, U. *Z. anorg. Chem.* **1934,** *218,* 161.
83. Briegleb, G. *Naturwiss.* **1941,** *29,* 420. *Z. phys. Chem.* **1941,** *B51,* 9. *Z. phys. Chem.* **1942,** *B52,* 368.
84. Long, R. W.; Hildebrand, J. H.; Morrell, W. E. *J. Am. Chem. Soc.* **1943,** *65,* 182.
85. Briegleb, G.; Strohmeier, W. *Z. Electrochem.* **1953,** *57,* 668. Strohmeier, W.; Briegleb, G. *Z. Electrochem.* **1953,** *57,* 662.
86. Yabroff, R. M.; Smith, S. C.; Lightcap, E. H. *J. Chem. and Eng. Data* **1964,** *9,* 178.
87. Shamir, J.; Netzer, A. *J. Sci. Instr.* **1968,** *1,* 770.
88. Hyman, H. H.; Katz, J. J. In "Non-Aqueous Solvent Systems"; Waddington, T. C., Ed.; Academic Press: New York, 1965, Chapter 2, p. 59.
89. Fredenhagen, K.; Cadenbach, G. *Z. anorg. Chem.* **1929,** *178,* 289.
90. Gillespie, R. J.; Moss, K. C. *J. Chem. Soc. (London) A* **1966,** 1170.
91. Kilpatrick, M.; Lewis, T. J. *J. Am. Chem. Soc.* **1956,** *78,* 5186.
92. Devynck, J.; Ben Hadid, A.; Fabre, P. L.; Tremillion, B. *Anal. Chim. Acta* **1978,** *100,* 343.
93. Ref. 11, p. 145.
94. Jones, J. G. *Ph.D. Dissertation, Illinois Institute of Technology,* 1964, p. 4.19; also Ref. 9, p. 54 and Ref. 39.
95. Perkins, A. S. *J. Phys. Chem.* **1964,** *68,* 654.
96. Kompa, K. L.; Pimentel, G. C. *J. Chem. Phys.* **1967,** *47,* 857.
97. Deutsch, T. F. *Appl. Phys. Letters* **1967,** *10,* 234.
98. Bougon, R.; Wilson, W. W.; Christe, K. O. *Inorg. Chem.* **1985,** *24,* 2286, and references therein.
99. Ref. 15, p. 120.
100. Gennick, I.; Harmon, K. M.; Potvin, M. M. *Inorg. Chem.* **1977,** *16,* 2033. Harmon, K. M.; Gennick, I. *J. Mol. Struct.* **1977,** *38,* 97. Coyle, B. A.; Schroeder, L. W.; Ibers, J. A. *J. Solid State Chem.* **1970,** *1,* 386. See also References 137 and 138.
101. Jenkins, H. D. B.; Pratt, K. F. *J. Chem. Soc. Faraday Trans A* **1979,** *73,* 812.
102. O'Donnell, T. A. *J. Fluorine Chem.* **1984,** *25,* 75, and references therein.
103. Kongpricha, S.; Clifford, A. F. *J. Inorg. Nucl. Chem.* **1961,** *18,* 270.
104. Jache, A. W.; Cady, G. H. *J. Phys. Chem.* **1952,** *56,* 1106.
105. Kongpricha, S.; Jache, A. W. *J. Fluorine Chem.* **1971,** *1,* 79.
106. Liang, J. J. Ph.D. Thesis, McMaster University, Ontario, 1976.
107. Gillespie, R. J.; Peel, T. E.; Robinson, E. A. *J. Am. Chem. Soc.* **1971,** *93,* 5083.
108. Gillespie, R. J.; Peel, T. E. *J. Am. Chem. Soc.* **1973,** *95,* 5173.
109. Engelbrecht, A.; Tshager, E. *Z. anorg. allg. Chem.* **1977,** *433,* 19.
110. Barraclough, C. G.; Cockman, R. W.; O'Donnell, T. A.; Schofield, W. S. J. *Inorg. Chem.* **1982,** *21,* 2519, and references therein.
111. Sheft, I.; Hyman, H. H.; Katz, J. J. *J. Am. Chem. Soc.* **1953,** *75,* 5221.
112. Seeley, R.; Jache, A. W. *J. Fluorine Chem.* **1972,** *2,* 225.
113. Cady, G. H. *J. Am. Chem. Soc.* **1934,** *56,* 1431.
114. Webb, K. R.; Prideaux, E. B. R. *J. Chem. Soc.* **1939,** 111.
115. Winsor, R. V.; Cady, G. H. *J. Am. Chem. Soc.* **1948,** *70,* 1500.
116. Cady, G. H.; Hildebrand, J. H. *J. Am. Chem. Soc.* **1930,** *52,* 3843.
117. Pawlenko, S. *Z. anorg. Chem.* **1964,** *328,* 133.
118. Ferriso, C. C.; Hornig, D. F. *J. Chem. Phys.* **1955,** *23,* 1464.
119. Viegweg, R. *Chem. Tech. (Berlin)* **1963,** *15,* 12.
120. Viegweg, R. *Chem. Tech. (Berlin)* **1963,** *15,* 734.
121. Vdovenka, L. N.; Lazarev, L. N.; Shirvinskii, E. V. *Radiokhim* **1965,** *7,* 46.
122. Vdovenka, L. N.; Lazarev, L. N.; Shirvinskii, E. V.; Gurikov, Yu. V. *Radiokhim* **1965,** *7,* 151.
123. Boinon, B.; Marchant, A.; Cohen-Adad, R. *J. Thermal Analysis* **1976,** *9,* 375. Boinon, B.; Marchant, A.; Cohen-Adad, R. *C. R. Acad. Sc. Paris* **1975,** *280,* 1413.

124. Tananaev, I. V. *Zhur. obschchei Khim.* **11,** 267.
125. Morrison, J. S.; Jache, A. W. *J. Am. Chem. Soc.* **1959,** *81,* 1821.
126. Vouillon, J. C. Docteur-Ingenieur Thesis, Universite De Lyon, 1968.
127. McGaw, B. L.; Ibers, J. A. *J. Chem. Phys.* **1963,** *39,* 2677, and references therein.
128. Frevel, L. K.; Rinn, H. W. *Acta Cryst.* **1962,** *15,* 286.
129. Higgins, T. L.; Westrum, E. F. *J. Phys. Chem.* **1961,** *65,* 830.
130. Cady, G. H. *J. Am. Chem. Soc.* **1934,** *56,* 1431.
131. Westrum, E. F.; Pitzer, K. S. *J. Am. Chem. Soc.* **1949,** *71,* 1940.
132. Tananaev, I. V. *Zhur. prikl. Khim.* **1938,** *11,* 214.
133. Davis, M. L.: Westrum, E. F. *J. Phys. Chem.* **1961,** *65,* 338, and references therein.
134. Ref. 11, p. 228.
135. Helmholz, L.; Rogers, M. T. *J. Am. Chem. Soc.* **1939,** *61,* 2590.
136. Kruh, R.; Fuwa, K.; McEver, T. E. *J. Am. Chem. Soc.* **1956,** *78,* 4256.
137. Ibers, J. A. *J. Chem. Phys.* **1964,** *40,* 402, and references therein.
138. Forrester, J. D.; Senko, M. E.; Zalkin, A.; Templeton, D. H. *Acta Cryst.* **1963,** *16,* 58.
139. Blinc, R.; Trontelj, Z.: Volavsek, J. *J. Chem. Phys.* **1966,** *44,* 1028.
140. Euler, R. D.; Westrum, E. F. *J. Phys. Chem.* **1961,** *65,* 1291.
141. Ruff, O.; Staub, L. *Z. anorg. Chem.* **1933,** *212,* 339.
142. Buettner, J. P.; Jache, A. W. *Inorganic Chemistry* **1963,** *2,* 19.
143. McDonald, T. R. R. *Acta Cryst.* **1960,** *13,* 113.
144. Plumb, R. C.; Hornig, D. F. *J. Chem. Phys.* **1955,** *23,* 1464.
145. Boinon, B.; Marchant, A.; Cohen-Adad, R. *J. Thermal Analysis* **1976,** *10,* 411, and references therein.
146. Opalovskii, A. A.; Fedetova, T. D. *Izv. Sib. Otd. Akad. Nauk. Ser. Khim. Nauk.* **1967,** *2,* 50.
147. Boinon, B. *C. R. Acad. Sc. Paris* **1975,** *280,* 657.
148. Webb, K. R.; Prideaux, E. B. R. *J. Chem. Soc.* **1939,** 111. Webb, K. R.; Prideaux, E. B. R. *J. Chem. Soc.* **1937,** 1.
149. Mathers, F. C.; Stroup, P. T. *Trans. Am. Electrochem. Soc.* **1934,** *66,* 245.
150. Boinon, M. J. Docteur-Es-Sciences-Physiques, L'Universite Claude Bernard. Lyon I, 1974, and references therein.
151. Hassel, O.; Kringstad, H. *Z. anorg. allgem. Chem.* **1930,** *191,* 36. Hassel, O.; Kringstad, H. *Z. anorg. allgem. Chem.* **1932,** *208,* 382.
152. Thomas, H. J.; Jache, A. W. *J. Inorg. Nucl. Chem.* **1960,** *13,* 54.
153. Buettner, J. P.; Jache, A. W. *J. Inorg. Nucl. Chem.* **1961,** *19,* 376.
154. Guntz, A.; Guntz, A. A., Jr. *Ann. Chim.* **1915,** *2,* 101.
155. Fredenhagen, K.; Cadenbach, G. *Z. phys. Chem.* **1930,** *A146,* 245.
156. Klatt, W. *Z. phys. Chem.* **1939,** *A185,* 306.
157. Clifford, A. F.; Pardieck, W. D.; Wadley, M. W. *J. Phys. Chem.* **1966,** *70,* 3241. See also Clifford, A. F.; Zamora, E. *Trans. Faraday Soc.* **1961,** *65,* 338.
158. Clifford, A. F.; Tulumello, A. C. *J. Chem. and Eng. Data* **1963,** *8,* 425.
159. Kurtenacker, A.; Finger, W.; Hey, F. *Z. anorg. Chem.* **1933,** *211,* 83, 281.
160. Clifford, A. F.; Sargent, J. *J. Am. Chem. Soc.* **1957,** *79,* 4041.
161. Ref. 11, p. 202.
162. Gillespie, R. J.; Hulme, R. *J. Chem. Soc. Dalton Trans.* **1973,** 1261.
163. Price, W. C.; Passmore, T. R.; Roessler, D. M. *Discussions Faraday Soc.* **1963,** *35,* 201.
164. Wilson, J. N. In "Advances in Chemistry Series No 54"; American Chemical Society: Washington, DC, 196, p. 30.
165. Tolberg, W. E.; Rewick, R. T.; Stringham, R. E.; Hill, M. E. *Inorg. Nucl. Chem. Letters* **1966,** *2,* 79. *Inorg. Chem.* **1967,** *6,* 1156.
166. Christe, K. O.; Guertin, J. P.; Pavlath, A. E. *Inorg. Nucl. Chem. Letters* **1966,** *2,* 83. Guertin, J. P.; Christe, K. O.; Pavlath, A. E. *Inorg. Chem.* **1966,** *5,* 1921.
167. Christe, K. O.; Wilson, W. W. *Inorg. Chem.* **1983,** *22,* 1950.
168. Malm, J. G.; Carnall, W. T. *J. Fluorine Chem.* **1985,** *29,* 26.

169. Bond, P. A.; Williams, D. A. *J. Am. Chem. Soc.* **1932,** *53,* 34.
170. Miller, G. R.; Gutowsky, H. S. *J. Chem. Phys.* **1963,** *37,* 198.
171. Gambardella, M. A.; Jache, A. W.; Kongpricha, S. United States Patent 3,446,592; 1969.
172. Jache, A. W.; Kongpricha, S.; Pitts, J. J. United States Patent 3,451,775; 1969.
173. Jache, A. W.; Gambardella, M.; Kongpricha, S.: Pitts, J. "Abstracts of Papers"; 8th
     International Symposium on Fluorine Chemistry, Kyoto, Japan; August 22–27, 1976.
174. Russell, J. L.; Jache, A. W. *J. Inorg. Nucl. Chem.* **1976,** Supplement (Hyman Memorial),
     81.
175. Pitts, J. J.; Jache, A. W. *Inorg. Chem.* **1967,** *7,* 1661.
176. Russell, J. L.; Jache, A. W. *J. Fluorine Chem.* **1976,** *7,* 205.
177. Knox, K.; Coffey, C. E. *J. Am. Chem. Soc.* **1959,** *81,* 5.
178. McCaulay, D. A.; Higley, W. S.; Lien, A. P. *J. Am. Chem. Soc.* **1956,** *78,* 3009.
179. Ruff, O.; Eisner, F. *Ber.* **1905,** *38,* 742.
180. Muetterties, E. L.; Castle, J. E. *J. Chem. Soc.* **1954,** 3922.
181. Lau, C., Passmore, J. *J. Fluorine Chem.* **1975,** *6,* 77.
182. Clifford, A. F.; Beachell, H. C.; Jack, W. M. *J. Inorg. Nuclear Chem.* **1957,** *5,* 57.
183. Pitts, J. J.; Kongpricha, S.; Jache, A. W. *Inorg. Chem.* **1965,** *4,* 257.
184. Jache, A. W.; White, W. E. United States Patent 3,337,295; Aug. 22, 1967.
185. Carre, J.; Germain, P.; Thourey, J.; Perachon, G. *J. Fluorine Chem.* **1986,** *31,* 241.
186. Christe, K. O. *Inorg. Chem.* **1986,** in press, personal communication.
187. Ruff, O. *Ber. Dtsch. Chem. Ges.* **1906,** *39,* 4310.
188. Bode, H.; Jensen, H.; Bandte, F. *Angew. Chem.* **1953,** *65,* 304, and references therein.
     Chaudhuri, M. K. *J. Inorg. Nucl. Chem.* **1981,** *43,* 85.
189. Ref. 15, p. 122.
190. *J. Fluorine Chem.* (J. H. Simons Memorial Issue) **1986,** *32,* 1.
191. Pearlson, W. H. *J. Fluorine Chem.* **1986,** *32,* 29.
192. Gamberetto, G. P.; Napoli, M.; Scipioni, A.; Armelli, R. *J. Fluorine Chem.* **1985,** *27,* 147.
193. Ref. 15, pp. 72, 232–238, 245, 313.
194. Patten, K. O., Jr.; Andrews, L. *J. Phys. Chem.* **1986,** *90,* 1073.
195. Ref. 15, p. 225.

CHAPTER 10

Structure and Bonding In N, O, F Compounds

Nancy J. S. Peters

Natural Science Division of Southampton College, Long Island University, Southampton, New York

Leland C. Allen

Department of Chemistry, Princeton University, Princeton, New Jersey

CONTENTS

1. INTRODUCTION

In this chapter we attempt to list and give an account of the geometry and electronic structure for all molecules containing nitrogen, oxygen, and fluorine only. These compounds are treated together because N, O, and F are the most electronegative atoms in the Periodic Table that form compounds and because it is on these three atoms that the octet rule imposes its most unique and stringent conditions. In addition, the collection of N, O, F containing molecules have a number of interrelated properties that are not brought out if these molecules are separately discussed under the usual textbook categories of oxides or fluorides. We start by stating a set of bonding rules that derive from our analysis of experimental data and *ab initio* electronic structure calculations and then present detailed case studies for eight species that illustrate these rules. We conclude with a list of all N, O, F compounds believed to exist, their measured or predicted geometries, and a simple pictorial representation of their electron distribution. We have found that a useful electronic structure description can be given within the framework of Linnett's double-quartet model[1] and/or even simpler extensions of Lewis Dot resonance structures.

Compared to other categories of molecules, the literature on N, O, F species is sparse. This is not surprising for several reasons: The compounds are generally difficult to synthesize and frequently explode. They are toxic and corrosive. Similarly, because of their high electron density, accurate ab initio wavefunctions have proved extremely difficult to obtain. During the 1960s there was a large-scale synthetic effort sponsored by the U.S. Defense Department. Three books resulting from this program[2,3,4] still represent some of the most important reference works.

Because of the problems noted above, practical uses for these compounds have been limited. The Defense Department program of the 1960s was directed toward rocket propellants, but none of the compounds studied has come into widespread use for this purpose. Currently, there is some research on N, O, F compounds as energy sources for high-powered lasers. Some are also employed as fluorinating agents. At this time, a greater range of practical applications is held up by lack of a systematic knowledge of their structure and bonding. The objective of this chapter is to advance such understanding.

2. BONDING RULES

1. The octet rule is rigorously obeyed for every atom in N, O, F compounds (except for two radicals, F_2N and NO_2, which each possess seven electrons around N). The octet rule is such a strong driving force that Lewis

Dot structures predict stability in a straightforward manner. The fact that neutral nitrogen cannot attach more than three covalent (i.e., non-dative) bonds, oxygen no more than two, and fluorine no more than one, results from two factors: (a) Extreme electron–electron repulsion resulting from the high density and small size of these three atoms, and (b) electronegativity so high that they form lone pairs rather than bond pairs, i.e., these molecules make poor Lewis bases. (These two factors may be identical when electronegativity is understood at a deeper level.)

2. The strength of NN, OO, and NO bonds is proportional to their bond order. By comparing the NN bond strengths of molecules with typical NN single, double, and triple bonds one can determine that the NN sigma bond is 40 kcal/mol, the first π bond 60 and the second π bond 125. Thus N≡N has a total bond strength of 225 kcal/mol. In a similar fashion, one can determine that the OO sigma bond is 34, and π bond 84, thus a total for O=O of 118 kcal/mol. The sigma bond in NO is 48 and the π increment is 97 to 114 kcal/mol corresponding to the range for a double to a double and a half bond giving a total bond energy for multiply bonded NO of 145 to 162 kcal/mol. It may also be noted that energies for the three possible single bonds to F are in the same range as those for the N, O combinations. F—N is 45 kcal/mol, F—O is 68, and F—F is 37. Thus there is a strong driving force to multiple bonding in N, O, F compounds. Lewis Dot resonance structures with N/O multiple bonds are often favored over single bonds in other parts of the molecule and, for many, "no bond" resonance is a major contributor to the electron distribution. CC multiple bonding shows an opposite trend, and because of the general familiarity most scientists have with carbon compounds, results for N, O, F molecules can appear counter intuitive. (The CC sigma bond is 83 kcal/mol, first π 61, second π 56, yielding a total for C≡C of 200 kcal/mol.)

3. Multiple bonds to F neutral compounds do not occur, with two exceptions: :F̈=N̈: and :F̈—Ö:. There is no other way to satisfy the octet rule for these two atom species. (This situation is in sharp contrast to some other main group compounds, eg, BF_3.)

4. The principle stable radicals are: NO, NO_2, NF_2, and CF, the first two having high stability and the latter two relatively lower stability. Compounds formed as dimers of radicals of high stability have unusually long NN bonds holding the monomers together because NO bonds are stronger than the corresponding NN bonds. Compounds formed between two low-stability radicals and between low- and high-stability radicals tend to have a shorter separation between the parent radicals.

5. O, N, F molecules with six or more atoms tend to separate into two, three, or four atom fragments. (There are only a very few exceptions and these are easy to rationalize.) The reason is that breaking of a bond pair permits a lone pair to be converted to multiple N/O bonds. For example,

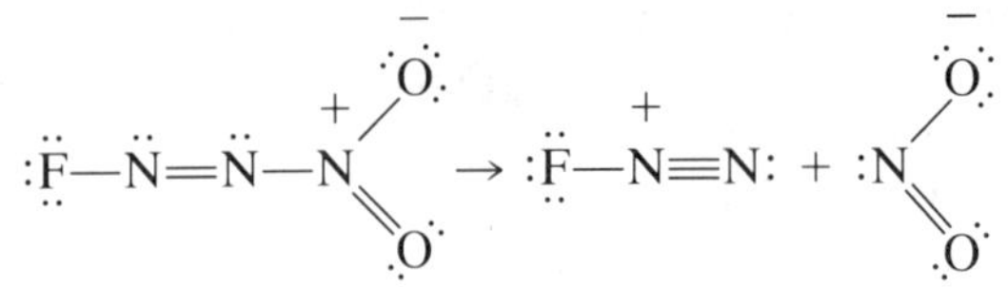

$$:F-N=N-N\begin{matrix}\overset{+}{\underset{}{}}\overset{\cdot\cdot}{O}:^{-}\\ \diagdown\\ \diagup\\ O\cdot\end{matrix} \rightarrow :F-N\equiv N: + :N\begin{matrix}\overset{\cdot\cdot}{O}:^{-}\\ \diagup\\ \diagdown\\ O\cdot\end{matrix} \qquad (10\text{-}1)$$

and

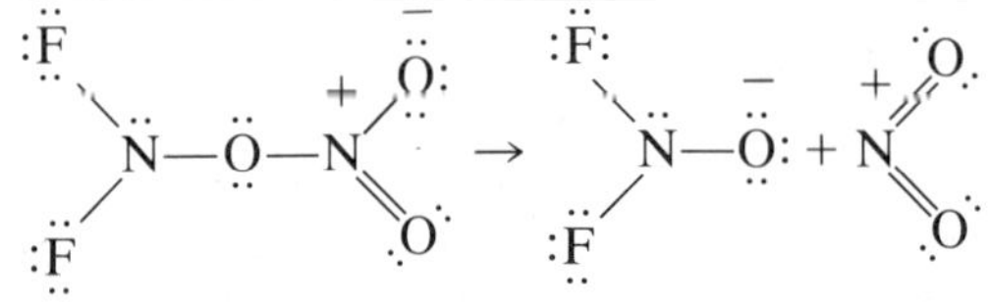

$$\begin{matrix}:F\\ \diagdown\\ N-\overset{+}{O}-N\\ \diagup\\ :F\end{matrix}\begin{matrix}\overset{\cdot\cdot}{O}:^{-}\\ \diagdown\\ \diagup\\ \cdot O\cdot\end{matrix} \rightarrow \begin{matrix}:F:\\ \diagdown\\ N-\overset{\cdot\cdot}{O}:^{-}\\ \diagup\\ :F\end{matrix}\begin{matrix}\overset{+}{\underset{}{}}\overset{\cdot\cdot}{O}\cdot\\ \diagup\\ \diagdown\\ \cdot O\cdot\end{matrix} \qquad (10\text{-}2)$$

Even though the number of bond pairs and the number of lone pairs are independently preserved for each pair of resonance structures above, the trade of an NN double bond for an NN triple bond or an NO single bond for an NO double bond strongly favors the separated fragment ions because of rule 2 above. The energy required to separate + and − charges to create ionic fragments is not appreciably greater than that required to generate intramolecular charge separation in the larger molecule. A counter example is the well-known existence of N_2O_5:

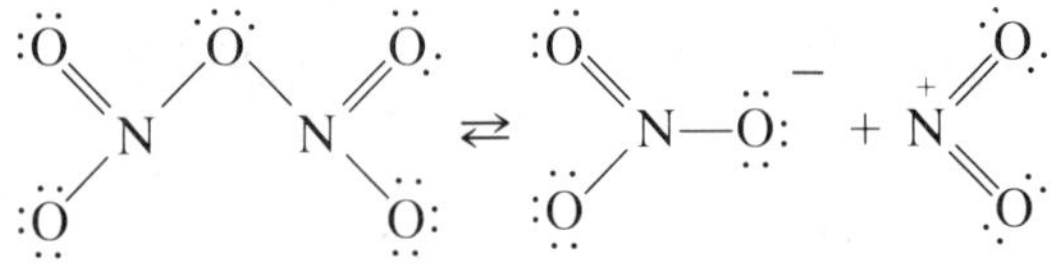

$$\begin{matrix}:\overset{\cdot\cdot}{O}\\ \diagdown\\ N\\ \diagup\\ :\overset{\cdot\cdot}{O}\end{matrix}\begin{matrix}\cdot\overset{\cdot\cdot}{O}\\ \diagdown\\ \diagup\\ \cdot O:\end{matrix}\begin{matrix}\overset{\cdot\cdot}{O}.\\ \diagdown N\\ \diagup\\ \cdot O:\end{matrix} \rightleftarrows \begin{matrix}:\overset{\cdot\cdot}{O}\\ \diagdown\\ N-\overset{\cdot\cdot}{O}:^{-}\\ \diagup\\ :\overset{\cdot\cdot}{O}\end{matrix} + \begin{matrix}\overset{+}{\underset{}{}}\overset{\cdot\cdot}{O}.\\ N\\ \diagup\\ \diagdown\\ \cdot O\cdot\end{matrix} \qquad (10\text{-}3)$$

The separation into asymmetrical charged fragments is prevented (in the gas phase, but not in the crystal where the Madelung potential greatly stabilizes the fragment ions) because of the planarity and high symmetry of N_2O_5. Thus its extensive delocalization lowers the energy more than the difference in energy between an NO single and double bond.

3. ELECTRONIC STRUCTURE DETERMINATION

Because of the exceptionally high electron density in N, O, and F, electronic wavefunctions that will describe all molecular properties to high accuracy must contain many instantaneous electron–electron correlation terms. Such high accuracy wavefunctions with full optimization for more than two atoms currently only exist for F_2O_2 and FNO, but even for these two molecules their necessary complexity is self-defeating because the results are so difficult to interpret. Direct use of information from these wavefunctions and others of slightly lesser accuracy are discussed in the case study on F_3NO. However, it is clear that we must devise another computational strategy to make progress toward a chemically useful electronic structure description even if it is not as complete or accurate as ultimately desired. We have, therefore, chosen to use a computational level that has proved to adequately describe most chemically interesting properties of molecules in categories other than N, O, F compounds. Thus we employ *ab initio* SCF calculations with a split s, p basis set (4-31G).[5] It is important that ab initio (exact Hamiltonian, all electrons included simultaneously) wavefunctions be used because it has been repeatedly shown that this is the only way an internally consistent set of results can be obtained. The computations were carried out using the GAUSSIAN series of computer programs[6] and geometry optimization utilized the Pulay force relaxation technique.[7]

The 4-31G basis set was chosen specifically because previous work on N, O, F compounds has shown it to yield reasonable geometries. This success derives from a compensation phenomenon wherein bond shortening errors introduced by the addition of d-polarization functions to the basis set are approximately cancelled by an elongation produced when an extended configuration interaction is subsequently carried out.[8] The 'compensation phenomenon' can be understood conceptually on the basis of Hartree-Fock perturbation theory. Thus the addition of d-polarization orbitals in the basis provides more opportunity for orbital overlap and consequently shortens bonds. Electron–electron correlation, on the other hand, forces an increase in the volumes available to electrons and therefore tends to increase bond lengths.

The effect of correlation on orbital shape can be estimated by applying a uniform field perturbation to the atoms. This perturbs s and p orbitals according to s → p', p → s', p → d' where the primes indicate the perturbed orbitals describing correlation.[9] It turns out that the radial probability distribution of p' is very close to that of p and s' is close to that of s so that the principle effect of introducing correlations is manifest in d'. The d', like s' and p', has a radial probability distribution similar to atomic d orbitals and thus similar to the d polarization functions used to improve basis sets. Polarization and correlation are therefore arising primarily in the same orbital shape change but with approximately equal and opposite physical driving forces.

In order to substantiate our computational procedure we have calculated optimized geometries for 30 N, O, F molecules, compared them with experiment (Table 10-1, Figure 10-1), and classified their bonding patterns. Four observations can be made from the results:

(a) the geometries of radicals are experimentally known to be preserved from molecule to molecule, eg, NO and NO_2 in FNO, HNO, FNO_2, $FONO_2$, $HONO_2$, and this is also found in our calculations (±0.02 Å, $\pm$ 4°).

TABLE 10-1. Ground State N, O, F Species: Experimental and
Theoretical Structures[a,b]

Species[c]	Geometry	Expt[d]	4-31G[e]
N_2^+	R_{NN}	1.12	1.105
N_2	R_{NN}	1.0976	1.098
N_2^-	R_{NN}	1.20	
NO^+	R_{NO}	1.06	1.048
NO	R_{NO}	1.151	1.155
NO^-	R_{NO}	1.29	1.282
O_2^+	R_{OO}	1.1227	
O_2	R_{OO}	1.2107	1.196
O_2^-	R_{OO}	1.28	
NF	R_{NF}	1.3173	1.371
OF	R_{OF}	1.417	(>2.3)
F_2^+	R_{FF}	1.305	1.327
F_2	R_{FF}	1.418	1.420
N_3	R_{NN}	1.189	1.124
N_3^-	R_{NN}	1.174	1.172
N_2O^+	R_{NO}	1.155	
	R_{NN}	1.185	
N_2O	R_{NO}	1.129	1.097
	R_{NN}	1.188	1.231
NO_2^+	R_{NO}	1.10	1.109
NO_2	R_{NO}	1.197	1.353, 1.389
	<ONO	134.3	108.3
NO_2^-	R_{NO}	1.236	1.256
	<ONO	115.4	117.0
O_3	R_{OO}	1.278	1.255
	<OOO	116.8	119.2
O_3^-	R_{OO}	1.19	
	<OOO	100°	
FNO	R_{NF}	1.512	1.468
	R_{NO}	1.136	1.137
	<FNO	110.0	110.2
NF_2	R_{NF}	1.363	1.378
	<FNF	102.5	102.9
FO_2	R_{FO}	1.649	
	R_{OO}	1.200	
	<FOO	111.2	
OF_2	R_{OF}	1.40	1.422
	<FOF	103.0	102.5
N_4^-	R_{NN}	(4 equivalent N's)	1.406 & 1.307 (rectangle)
N_2O_2	R_{NO}	1.161	1.166 trans)
	R_{NN}	2.237	1.612
	<ONN	99.6°	
NO_3^-	R_{NO}	1.218	1.256
N_2F_2	R_{NN}	1.2139	1.198
	R_{NF}	1.410	1.403
	<FNN	114.4	114.7
NF_3	R_{NF}	1.37	1.383
	<FNF	102.2	102.5

TABLE 10-1.   Cont.

Species[c]	Geometry	Expt[d]	4-31G[e]
FNO_2	R_{NF}	1.47	1.407
	R_{NO}	1.18	1.186
	<FNO	112.0	113.4
F_2O_2	R_{FO}	1.575	1.432
	R_{OO}	1.217	1.395
	<FOO	109.5	104.7

N_2O_3

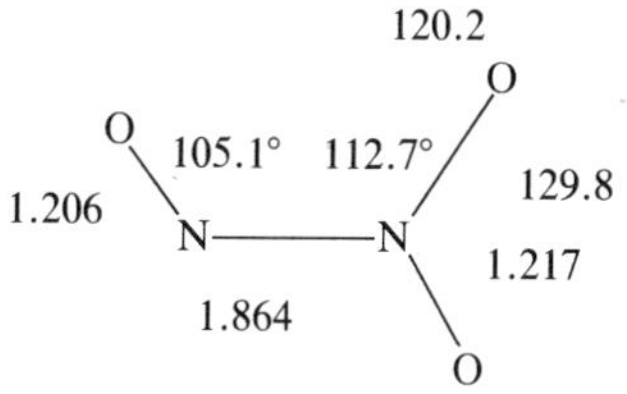

F_3NO	R_{FN}	1.432	1.412
	R_{NO}	1.159	1.190
	<FNO	117.4	117.2
NF_4^+	R_{NF}	1.24–1.40	1.340
N_2O_4	R_{NO}	1.18	1.188
	R_{NN}	1.75	1.631
	<NNO	113.2	113.4

$FO'NO_2$

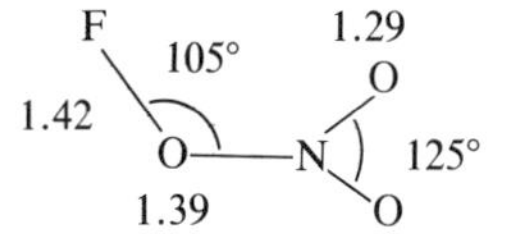

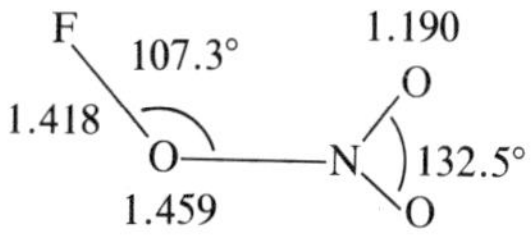

N_2F_4	R_{NF}	1.372	1.387
	R_{NN}	1.492	1.470
	<FNF	103.1	102.9
	<FNN	101.4	101.4

N_2O_5

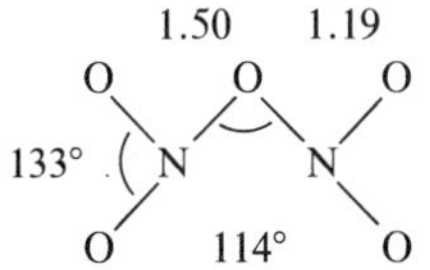

[a] Bond lengths in Angstroms, angles in degrees.

[b] We have limited ourselves to known structures of stable neutral species and stable monopositive and mononegative ionic species.

[c] The following have been detected but there is little or no structural data available: NF^+, NF^-, N_2F^+, N_3F, F_2NO^+ (planar with NO bond in range of $N\equiv O$ in IR spectrum), F_3NO^-, F_2O_4, O_3^+, F_2^-, N_5^-, N_6, N_3^+ and the following are of theoretical interest, but have not yet been detected: N_4, O_4, OF^-, FON, NF_4, NF_5.

[d] Taken from Reference 32 and N_2^+, O_2^+ from G. Herzberg, "Infrared Spectra of Diatomic Molecules", 2nd Ed., Van Nostrand Co., New York, 1950; N_3, N_2O^+ from G. Herzberg, "Infrared Spectra of Polyatomic Molecules", 2nd Ed., Van Nostrand Co., New York, 1950; N_2^- from *Rev. Mod. Phys.*, 1973, *45*, 423; O_2^- from *Acta Cryst.*, 1955, *8*, 503; NO^+, NO^- from *Prog. Inorg. Chem.*, 1980, *27*, 465; N_3^- from *Acta Chem. Scand.*, 1963, *17*, 1444;

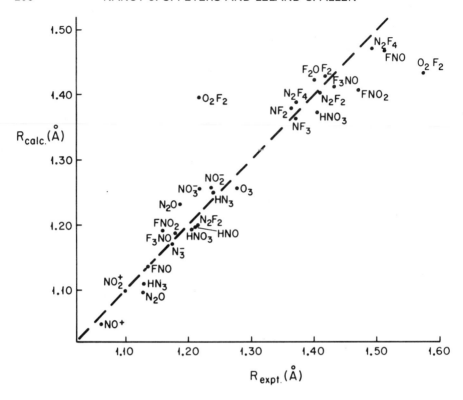

Figure 10-1. Calculated versus experimental bond lengths (Å). (Several points corresponding to the same molecule refer to different bonds.) $-r_{calc} = r_{expt}$.

(b) All normal single bond lengths are equal to experimental values or slightly longer (± 0.02 Å).

(c) Multiple bonds are well represented (± 0.02 Å).

(d) All long NF Bonds (those generally rationalized by hyperconjugation or 'no bond' resonance, eg, FNO_2, cis F_2N_2, FNO) are computed to be significantly longer than normal, but slightly (0.01–0.06 Å) too short.

NF_4^+ from Ref. 45 and 46; NO_2^+ from *Acta Cryst.* , 1960, *13*, 855; NO_2^- from ibid., 1961, *14*, 56; NO_3^- from ibid., 1957, *10*, 567; O_3^- from *Proc. Nat. Acad. Science*, 1963, *49*, 1; N_2O_2 from *J. Amer. Chem. Soc.*, 1982, *104*, 4715; N_2O_5 from *Prob. Inorg. Chem.*, 1980, *27*, 465; F_3NO from *J. Chem. Phys.*, 1970, *53*, 3488; N_2F_4 from ibid., 1975, *56*, 1961; and N_2O_4 from ibid., 1974, *61*, 1248.

[e] This lab except F_2^+ (3-21G basis set with inversion symmetry enforced) from *J. Phys. Chem.*, 1983, *87*, 79; N_3 from *Chem. Phys. Lett.*, 1976, *44*, 569; N_2 from *J. Amer. Chem. Soc.*, 1974, *96*, 4753; N_3^+ from ibid., 4753; F_3NO from Ref. 20; N_2O_4 from Ref. 56; F_2O_2 from *J. Chem. Phys.*, 1978, *68*, 2507; NO_2^+, NO_2^+, N_2O, NO from *Chem. Phys. Lett.*, 1981, *80*, 83; NO_3^-, NO_2^- from *Australian J. Chem.*, 1976, *29*, 1635; NO^+, NO^- from *Progr. Phys. Org. Chem.*, 1974, *11*, 175; F_2O from *J. Am. Chem. Soc.*, 1971, *93*, 289.

We can conclude from these observations that optimized geometries for species whose structures are unknown are accurate to ± 0.02 Å and $\pm 4°$. We expect other one-electron properties of these molecules, particularly the charge distribution, to be equally well represented at the 4-31G level because of the Extended Brillouin Theorem.[10]

The molecules we have studied in considerable detail are: F_3NO, NF_4^+, FN_3, $FONO_2$, F_2NNO_2, F_2NNO_2, NF_4, and NF_5. In the paragraphs below we give some of the facts and questions concerning these species and thus our motivation for selecting these particular species for our case studies.

4. F₃NO

Background

F_3NO is the best-characterized molecule of those to be discussed and the most unusual. It has only been known for a relatively short time and its existence, as well as its synthesis from NF_3,[11,12] was unexpected. NF_3 generally reacts only in the presence of very strong Lewis acids, e.g. $AlCl_3$, AsF_5, SbF_5, and amine oxide formation almost always occurs with the most basic amines. The isomer F_2NOF, analogous to hydroxyl amine, is unknown. Thus it is not surprising that F_3NO displays novel properties.

Its unusual bonding was first suggested when several papers[11-15] reported NO stretching frequencies in the range of typical NO double bonds rather than typical amine oxide bonds. An x-ray diffraction study[16] confirmed the existence of exceptionally short NO bonds (1.159Å as compared to 1.211Å in HNO[17] and 1.54Å in Me_3NO[18]) and exceptionally long NF bonds (1.432Å as compared to 1.37Å in NF_2, NF_3, and N_2F_4[19]).

The question of whether or not the bonding in this molecule represents a violation of the octet rule around the nitrogen was not clearly decided. The recent detection of F_3NO^- by ESR spectroscopy[20] further increases interest in F_3NO bonding.

F_3NO is one of a class of unusual N, O, F compounds, e.g. FNO, F_2O_2 and N_2F_2. In these molecules OF and NF bonds are significantly elongated relative to the reference species NF_3 and F_2O, and the attendant NO, OO, and NN bonds are significantly shortened relative to representative bonds in standard reference species. Fluorine no-bond resonance structures are often invoked to explain these results, as is the exceptional ability of N, O, F compounds to form multiple bonds.

A number of theoretical studies of F_3NO were done, a CNDO/2 study failing to reproduce any of the atypical behavior.[21] *Ab initio* studies by Olsen and Howell[22-25] were able to reproduce the experimental geometry reasonably well. They analyzed the bonding and made comparisons with CF_3O^- and H_3NO.

TABLE 10-2.   Experimental Geometries, Dipole Moments, and E_T Computed at This Geometry

Molecule	$R_{on}(A)$	$R_{nx}(A)$	$<ONX$	$\mu(expt)$	$\mu(cal(4\text{-}31G)$	$\mu(cal(6\text{-}31G^*))$	$E(4\text{-}31G)$	$E(6\text{-}31G^*)$
				Dipole moment[d]				
FNO	1.136	1.512	110.0	1.81	1.59	1.61	−228.29167	−228.51254
F_3NO	1.158	1.431	117.1	0.04	0.55	0.93	−426.72703	−427.30700
Me_3NO	1.36	1.54	110.0	5.02	5.41	5.11	−247.67502	−248.03609

[a] Reference 28
[b] Reference 16
[c] Reference 18
[d] in Debyes

An extensive study by Lawlor and Grein[26] used polarization functions and configuration interaction methods. The Hartree-Fock wavefunctions suggest that F_3NO can be thought of as a donor-acceptor product of NF_3 and O with type back donation from O to NF_3 acceptor orbitals. However, the limited post-Hartree-Fock optimization yielded results that were difficult to interpret within this simple model.

We feel that it is important to reexamine ab initio molecular orbital methods including a full s, p, d orbital basis, Mulliken populations, molecular orbital diagrams, and a detailed comparison with Me_3NO and FNO (Me_3NO is considered to be a typical amine oxide with a well-characterized dative bond and formal charges of $+1$ on nitrogen and -1 on oxygen and FNO has a characteristic π bond and exceptionally long NF bonds).[27]

A. Geometry

Table 10-2 gives the experimental geometries, Hartree-Fock total energies, and dipole moments for the different basis sets. Table 10-3 gives the optimized geometry and dipole moment for F_3NO for the different basis sets.

The influence of d orbitals is evident in the 6-31G* results. The geometry optimization (Table 10-3) gives a small improvement in the NO bond but the NF bonds are given as normal, not exceptionally long. This is not uncommon in N, O, F containing molecules where exceptionally long bonds are present.[28]

The inclusion of correlation effects corrects this foreshortening at the 6-

TABLE 10-3.   Optimized F_3NO Geometries

Basis set	$R_{NF}(A)$	$R_{NO}(A)$	$<ONF$	$\mu^a(D)$	$E(hartrees)$
4-31G[b]	1.412	1.187	117.2	0.86	−426.7287
4-31G(d)[b]	1.41	1.16	116.0		−426.8169
6-31G*[c]	1.3496	1.1717	116.32	1.31759	−427.32057

[a] Experimental value: $\mu = 0.0390$ D.
[b] Reference 22, 4-31G(d) = d orbitals on N.
[c] This work.

31G* level and we discuss the anticipated results of these corrections in a later section.

The dipole moments obtained for F_3NO at the experimental and 4-31G optimized geometries (Table 10-3) and for FNO and Me_3NO (Table 10-2) are quite reasonable and in excellent agreement with experimental values, although somewhat fortuitously. (See discussion on F_2NNO_2, for example.) The larger value for the 4-31G optimized F_3NO, as compared with the experimental, structure (0.86 versus 0.55) reflects the slight changes in geometry, ie 4-31G optimization gives a slightly shorter NF bond and a slightly longer NO bond in comparison to the experimental geometry. The significantly shorter NF bonds obtained at the 6-31G* level of optimization (1.35 versus 1.43) cause a significant increase in the dipole moment as well (to 1.32 D).

B. Bonding

Molecular Orbital Diagram

A molecular orbital diagram of the NX_3 group (NF_3 or NMe_3) with singlet oxygen (see Figure 10-2) shows the principle parentage of the resulting F_3NO or Me_3NO molecular orbital, not the entire interaction.

The 6-31G* wavefunction has a considerably lower energy than the 4-31G wavefunction (E_T is 300 kcal/mol lower than the 4-31G result), but the one-electron properties are of significantly poorer quality for the 6-31G* optimized structure, e.g., the NF bond length and the dipole moment. Much of the following discussion, therefore, relies on the 4-31G wavefunction, as other one-electron properties, e.g., charge density, should be represented as well as the geometry and dipole moment.

In the 6-31G* results, d orbital contributions are an order of magnitude smaller than those for the s and p orbitals. Their orientations are such as to strengthen the NO and NF overlaps, thus accounting for the shorter bond lengths of the 6-31G* optimized geometry. Generally the largest d orbital contributions are on the nitrogen atom, consistent with Olsen and Howell's suggestion that d orbitals are most significant on the central, rather than ligand, atom.

The MO diagram (Figure 10-2) clearly illustrates the differences between F_3NO and Me_3NO: the energy levels of both NMe_3 and Me_3NO are significantly higher in energy than the fluorinated analogs; and, although the 5e, $8a_1$, and $1a_2$ orbitals are close in energy for both molecules, the ordering is different. The $6a_1$, 5e, and $1a_2$ orbitals of NX_3 are all nominally nonbonding orbitals, but interaction with oxygen significantly lowers the $6a_1$ relative to the others. The effect is due to the introduction of some NO bonding features into the $8a_1$ of the product (see Figure 10-3) while the others remain essentially unchanged. The 4, 5, and 6 e orbitals of F_3NO remain essentially fluorine lone pairs (4e and 5e) and an oxygen lone pair (6e), as shown in Figure 10-3.

The most significant difference between the methyl and the fluoro com-

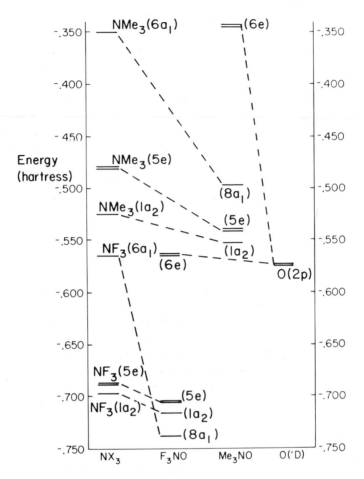

Figure 10-2. Molecular orbital correlation diagram for NX_3, $O(^1D)$ forming X_3NO (obtained using the 4-31G basis). One-electron energies in hartrees.

pounds, however, is the energy of the $6a_1$ orbitals of NF_3 and NMe_3. For the trifluoro case, this orbital is approximately the same energy level as the 2p orbitals of oxygen. This result is consistent with the formation of a more covalent NO bond in F_3NO than in Me_3NO, ie the electronegativity of N in NF_3 appears to be comparable to that of oxygen and results in electrons shared more equally than in a typical amine oxide.

π Bond

It has been suggested[23] that the origin of the π bond in F_3NO is the interaction of the unoccupied 6e (antibonding) orbitals of NF_3 with the oxygen p atomic orbitals (see Figure 10-4). This interaction is shown by the participating atomic orbital coefficients (Table 10-3) for the HOMO, 6e orbitals, of F_3NO and Me_3NO. The interaction between oxygen and nitrogen in the case of Me_3NO is antibonding, while that of F_3NO is bonding. The NF interaction

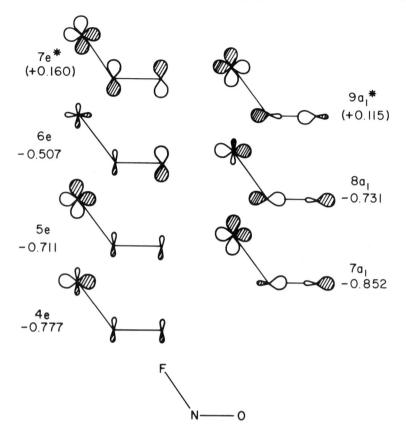

Figure 10-3. Schematic showing the principle atomic orbital contributions to the valence orbitals of F_3NO in the ONF_1 plane.

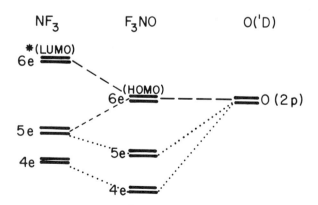

Figure 10-4. π molecular orbital correlation diagram for NF_3 and $O('D)$ forming F_3NO. Line weights refer to magnitude of NO π mixing coefficients: ————, very strong (0.9); ———, strong (0.3–0.9); ---------, medium (0.13–0.3); ...., weak (0.02–0.13).

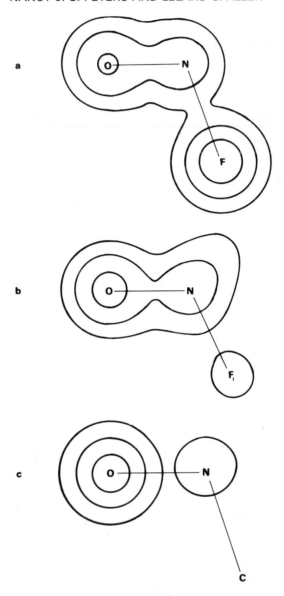

Figure 10-5. π charge density summed over the principle π molecular orbitals of FNO, F_3NO, and Me_3NO in a plane 0.5 Å above the FNO and CNO planes. Contours are: 0.2828, 0.08936, 0.02828, and 0.008936 e/A^2. a. The $1a''$ and $2a''$ of FNO. b. The 5e and 6e of F_3NO. c. The 5e and 6e of Me_3NO.

is antibonding (note that the fluorine is in the positive x direction with re-spect to the nitrogen, so the positive coefficient is actually out of phase with that on the nitrogen), as suggested by Olsen and Howell, while the NC interaction of Me_3NO is bonding.[29]

A further indication of the π bond formation is given in Figure 10-5, a charge density plot of the 5e + 6e molecular orbitals of F_3NO and Me_3NO, as well as the 1a″ + 2a″ orbitals of FNO, 0.5 Å above the xz plane, the z-axis being the C_{3v} axis. Several features stand out:

1. The FNO plot indicates the presence of the π bond, with charge buildup between the oxygen and nitrogen.
2. The dative bond of Me_3NO is also shown with the break in charge density along the NO bond, illustrating the ionic nature of the bond.
3. The F_3NO plot clearly resembles the typical π bond with charge buildup between the oxygen and nitrogen.

These plots indicate that the NO bond of F_3NO has well-developed multiple bond character.

The existence of π bonding across the NO bond does not, in fact, indicate a violation of the octet rule around the nitrogen atom. In a molecular orbital sense, π back bonding from an oxygen lone pair into an NF antibonding orbital takes place. In the Lewis sense, both structures shown below contribute to the overall F_3NO.

Isoelectronic Comparisons

In the paragraphs above a quantitative molecular orbital diagram for the formation of the NO π bond in F_3NO is given. An understanding of the bonding in F_3NO can be enhanced by a qualitative description of the change in bonding through the isoelectronic sequence F_3NF^+, F_3NO, F_3NN^-. NF_4^+ is discussed later and $F_3N_2^-$ may or may not be a stable species, but its properties help establish trends.

The principal Lewis structures for these three are:

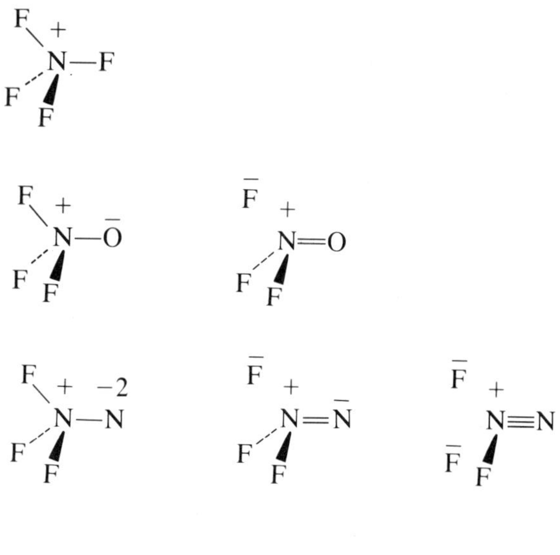

Relative to the normal reference molecule, NF_3 (1.37 Å), the addition of F^+ to form NF_4^+ results in four equal, single bonds that are somewhat more covalent than in NF_3 and have an NF bond length of 1.32 Å. The electronegativities and HOMO energies of NF_3 and O are comparable, thereby allowing two (or slightly more) oxygen lone-pair electrons to participate in the NO and NF bonds. Because all NF bonding orbitals are occupied, the charge acquired by the fluorines must go into antibonding, unoccupied orbitals, thus elongating the NF bond to 1.43 Å. In $F_3N_2^-$, the higher electronegativity of NF_3 relative to N^- will lead to a great increase in NN bonding, therefore to a very short NN separation and to a large decrease in NF bond strengths with concommitant long NF bonds. The LUMOs will be NN bonding and NF antibonding in both xz and yz planes. The bonding in $F_3N_2^-$ bears resem-

$$\underset{N=O}{F\diagdown} \qquad\qquad \overset{\overline{F}}{\underset{N\equiv\overset{+}{O}}{}}$$

blance to that in FNO and suggest why this species would have even more extreme bond lengths than F_3NO.

Mulliken Population Analysis

A comparison of Mulliken charges[30] for oxygen and nitrogen for the 4-31G, 4-31G(d), and 6-31G* basis sets for F_3NO and Me_3NO is given in Table 10-5. To account for the formation of the NO π bond and the increase in NO overlap that they obtain with the 4-31G(d) basis relative to the 4-31G result, Olsen and Howell cite the loss of charge around the oxygen and the gain of charge around the nitrogen in the 4-31G(d) basis relative to the 4-31G result. However, the more equal representation of the N, O, and F atoms at the 6-31G* level shows that, while NO overlap is increased, so is the charge on both oxygen and nitrogen (this charge increase being even greater than that for Me_3NO), reversing the 4-31G(d) analysis. Both basis sets show the importance of d orbitals in strengthening the NO and NF bonds, the 4-31G(d) overestimating the NF bond and underestimating the NO bond, relative to the 6-31G* results. The increase in bond strength is then at the cost of nitrogen electron density, not oxygen or fluorine density. This result is consistent with the atomic orbital coefficients for the 6e orbital (Table 10-4). The

TABLE 10-4. Atomic Orbital Coefficients of the HOMO 6e

Atomic orbital	4-31G		6-31G*	
	F_3NO	Me_3NO	F_3NO	Me_3NO
N $2p_{XI}$	0.010	0.138	0.044	0.137
N $2p_{XO}$	0.133	0.053	0.081	0.073
O $2p_{XI}$	0.558	−0.582	0.584	−0.594
O $2p_{XO}$	0.409	−0.518	0.411	−0.500
F(C) $2p_{XI}$	0.201	−0.133	0.106	0.000
F(C) $2p_{XO}$	0.214	−0.113	0.090	0.020

TABLE 10-5. Mulliken Populations

	$F_3NO^{a,b}$				Me_3NO^c			
	q_N	q_O	NO overlap	NF overlap	q_N	q_O	NO overlap	NC overlap
4-31G	+0.872	−0.306	0.105	0.052	−0.440	−0.590	−0.136	0.133
4-31G(d)	+0.542	−0.083	0.399	0.243	—	—	—	—
6-31G*	+1.262	−0.453	0.456	0.210	−0.202	−0.658	0.010	0.233

[a] 4-31G and 6-31G* results, this work, optimized geometry.
[b] 4-31G(d) results of Olsen and Howell, Reference 22.
[c] This work, experimental geometry.

nitrogen $2p_x$ coefficient for F_3NO is only about $\frac{1}{7}$ that of the oxygen $2p_x$ and $\frac{3}{4}$ that of the nitrogen $2p_x$ for Me_3NO.

Comparison of the NF and NO overlaps with other N, O, F molecules shows the limitations of Mulliken analysis. While the NF overlap of F_3NO (0.052) at 4-31G is less than that of NF_3 (0.060), consistent with the longer, i.e., weaker, bond in F_3NO, it is also less than that of FNO (0.057) which should have the weaker bond of the two (the N—F force constants are compared in discussion of $NF_4{}^+$ and in Figure 10-8). The NO overlaps of FNO (0.088), with $1\frac{1}{2}$ π bond, HNO (0.096), with one π bond, and F_3NO (0.105), with $\frac{1}{2}$ π bond, are inconsistent with decreasing bond orders and anticipated decreasing density. The 6-31G* results are comparably inconsistent: NF overlaps for NF_3 (0.127), F_3NO (0.210), and FNO (0.085) are inconsistent with bond lengths, as the NO overlaps are for bond order for FNO (0.361), HNO (0.286), and F_3NO (0.456).

Comparison with the NO overlaps of Me_3NO, however, is consistent as it clearly suggests the strength of the NO bonding in the trifluoro case and the antibonding nature of the NO interaction in the trimethyl case. The influence of d orbitals in the trimethyl compound is as great as in the trifluoro, the NO overlaps becoming bonding at the 6-31G* level.

Heat of Formation

Energies at the 6-31G* level and experimental heats of formation were used to estimate the heat of formation of F_3NO using the following isodesmic reaction:

$$NF_3 + N_2O \rightarrow F_3NO + N_2 \tag{10-4}$$

Isodesmic reactions minimize the effect of electron correlation by incorporating the same number and type of bonds on both sides of the equation. They yield good estimates of the heats of formation.[31] The NO bond in N_2O and F_3NO has Lewis resonance structures with single and double bonding. The NN bond in N_2O has one resonance structure with a triple bond like that of N_2. The other resonance structure is a double bond, which would make the NN bond in N_2O slightly too long compared to that in N_2. However, the

TABLE 10-6. Energies Used to Calculate the
Heat of Formation of F_3NO

Molecule	6-31G* energy[a]	$\Delta H_f^{\circ b,c}$
NF_3	-352.54005	-29.8
N_2O	-183.68012	$+19.61$
N_2	-108.94395	0
F_3NO	-427.32057	-39.0 ± 5^c
		$(-38.02)^d$

[a] Hartrees
[b] kcal/mol
[c] Same as Reference c on Table 10-12
[d] 6-31G* calculation

FN bond in F_3NO is clearly longer than the FN bond in NF_3. These two differences between the left-hand side and the right-hand side should cancel and therefore permit the theoretical energies to give good results.

As shown in Table 10-6, the results of our 6-31G* calculations are in excellent agreement with the experimental data.

Correlation Effects

Because of the large number of basis functions in F_3NO (75 at 6-31G*) we have not undertaken calculations to fully optimize the structure at the MP2 level. (This calculation requires in excess of 425,000 blocks of memory on the VAX 11/780.) The results of Lawlor and Grein[26] suggest that a full optimization would yield a structure in excellent agreement with experiment. However, the dipole moment is still an order of magnitude too large. It may be necessary to include high-order angular momentum functions, single excitations, and excitations from the core to correct the dipole moment. Inclusion of these terms may yield to a simple analysis analogous to the FNO situation. The present results for F_3NO do not show the straightforward pattern of replacing bonding NF orbitals with antibonding NF orbitals.

5. FN₃

FN_3 is difficult to work with (it explodes), and contains a potentially interesting NF bond. In similar species, with one fluorine bound to nitrogen, the NF bond is exceptionally long. This elongation is explained by invoking fluorine no-bond resonance structures, but these structures cannot be considered for FN_3 without some difficulty.

A. Geometry

There are two classes of NF bonds: those of normal length (1.37 Å) and those unusually long (1.41–1.51 Å). The challenge in FN_3 is to ascertain which of these cases applies. Table 10-7 gives the list of bond lengths in known NF compounds and other relevant molecules. The NF separation is unusually large in all known polyatomic species containing a single fluorine atom bonded to nitrogen (ie, FNO_2, FNO, and F_2N_2) thus suggesting a similar assignment for FN_3. A second line of reasoning leading to the same conclusion comes from the extensive theoretical investigation of 30 N, O, F compounds which yielded excellent agreement with experimental data. N_2, O_2, O_2^+, F_2, NO, NO^+, NO^-, O_3, N_3^-, NF_2 N_2O, NO_2^+, F_2O, N_2F_2, NF_3, FNO_2, NO_3^- N_3, HNO, N_2F_4, and HNO_3 were calculated ab initio with a flexible s, p basis (4-31G). *Ab initio* molecular orbital calculations predicted an NF bond length in FN_3 of 1.46 Å.

On the other hand, a long NF bond is clearly inconsistent with simple Lewis or Linnett structure predictions. Below we give the expected Linnett structures because they embody more physical content[33] and because they have been shown to give useful rationalizations for numerous N, O, F compounds.[1,2] Linnett theory treats the valence electrons as two spin sets, a double quartet. Thus electrons are not constrained to be grouped in pairs. One of the consequences of this separation of α and β electrons is the possibility of L strain; that is, the two electrons involved in a given bond may not lie along the internuclear axis and hence weaken the bond.

Obviously structure **1** has the lowest energy. It has no L strain, good electron correlation, and formal charges that are the most favorable of the four. Structures **2** and **3** are ruled out because they have one less binding

TABLE 10-7. Experimental Geometries[a]

NF_3	r_{NF}	1.365 Å
F_2N_2	r_{NF}	1.410
FNO_2	r_{NF}	1.467
FNO	r_{NF}	1.512
NCl_3	r_{NCl}	1.754
$ClNO_2$	r_{NCl}	1.840
ClNO	r_{NCl}	1.975
HN_3	$<HN_1N_2$	114.1°
	$r_{N_1N_2}$	1.237
	$r_{N_2N_3}$	1.133
ClN_3	$<ClN_1N_2$	108.7°
	$r_{N_1N_2}$	1.252
	$r_{N_2N_3}$	1.133
	r_{ClN}	1.745

[a] Reference 32

$$-\overset{\underset{|}{|}}{F}-\overset{\underset{|}{\cdot\cdot}}{N}\overset{-\frac{1}{2}}{-}\overset{+1}{N}\overset{-\frac{1}{2}}{=}\overset{\cdot\cdot}{N}-$$

1

$$-\overset{\underset{|}{|}}{F}\cdot\overset{\overset{+\frac{1}{2}-\frac{1}{2}}{\cdot\cdot}}{N}\overset{+1}{=}N\overset{-1}{=}N-$$

2

$$-\overset{\underset{|}{|}}{F}\cdot\overset{\overset{-\frac{1}{2}-\frac{1}{2}}{\cdot\cdot}}{N}\overset{+1}{=}N\overset{\cdot\cdot}{=}N-$$

3

$$-\overset{\underset{|}{|}}{F}\overset{+\frac{1}{2}}{-}\overset{-\frac{1}{2}}{N}-\overset{+\frac{1}{2}}{N}\overset{-\frac{1}{2}}{=}\overset{\cdot\cdot}{N}-$$

4

electron. Structure **4** will make a relatively small contribution because of its unfavorable NN bond, L strain and unfavorable charge separation. Therefore Lewis or Linnett structures imply a normal or slightly shorter NF bond length. The ultimate reason why elongation of the NF by "no-bond" resonance is energetically unfavorable is lack of lone-pair electrons on the middle nitrogen atom, thus preventing such electrons from adding compensatory binding electrons to the first NN bond when NF is weakened.

A normal length bond is also expected by analogy with the known molecule ClN_3 which has an NCl bond very close (slightly shorter) to that in NCl_3. By itself, this observation is not fully creditable because Cl, Br, and I often act normal against an abnormal behavior for F. In the present case, however, high resolution microwave spectroscopy values for the ClN bond lengths in ClNO and $ClNO_2$ are known and these show the same bond length ratio with NCl_3 as the fluorine analogs.

Further exploration with the same computational approach applied to the N, O, F compounds revealed that ClN_3 yields the same incorrect elongation of the ClN bond found previously for the NF bond in FN_3. The poor representation of these species compared to the success with other species arises from a greater need for inclusion of instantaneous electron–electron correlation in these two molecules than the others. This fact is brought out by comparing the Mulliken net charges[30] on fluorine in FN_3 and representative molecules with normal NF bond lengths. Thus the charge on F is -0.32 e in FN_3, -0.26 in NF_3, and -0.25 in N_2F_4.

The N_3 moiety is well preserved in HN_3 and ClN_3 (see Table 10-7) and one expects FN_3 to show a similar result. Thus the predicted structure of FN_3 is planar with the angles and bond lengths given below based on analogy with HN_3 and ClN_3 rather than on the calculational results.[33a]

$$\underset{1.37}{F}\diagdown\overset{110°}{\underset{\underset{1.25\quad1.13}{N——N——N}}{}}$$

B. Bonding

In order to understand the bonding in this molecule we have computed and plotted the valence molecular orbitals at this geometry. The orbital energies are specified with each orbital plot (Figure 10-6).[34] The lowest four orbitals,

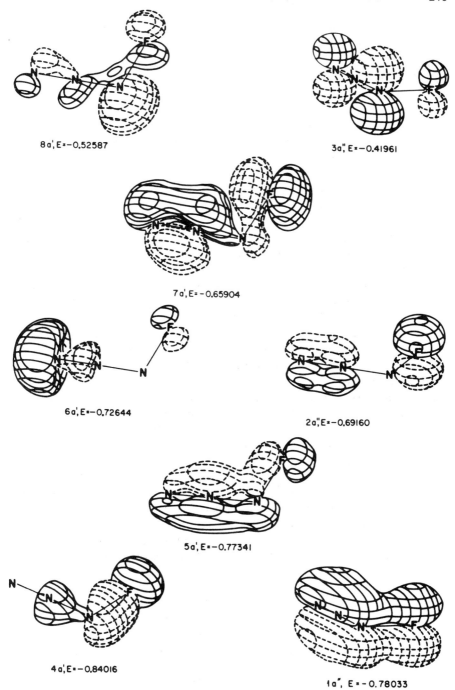

Figure 10-6. Contour plots of the valence molecular orbitals for FN_3 calculated with the 4-31G basis set at the predicted geometry (orbitals, 1a′, 2a′, and 3a′, not shown). One-electron energies for each orbital in hartrees.

not shown in Figure 10-6, are quite simple. The lowest is NF bonding, with some small NN overlap: it is pear-shaped with fluorine at the large-diameter end, similar to that found in NF_3. The next two π orbitals preserve the N_3 moiety; that is, two NN bonds are indicated, the lower orbital with no node, the next with one node.

For the π orbitals there is a common in-plane node. In addition, the out-of-plane nodal pattern in the 1a″, 2a″, and 3a″ is: no nodes, one node, and two nodes, respectively. The first node is on the nitrogen attached to the fluorine, i.e., the weakest of the three bonds (NF). The 4a′ through 8a′ orbitals mix bonding and nonbonding features. The lowest, the 4a′, has a large fraction of its density in the NF region. The 5a′ is an "in-plane π." It is "π-like" bonding over all four atoms, in essence a four-center, two-electron bond. The 6a′ is obviously a nitrogen lone pair. It lies lower in energy than other nonbonding orbitals because it lacks an appreciable antibonding contribution. It is additionally interesting to note that the 4a′ and the 6a′ can be viewed as a nonbonding N_3 orbital split by a perturbing fluorine atom. The lower energy 4a′ acquires an NF overlap while the 6a′ remains nonbonding. The principal organizing feature of the 7a′ and 8a′ orbitals is the increasing number of nodes and their concommitant antibonding character. Nonbonding portions are also prominent features. Although most molecular orbitals mix in contributions from atoms in different functional groups, and many MOs contain parts of both bond pairs and lone pairs, it is nevertheless true that, to a good approximation, the molecular orbitals for a given functional group (unnormalized) are just those contributions around the nuclear centers defining the functional group. These functional group orbitals are themselves largely transferable from one molecule to another and therefore the NF bond given in Figure 10-6 is very similar to that in NF_3 and the bonding in the N_3 fragment very similar to that in N_3^-.

Quantum chemistry has traditionally relied on the use of Mulliken overlap populations to describe bond strengths and in particular to establish simple bond length—bond order (overlap population) relationships among similar bonds. Thus it is important to note that such a relationship does not exist for normal NF bonds. Total NF overlap populations in species with representative normal NF bonds (ie, all with NF bond lengths very close of 1.37 Å) are: 0.033, 0.039, 0.060, and 0.109 electrons in NF_2, N_2F_4, NF_3, and FN_3, respectively. A counter-intuitive relationship also exists in N, O, F compounds for species with unusually long NF bonds, thus a larger overlap, 0.09 electron, is present in FNO_2 compared to 0.07 in N_2F_2, even though the latter possess the stronger bond. The failure of Mulliken population analysis in these species appears to be related to the extremely high electron density of N, O, F atoms with the consequent difficulty in achieving the high accuracy separation of lone-pair and bond-pair MO coefficients required for successful use of Mulliken overlaps in molecules made up of these atoms. It could be argued that a higher level basis (eg, 6-31G*) might alleviate this problem, but calculations at this level on other N, O, F compounds (eg, NF_4^+, NF_3) indicate that this is not the case. It seems more likely that an extensive configuration interaction treatment is required, but with such a wavefunction population analysis loses much of its meaning. Fortunately, other methods of extracting bonding information from *ab initio* molecular wavefunctions, e.g., the orbital contour plots given in this chapter, are available.

6. FONO₂

A low-precision electron diffraction structure is available for $FONO_2$, but we question the bond lengths and angles assigned to the NO_2 moiety. Again, this chemical is difficult to work with in the laboratory because it is shock sensitive.

A. Geometry

The experimental measurements upon which the currently accepted NO bond lengths are based is a 1937 electron diffraction study carried out to decide a qualitative question concerning the importance of resonance structures with like charges on adjacent atoms.[35] In the intervening years ample experimental evidence has accumulated to establish NO_2 as a very tightly bound unit which retains its NO bond length at 1.18–1.22 Å regardless of its attachment to other atoms, eg, FNO_2, N_2O_4, $HONO_2$, and N_2O_3. See Table 10-8.

The $FO'NO_2$ geometry assignment (O' = oxygen attached to F) in Reference 33 was with an FO'NO dihedral angle of 90°. Some controversy has existed about this assignment[36–40] and the most definitive spectroscopic study to date[41] indicates a planar molecule.

TABLE 10-8. Experimental Geometries
Relevant to OF and NO_2 in $FONO_2$[a]

	$R_{ON}(Å)$	$R_{FO}(Å)$	$<ONO(°)$
FNO_2	1.180	—	136.0
HNO_3	1.206	—	130.3
N_2O_3	1.211[b]	—	129.8
N_2O_4[c]	1.211[b]	—	134.0
F_2O	—	1.405	—
FOH	—	1.442	—

[a] Reference 32, except as noted.
[b] NO length in the NO_2 group.
[c] Bibart, C. H. and G. E. Ewing, *J. Chem. Phys.*, 1974, *61*, 1248; Snyder, R. G. and I. C. Hisatsumi, *J. Mol. Spec.*, 1957, *1*, 139.

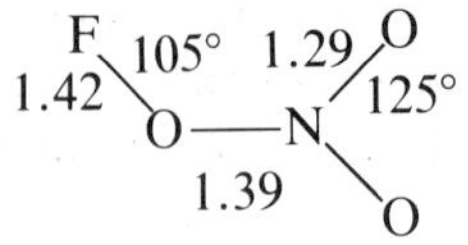

$$\text{F} \overset{105°}{\underset{1.42}{\diagdown}} \overset{1.29}{\underset{\text{O}-\text{N}}{\diagup}} \overset{\text{O}}{\underset{1.39}{\diagup}} 125°$$

The O'F bond is consistent with other molecules, eg, F_2O and FOH (Table 10-8). The NO' bond is approximately the same length as that in HO'NO$_2$, but single NO bonds can range as long as 1.46 Å (as in H$_2$NOH). Fluorine, substituting for hydrogen, acts as a σ acceptor and can induce a lengthening of an adjacent bond. However, an examination of several molecules for which both hydrogen and fluorine (or choline) analogs are known (see Table 10-9) suggests that the effect of fluorine on the NO' bond is minimal. The correct structure for FONO$_2$ is then predicted to be

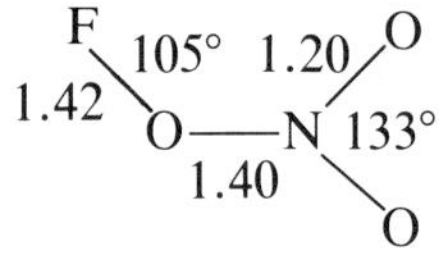

$$\text{F} \overset{105°}{\underset{1.42}{\diagdown}} \overset{1.20}{\underset{\text{O}-\text{N}}{\diagup}} \overset{\text{O}}{\underset{1.40}{\diagup}} 133°$$

The 4-31G optimized structure supports this assignment and is

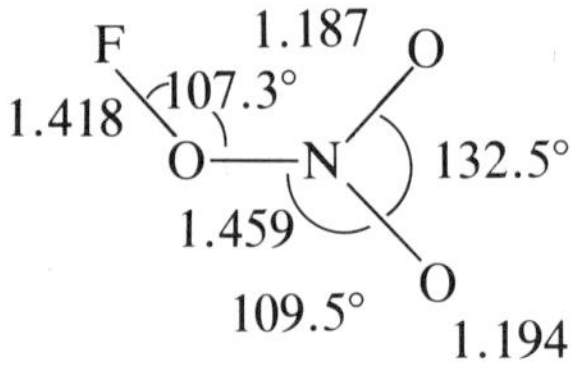

$$\text{F} \overset{107.3°}{\underset{1.418}{\diagdown}} \overset{1.187}{\underset{\text{O}-\text{N}}{\diagup}} \overset{\text{O}}{\underset{1.459}{}} 132.5°$$
$$109.5° \quad \overset{\text{O}}{1.194}$$

B. Molecular Orbitals and Bonding

Molecular orbitals for the predicted geometry have been calculated and are shown in Figure 10-7. The orbital energies are identified with each plot. The σ orbitals, not shown in Figure 10-7, can be described as follows: the 1a' is

TABLE 10-9. Bond Lengths Relevant to ON Bond Length in FNO$_3^a$

	Bond	Length
HOH	HO	0.965
HOF	HO	0.966
H$_3$COH	CO	1.425
H$_3$COCl	CO	1.418
H$_3$CNH$_2$	CN	1.471
H$_3$CNF$_2$	CN	1.450

^{a} Reference 32

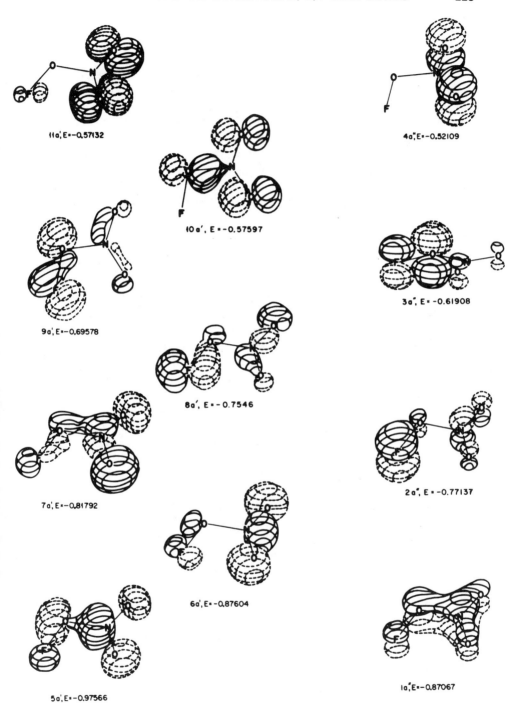

Figure 10-7. Contour plots of the valence molecular orbitals for $FONO_2$ calculated with the 4-31G basis set at the predicted geometry (orbitals, 1a′, through 4a′, not shown). One-electron energies for each orbital in hartrees.

σ bonding, encompassing all five atoms, that is, a two-electron, five-center bond. The 2a' is σ bonding over the NO_2 and O'F fragments, with a node in the NO' bond. The 3a' is a NO_2 bond with a nodal plane passing through the nitrogen. The 4a' has contributions which include the N, O', F atoms. It is bonding NO' and antibonding O'F. The charge in the antibonding O'F fragment is significant and takes charge away from the bonding in the NO' region. The remainder of the charge needed to complete the NO' bond is obtained from the 5a' and 10a'.

Since all the a'' orbitals are π orbitals, they have a common nodal plane in the plane of the molecule. Thus we describe these orbitals neglecting the molecular plane: the 1a'' has no nodes. The 2a'', 3a'', and 4a'' all have one node. The 3a'', in effect, has a single node because of the small amplitudes on the two oxygens and the absence of any contribution on the nitrogen itself. Returning to the 2a'', the node is NO' antibonding, thereby separating the two functional groups, NO_2 and O'F. It is the lowest of the three single-node orbitals (each functional group retains its own bonding characteristics). In the 3a'' we have an O'F antibond and in the 4a'', a classic NO_2 nonbonding orbital; that is, a π lone pair on the oxygens. Taking the 1a'', 2a'', and 3a'' orbitals together (that is, adding the charge density contours for all three of these orbitals together), it is apparent that the contributions from the O'F fragments cancel, as do the contributions in the NO' region. On the other hand, it is also apparent that the NO_2 fragment sums to one π bond. In net, the four a'' orbitals represent one π bond (as required by the Lewis structure for $FO'NO_2$) and three lone pairs.

The remaining seven a' orbitals are in fact lone pairs except for the "filling in" of the NO' bond by the 5a' and 10a' orbitals (noted above). For the NO_2 unit, orbitals 6a', 7a', and 11a' each contribute a lone pair of electrons. The 5a' and the 10a' contribute another between them, thereby giving four lone pairs. When added to the 4a'' orbital, these yield the five lone pairs on NO_2 required by the Lewis structure. The detailed assigning of specific lone pair contributions between the seven nonbonding σ orbitals and the three nonbonding π orbitals is somewhat ambiguous for the O'F fragment. One assignment could be the 8a' as a fluorine lone pair and the 9a' as half a lone pair, implying that the remaining one and a half lone pairs on fluorine come from the π set. For O', one-half a lone pair coming from each of the 5a', 9a', and 10a' orbitals would require the other one-half coming from the π set. Another division, assigning small contributions of fluorine orbitals among the 5a', 6a', and 7a' orbitals is possible, but this assignment is complex and probably more confusing than clarifying.

As discussed for the NF bond in FN_3, the Mulliken overlap populations for OF prove not to be particularly useful. Thus we know from its length that the OF bond is to be classified as normal and its overlap is 0.028 electron in $FO'NO_2$ compared to 0.033 in the reference molecule F_2O. The problem is that the erratic variation in other N, O, F bonds makes it uncertain whether or not the 15% difference between these two overlaps is to be regarded as significant. Mulliken net charges also suffer well-known defects that are apparent in molecules made from atoms with less extreme densities than N, O, F, but there are always selected cases where useful insight is gained. This occurs in the sequence $HONO_2$, $FONO_2$, and FNO_2 where the NO_2 group charges are -0.013, $+0.110$, and $+0.225$ electrons, respectively. The increasingly positive NO_2 charge reflects the increasing group electronegativity of OH, OF, and F, respectively.

7. NF_4^+

The synthesis of NF_4^+ salts by Christe et al.[42–48] has been an unexpected and useful development that has led to important applications, such as high-energy oxidizers. Their existence and the nature of their bonding is a challenge to chemical bonding theory because of the low basicity of difluoramine and nitrogen trifluoride, the anticipated nonexistence of the conjugate base, NF_5, and their predicted instability.[49] Reaction mechanisms proposed for their formation invoked exceptional NF compounds[50,51] which may violate the octet rule.

There have been several estimates of the NF bond length in NF_4^+, varying from 1.24 Å to 1.30 Å, based on extrapolation of force constants,[52] and from 1.30 Å to 1.40 Å, based on electron diffraction.[53] A heat of formation for gaseous NF_4^+ has also recently been reported.[54] Christe has recently reported the bond length to be 1.299 Å.[55]

A. Geometry

The experimental and computed bond lengths optimized at the 4-31G basis set level for NF, NF_2, NF_3, N_2F_2 (cis), N_2F_4 (trans), and NF_4^+ are given in Table 10-10. For the NF compounds in Table 10-10 the average deviation from experiment is 0.019 Å (this also turns out to be the deviation for our principle reference, NF_3). Applying this deviation gives a predicted bond length for NF_4^+ of 1.321 Å. Optimization at the 6-31G* level yields a bond length of 1.286 Å which is too short, as expected. However, calibration of this value against the 6-31G* and experimental bond length for NF_3 leads to an independent estimate of the NF_4^+ bond length which is 1.323 Å.

Bond length estimates are often made by interpolating or extrapolating relationships between force constants and bond lengths. Such a correlation for NF bonds using the available experimental data on NF, NF_2, NF_3, FNO, and F_3NO (Table 10-11) is shown as Figure 10-8. The relationship has only limited quantitative value and the force constant for NF_4^+ fits rather crudely. The corresponding NF bond length is 1.3 ± 0.03 Å which is at least consis-

TABLE 10-10. Experimental and Calculated NF
Bond Lengths (Å) and Total Energies (Hartrees)

Molecule	Expt[a]	Calc[b]	Energy[b]
NF	1.3173	1.371	−153.58312
NF$_2$	1.363	1.378	−252.82918
NF$_3$	1.365	1.383	−352.07605
		1.328[c]	−352.54005[c]
N$_2$F$_2$(cis)	1.410	1.403	−307.14607
N$_2$F$_4$(trans)	1.372	1.387	−505.62331
NF$_4^+$	1.299	1.340	−450.88237
		1.286[c]	−451.51971[c]

[a] Geometry for NF: Douglas, A. E. and W. E. Jones,
Can. J. Phys., 1966, 44, 2251; NF$_2$: Bohn, R. K. and
S. H. Bauer, Inorg. Chem., 1967, 6, 304; NF$_3$: Sheridan, J. and W. Gordy, Phys. Rev., 1950, 79, 513;
N$_2$F$_2$: Bohn, R. K. and S. H. Bauer, Inorg. Chem.,
1967, 6, 309; N$_2$F$_4$: Gilbert, M. M., G. Gundersen,
and K. Hedberg, J. Chem. Phys., 1972, 56, 1961;
NF$_4^+$: Reference 52.
[b] 4-31G optimized values unless specified otherwise.
[c] 6-31G* optimized values.

TABLE 10-11. Experimental NF Force
Constants (mdynes/Å) and Bond Lengths (A)

Molecule	Force constant[a]	Bond length[b]
NF	5.90	1.317
NF$_2$	5.10	1.363
NF$_3$	4.31	1.365
ONF	2.15	1.512
ONF$_3$	4.25	1.432

[a] NF: Milligan, D. E. and M. E. Jacox, J.
Chem. Phys., 1964, 40, 2461; NF$_2$: Harmony,
M. D. and R. J. Meyers, ibid., 1962, 37, 636;
NF$_3$ and ONF: Sawodny, W. and P. Pulay, J.
Mol. Spectrosc., 1979, 51; ONF$_3$: R. P.
Girshmann, D. F. Harnish, J. R. Holmes, J.
S. MacKenzie, and W. B. Fox, Appl. Spec.,
1969, 23, 333.
[b] NF, NF$_2$, NF$_3$: see footnote a of Table 10-11;
ONF: Millen, D. J., K. S. Buckton, and A. C.
Legon, Trans. Farad. Soc., 1969, 65, 1975;
ONF$_3$: Plato, V., W. D. Hartford, and K.
Hedberg, J. Chem. Phys., 1970, 53, 3488.

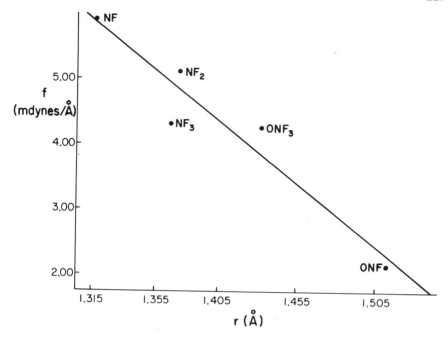

Figure 10-8.   Plot of NF force constant in mdynes/Å versus NF bond length in Å for NF, NF_2, NF_3, FNO, and F_3NO.

tent with our other estimate, but offers only a small degree of discrimination between previous predictions.

B. Bonding

Bond Energies

Energies at the 6-31G* level, as well as the experimental heats of formation for NF_3, NH_3, and NH_4^+, were used to estimate the heat of formation of NF_4^+ according to the following isodesmic reaction:

$$3NH_4^+ + 4NF_3 \rightarrow 3NF_4^+ + 4NH_3 \qquad (10\text{-}5)$$

As shown in Table 10-12, the results from our 6-31G* calculations are in good agreement with a recent experimental determination by Christe.[54] Using the experimental estimate of the heat of formation of NF_4^+, the bond energy can be obtained from Equation 10-6 as 329 kcal/mol, an average of 82.3 kcal/mol/bond.

$$NF_4^+ \rightarrow N^+ + 4F \qquad (10\text{-}6)$$

TABLE 10-12.   Energies for NH_3, NH_4^+, NF_3 and NF_4^+

Molecule	$E_T(4\text{-}31G)^a$	$E_T(6\text{-}31G^*)^a$	$\Delta H_f^\circ(298)^{b,c}$	Binding $E^{b,d}$	Avg. NF energy^{b,d}
NH_3	-56.10669^e	-56.18434^f	-11.02	—	—
NH_4^+	-56.45888^e	-56.53077^f	$+150.00$	—	—
NF_3	-352.07605	-352.54005	-29.8	199.2	66.4
NF_4^+	-450.88237	-451.52010	$+195.6^g$	329.5	82.4
			$(220.1)^h$	$(305.0)^n$	$(76.3)^h$

^{a} In Hartrees
^{b} In kcal/mol
^{c} Wagman, D. D., W. H. Evans, I. Halow, V. B. Parker, S. M. Bailey, and R. H. Schumm, *Selected Values of Chemical Thermodynamic Properties*, National Bureau of Standards.
^{d} Calculated from the $\Delta H_f^\circ(298)$ using the $\Delta H_f^\circ(298)$ of N, N$^+$, F from footnote C.
^{e} Lathan, W. A., W. J. Hehre, L. A. Curtiss, and J. A. Pople, *J. Am. Chem. Soc.,* 1971, *93,* 6377.
^{f} Hariharan, P. C. and J. A. Pople, *Mol. Phys.,* 1974, *27,* 209.
^{g} Reference 52.
^{h} Numbers in parentheses from 6-31G* calculations.

This latter value is to be compared with the calculated value of 76.3 kcal/mol/bond (Table 10-13). It may be noted these bond energies are significantly larger than the 66.4 kcal/mol in NF_3.

Another interesting bond energy question that can be answered by our calculations is the useful hypothesis put forth by Christe et al., that removal of a fluorine from NF_4^+ requires approximately the same energy as removal of a fluorine from NF_3. We exploit the fact that the increasing energy required to remove successive fluorine atoms from NF_3 is paralleled by an increasing NF force constant. The force constant for NF_4^+ turns out to be even greater than that for NF and extrapolation yields an NF_4^+/NF_3 fluorine removal energy ratio nearer to 1.2–1.3 than Christe's estimate of unity.

C. Comparison with NF_3 and F^+

Christe[51] has rationalized the bonding in NF_4^+, relative to NF_3, in the following manner: the polarity of the NF bonds in NF_3 is N^+F^-; therefore the addition of F^+ to NF_3 should reduce the polarity and induce a stronger, more

TABLE 10-13.   Population Analysis

	$q(F)$	$q(N)$	NF_4^+ NF overlap	$q(F)$	$q(N)$	NF_3 NF overlap
4-31G	-0.026	$+1.103$	0.092	-0.256	$+0.769$	0.060
6-31G*	-0.121	$+1.485$	0.281	-0.284	$+0.851$	0.127
6-31G*/4-31G		1.35	3.05		1.11	2.11

covalent bond. This model is consistent with the increase in the NF force constant from NF_3 to NF_4^+ and with Mulliken population data collected in Table 10-13. For both basis sets the fluorine is more negative in NF_3 and the NF overlap less. The ratio of 6-31G*/4-31G values in Table 10-13 indicated the population analysis is yielding results of comparable quality for NF_4^+ and NF_3. However, we put much greater emphasis on the molecular orbital contour maps discussed below.

A valence molecular orbital correlation diagram for NF_3 and NF_4^+ is shown in Figure 10-9, with the NF_4^+ orbitals labelled as C_{3v} to match NF_3. A dominating feature of the diagram is the nearly constant energy lowering of all the NF_4^+ levels relative to NF_3 due to the overall increase in the strength of the central field around nitrogen imposed by the positive charge. Likewise

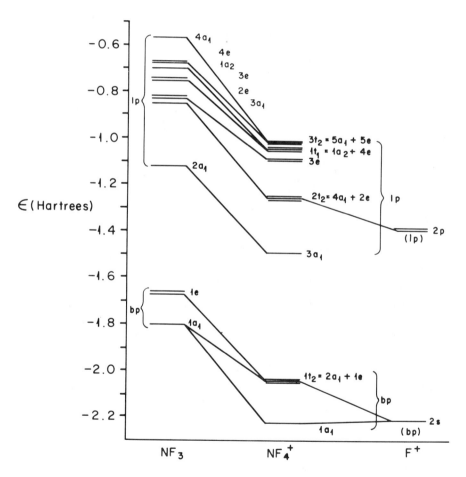

Figure 10-9. Valence molecular orbital correlation diagram for NF_3, NF_4^+, and F^+ (obtained with the 4-31G basis set). NF_4^+ orbitals labelled as C_{3v} to match NF_3.

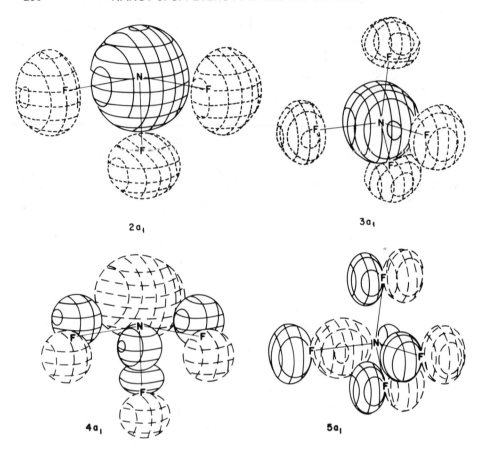

$2a_1$

$3a_1$

$4a_1$

$5a_1$

Figure 10-10.   Contour plots of valence molecular orbitals for NF_3 and NF_4^+ (4-31G basis set). Top: The $2a_1$ of NF_3 and the $3a_1$ of NF_4^+. Bottom: The $4a_1$ of NF_3 and the $5a_1$ of NF_4^+.

the shapes of the orbitals in the two species are quite similar (see Figure 10-10[34]). As an example, the $2a_1$ of NF_3 and the $3a_1$ of NF_4^+ (Figure 10-10) are seen to be identical except for the consequence of adding fluorine in NF_4^+: the nitrogen atom has a large s-type orbital that is out of phase with the s-type orbitals on the fluorines.

Consider now the HOMO of our principal reference species, NF_3 compared with that of NF_4^+ (Figure 10-10). The traditional lone pair on NF_3 is accompanied by three smaller, tangential p orbitals on the fluorines. When this orbital becomes the a_1 component of the NF_4^+ HOMO, the three tangential orbitals are retained, but the nitrogen lone pair has become partly an NF bond and partly a fluorine lone pair. This orbital, as well as the eleven other highest occupied orbitals, is classified as nonbonding but nevertheless contains significant NF bonding—a characteristic mixing of bonding and nonbonding features found in many molecular orbitals. Another instructive example of orbital mixing occurs in the NF_4^+ $2t_2$ orbital (Figure 10-11). This

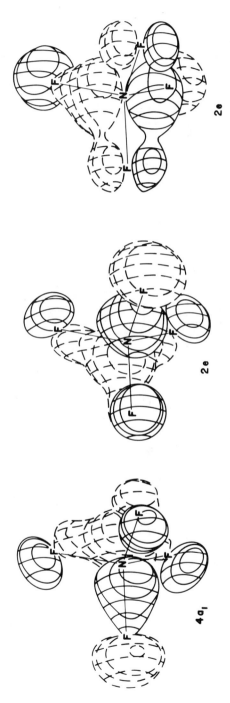

Figure 10-11. Contour maps of $2t_2$ valence MOs of NF_4^+ (4-31G). Orbitals are composed of three tangential lone-pair p's plus one radial bonding p. Radial bonding p along left NF bond in first panel; right NF bond, second panel; upper NF, third panel.

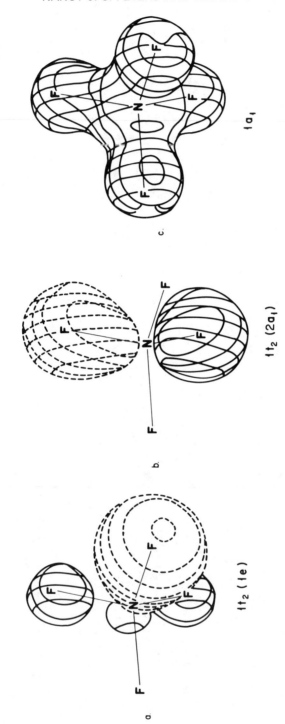

Figure 10-12. Contour maps of NF_4^+, $1t_2$, and $1a_1$ (4-31G). a. One of the 1e set in $1t_2$ b. The $2a_1$ in $1t_2$ c. The $1a_1$

orbital is of special interest because it mixes the NF_3 $3a_1$ with the 2p of F^+. Each t_2 is composed of three medium-sized tangential fluorine p orbitals with a fourth somewhat larger fluorine radial p overlapping a nitrogen p along an NF line. This again illustrates the mixture of a bonding component into nominally lone-pair orbitals.

Finally, a particularly interesting illustration of the molecular orbital representation of bonding is given by the four sigma bonds in NF_4^+. These arise from a mixing of the $1a_1$ and $1e$ of NF_3 with the 2s of F^+, resulting in the $1a_1$ and $1t_2$ of NF_4^+. The contributing NF_3 orbitals follow the well-known triangle phase pattern: the $1a_1$ consists of in-phase atomic orbital contributions from all three fluorine atoms; one of the $1e$ orbitals consists of medium-sized in-phase atomic orbitals from two of the fluorines with a larger out-of-phase component around the remaining fluorine atom and also the NF bond, the other consisting of two rather large out-of-phase atomic orbitals around two of the fluorines, extending along the NF bonds. The result of the mixing with the F^+ 2s is the three orbitals of $1t_2$: two as in Figure 10-12a, the other like Figure 10-12b. The $1a_1$, of course, is totally symmetric (Figure 10-12c). These four orbitals (eight electrons) together comprise the four sigma bonds.

The MO contour maps shown here (as well as the MO energy level diagram, bond lengths, force constants, binding energies, and Mulliken population analysis) support Christe's simple model that addition of F^+ to NF_3 produces NF bonds that are more stable and more covalent. Moreover, examination of the shapes and energies of the highest occupied MOs helps explain the bonding in NF_4^+ salts. Thus the top eight molecular orbitals are nonbonding, rather close is energy and separated by an energy gap (Figure 10-9) from those of lower energy. Together they are spherically symmetric and form a closed shell of radius slightly greater than the NF bond length. By Gauss's Law they act electrostatically as if all of their charge were concentrated at the nitrogen nucleus.

8. F_2NNO_2

F_2NNO_2 and F_2NNO are known from synthetic efforts,[2] but their structure and bonding have not heretofore been reported. These may be thought of as complexes of the radicals NF_2 with NO_2 or NO. Among N, O, F compounds, NO, NO_2, NF_2, and OF are the principal radicals, and they have stabilities ranging from high to low in the order given. When pairs of these radicals form complexes, unexpected properties arise: thus, those dimers made of monomers with the most tightly bound atoms, N_2O_2, N_2O_3, N_2O_4, possess exceptionally long NN bonds and low dissociation energies (Table 10-14). On the other hand, the dimer between the most weakly bound radical and a strongly bound one, $FONO_2$, is connected by a normal NO single bond, while the OF dimer, FOOF, contains an exceptionally short central bond.

TABLE 10-14. NN Bond Energies (kcal/mol) and Bond Lengths (Å) for Molecules with NF_2, NO, and NO_2 Fragments

Molecules	Bond	Energy[a]	R_{expt}[b]	R_{calc}[c]
N_2F_4	NN	21.0	1.492	1.470
	NF		1.372	1.387
N_2O_2	NN	—	2.18 (cis)	1.162 (trans)[d]
N_2O_3	NN	9.5	1.864	
N_2O_4	NN	12.9	1.75	1.631[e]
NF_3	NF	—	1.365	1.383
FNO	NF	—	1.512	1.468[f]
FNO_2	NF	—	1.467	1.407
H_2NNO_2	NN	—	1.427	—

[a] N_2F_4: Darwent, B. de B., *Bond Dissociation Energies in Simple Molecules*, NSRDS-NBS, 1970, *31*; N_2O_3 and N_2O_4: Kerr, J. A., *Chem. Rev.*, 1966, 465.

[b] N_2F_4: Gilbert, M. M., G. Gunderson and K. Hedberg, *J. Chem. Phys.*, 1972, *56*, 1691; N_2O_2: Reference 64; N_2O_3: Brittain, A. H., A. P. Cox, and R. L. Kuczkowski, *Trans. Farad. Soc.*, 1969, *65*, 1963; N_2O_4: Bibart, C. H. and G. E. Ewing, *J. Chem. Phys.*, 1974, *61*, 1248 and R. G. Snyder and I. C. Hisatsune, *J. Mol. Spectrosc.*, 1957, *1*, 139; NF_3: Sheridan, J. and W. Gordy, *Phys. Rev.*, 1950, *79*, 513; FNO: Millen, D. J., K. S. Buckton, and A. C. Legon, *Trans. Farad. Soc.*, 1969, *65*, 1975; FNO_2: Legon, A. C., and D. J. Miller, *J. Chem. Soc. A*, 1968, 1736; $H_2N_2O_2$: Tyler, J. K., *J. Mol. Spectrosc.*, 1963, *11*, 39.

[c] This work except as noted.

[d] Dykstra et al. with a basis set roughly comparable to 4-31G, obtain NN separation of ^1A_g(trans) = 1.542 A; $^1A_1$(cis) = 1.615 A.

[e] Reference 51.

[f] Ditchfield, R., J. A. Pople, and J. Del Bene, *J. Am. Chem. Soc.*, 1972, *94*, 4806.

The FONO and $FONF_2$ isomers do not exist (they exist as FNO_2 and F_3NO), but calculations[56] show $FONF_2$ to be nonplanar with the same NO bond length as $HONF_2$ (1.35 Å) but shorter than that in $HONH_2$ (1.45 Å). F_2NNF_2, the nonplanar dimer of moderately stable NF_2, had a normal NN single bond length (1.49).

The conformations of the radical complexes likewise are of interest, e.g., the planarity of N_2O_4 and its justification have been the subject of much study and many hypotheses.

The only study of the properties of F_2NNO_2 and F_2NNO is a calculation using the semi-empirical MINDO techniques.[57] Calibration of its adjustable parameters against heats of formation in related N, O, F compounds, plus the assumption of normal NN single bond lengths, enabled this scheme to approximately reproduce the experimental heat of formation in F_2NNO, but failed for F_2NNO_2. *Ab initio* calculations are needed to provide optimized structures to distinguish between weak and normal or normal and strong bonds.

A. Geometry

The predicted minimum energy structure for F_2NNO_2 is planar and given below:

$$
\begin{array}{ccc}
\text{F} \ 1.350 & 1.215 & \text{O} \\
115.8° \quad \diagdown & \diagup & \\
\text{N} \text{---} \text{N} & & 128.5° \\
\diagup & 1.353 \quad \diagdown & \\
\text{F} & & \text{O}
\end{array}
$$

The NF bond lengths are normal (1.365 Å, NF_3 as reference) and very close to those in the free NF_2 radical (1.36 Å), while the NN bonds are notably shorter than those in the NN single bond reference molecules N_2H_4 (1.45 Å) and N_2F_4 (1.49 Å), but longer than a full NN double bond (1.25 Å in H_2N_2).

The conventional Lewis resonance structures are:

$$
\begin{array}{ccc}
\text{F} \quad \text{O} & \text{F} \quad \text{O}^- & \text{F} \quad \text{O}^- \\
\diagdown \ _+ \ \diagup\!\!\diagup & \diagdown \ _+ \ \diagup & \diagdown + \ + \ \diagup \\
\text{N}\text{---}\text{N} & \text{N}\text{---}\text{N} & \text{N}\!=\!\text{N} \\
\diagup \quad \diagdown & \diagup \quad \diagdown\!\!\diagdown & \diagup \quad \diagdown \\
\text{F} \quad \text{O}^- & \text{F} \quad \text{O} & \text{F} \quad \text{O}^-
\end{array}
$$

As will be apparent from the results in the sections below, it is remarkable how much insight into the bonding these structures yield, although this success must be regarded as a posteriori since the formal charges do not give an adequate guide as to the weight to be assigned the several structures. Fluorine and oxygen no-bond structures might also be added to the set above, but prove to have no weight in the actual molecules.

B. Bonding

Conjugation

In his recent comprehensive treatment of 34 valence electron A_2X_4 species Schleyer et al.[58] lists and discusses four effects responsible for the stabilization of the D_{2h} (planar) conformation versus the D_{2d} (perpendicular): π conjugation, electrostatic effects, 1,4 interactions, and hyperconjugation. We likewise adopt these categories to structure the analysis of our 36 valence electron $X_2A_2Y_2$ system. The energy difference between optimized planar and perpendicular F_2NNO_2 is 19.2 kcal/mol and favors the planar form. π conjugation is the principal driving force for achieving the planar conformation in F_2NNO_2 but, as we will describe in the section on electrostatics, the actual stabilization energy is about equally divided between π conjugation and net attractive electrostatic interactions. The molecular orbital correlation diagram for the occupied MOs of planar and perpendicular F_2NNO_2 is shown in Figure 10-13. The MOs for the two radicals which combine to form

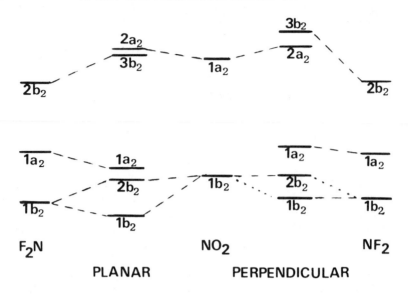

Figure 10-13. π molecular orbital correlation diagram for F_2NNO_2 in the planar and perpendicular conformations. Orbitals labelled in C_{2v} symmetry.

F_2NNO_2 can be qualitatively represented by the familiar three-center, three-orbital, bonding, nonbinding, antibonding schematic shown in Figure 10-13.

In the NF_2 fragment all three levels are doubly occupied, while in the NO_2 fragment the $2b_2$ is an unoccupied LUMO and is not shown in Figure 10-13. The a_2 levels are lone pairs and are shown in Figure 10-13 for completeness, but they are not germane to the analysis and are not discussed further. The $1b_2$ in planar F_2NNO_2 is the bonding combination of the $1b_2$ from NF_2 and the $1b_2$ from NO_2, while the $2b_2$ in the complex is the antibonding combination of these same radical orbitals with the phases of NO_2 reversed. The $3b_2$ arises largely from the NF_2 $2b_2$ and the NO_2 $2b_2$ LUMO and is therefore bonding NN, but antibonding in the NF and NO bonds. The positions of the F_2NNO_2 $1b_2$, $2b_2$, and $3b_2$ levels relative to the $1b_1$ and $2b_2$ in NF_2 and in NO_2 measure the π stabilization and they are governed by the familiar balance between the repulsion of doubly occupied pairs of orbitals against the energy lowering of unoccupied–occupied level interactions. The two NF_2 levels and two NO_2 levels result in four F_2NNO_2 levels (the $4b_2$ LUMO is not shown in Figure 10-13) with the usual strong stabilization of the lowest level ($1b_2$). The $2b_2$ of the complex is in effect lowered by its interaction with the NO_2 $2b_2$ LUMO. The orbital amplitudes on N are out-of-phase between the $1b_2$ and $2b_2$ LUMO of NO_2 and their mixing in the complex reduces the antibonding interaction across NN (and enhances the NO bonding). The $3b_2$ is NN bonding and therefore also contributes to π conjugation. Rotation from the planar to perpendicular conformation turns off the NN π bond and the level spacings on the right side of Figure 10-2 clearly show the consequences: there is

TABLE 10-15. Mulliken Charges and Overlap Population for N, F Molecules

Molecule	q_N	q_N	q_E	NF overlap	NN overlap
F_4N_2	+0.280	—	−0.280	0.039	−0.002
F_2N_2	+0.486	—	−0.243	0.068	0.076
$F_2NN'O$ (planar)	+0.409	+0.382	−02.10	0.075	−0.110
$F_2NN'O$ (perpendicular)	+0.331	+0.300	−0.217	0.085	−0.102
$F_2NN'O_2$ (planar)	+0.529	+0.521	−0.191	0.086	−0.163
$F_2NN'O_2$ (perpendicular)	+0.463	+0.463	−0.196	0.097	0.035

little mixing between the NF_2 $1b_2$ and NO_2 $1b_2$, thus the F_2NNO_2 $1b_2$ levels are close to those of the NF_2 $1b_2$ and NO_2 $1b_2$, respectively. The $3b_2$ is also less stable. The effect of π conjugation is manifest in the NN bond length change from 1.353 Å in the planar to 1.40 Å in the perpendicular conformation.

Electrostatic Effects

It is a trivial consequence of trigonometry that the terminal atoms in $A_2X_2Y_2$ and A_2X_4 systems are farther apart in the perpendicular conformation than in its planar rotomer. Therefore when X and Y are electronegative relative to A, electrostatic repulsion will favor the perpendicular structure and this has been an implicit or explicit assumption in most explanations to date. This assumption is true only if the atomic charges in both configurations are nearly the same. In order to investigate the effect of conformation-dependent atomic charge change we first calibrate the Mulliken net charges (Tables 10-15 and 10-16 give Mulliken populations for N, O, F compounds). One simple test is to use the net charges directly and calculate the molecular

TABLE 10-16. Mulliken Charges and Overlap Population for N, O Molecules

Molecule	q_N	$q_{N'}$	q_O	q_{NO}	q_{NO}	NO overlap
F_3NO	+0.872	—	−0.306	+0.566	—	0.105
FNO	+0.562	—	−0.093	+0.369	—	0.088
$F_2NN'O$ (planar)	+0.409	+0.482	−0.369	+0.013	—	0.161
$F_2NN'O$ (perpendicular)	+0.331	+0.300	−0.177	+0.23	—	0.069
$FO'NO_2$	+0.686	—	−0.288	—	+0.110	0.070
FNO_2	+0.790	—	−0.283	—	+0.234	0.089
$F_2NN'O_2$ (planar)	+0.529	+0.521	−0.334	—	−0.147	0.070
$F_2NN'O_2$ (perpendicular)	+0.463	+0.463	−0.267	—	−0.071	0.079

dipole moment. The result is 2.17 Debye compared to the dipole operator expectation value of 2.03 D calculated directly from the 4-31G molecular wavefunction. Although encouraging, it is clear that this indicates that the 4-31G Mulliken charges are too large because the contribution of the F and O lone pairs is inadequately taken into account: if they were, the dipole moment would be considerably increased. A similar comparison holds for other N, O, F molecules. Similar calculations on the dipole moments of F_3NO and FNO give values of 0.01 D and 2.51 D, respectively, in reasonable agreement with the experimental values of 0.04 D and 1.81 D, respectively. Another estimate can be obtained by comparing STO-3G and 4-31G charges, since it has been frequently argued that the short radial tails of the STO-3G basis largely eliminates the inaccuracy of the arbitrary Mulliken overlap division. STO-3G charges are almost uniformly half the 4-31G values, therefore suggesting that electrostatic interaction energies obtained from 4-31G charges may be up to four times too large.

We now carry out a set of calculations to separate the electrostatic effects due purely to geometry change from those of conformation-dependent atomic charge changes. 1) We calculate the electrostatic energy for the equilibrium perpendicular conformation, then rigidly rotate to the planar form and recompute the electrostatic energy using the same charges found for the perpendicular conformation. The energy difference is 1.2 kcal/mol favoring the perpendicular conformer as expected. 2) Repeat 1) but using the equilibrium planar geometry in the planar form. Again the perpendicular form is favored, now by 5 kcal/mol. 3) As a check, 2) is repeated using the planar charges for both conformations. The result is essentially identical: 6 kcal/mol favoring the perpendicular conformer. 4) We calculate the electrostatic energy difference between the perpendicular form in its equilibrium geometry with its equilibrium charges and the planar conformation at its equilibrium geometry with equilibrium charges. The result favors the planar form by 20 kcal/mol, thereby indicating that the atomic charge redistribution in the planar form is responsible for $20 + 5 = 25$ kcal/mol stabilization energy favoring the planar form. The calculations are summarized in Table 10-17. Since Mulliken net charges from the 4-31G basis set were used to generate the energy values quoted, we can expect that the actual stabilization of the planar form due to electrostatic effects ranges from perhaps 6–10 kcal/mol. Recalling that the total energy difference between the planar and perpendicular conformations is 19.2 kcal/mol, it is apparent that π conjugation and electrostatic attraction are comparable in magnitude.

The origin of these attractive electrostatic interactions are the increases in negative charge on the oxygens and positive charge on the nitrogens in the planar conformer. The fluorine charges remain unchanged and there is only a very small elongation of the NF bond length (1.350 to 1.362 Å) and NO bond lengths (1.208 to 1.215), with a corresponding very small increase in NF and NO overlaps, in going from the perpendicular to planar conformation. In the planar conformation the NO_2 $2b_2$ LUMO makes an important contribution to

TABLE 10-17. F_2NNO_2 Electrostatic Energy Differences (kcal/mol)[a]

Configurations	$r(N{-}N)$ perpendicular	$r(N{-}N)$ planar	Charges perpendicular	Charges planar	Preferred configuration	ΔE
1. Rigid rotation, fixed charges	1.404Å	1.404Å	perp.	perp.	perp.	1.2
2. Equil. geometries, fixed charges (perpendicular)	1.404	1.353	perp.	perp.	perp.	5.0
3. Equil. geometries, fixed charges (planar)	1.404	1.353	planar	planar	perp.	6.0
4. Equil. geometries, equil. charges	1.404	1.353	perp.	planar	planar	20.0

[a] Energies determined using Mulliken charges generated by the 4-31G basis set. E_T(perpendicular) = -456.4149 hartrees, μ(perpendicular) = 1.091D E_T(planar) = -456.44560 hartrees, μ(planar) = 2.034 D

the oxygen charge increase by increasing the oxygen π orbital coefficient and the positive charge increase of the NO_2 nitrogen by decreasing its π orbital coefficient. The reduction in NN bond lengths also causes a σ bond change that makes the nitrogens more positive and the oxygens more negative. The overall driving force is the formation of the NN π bond with the consequences enumerated above. Thus π conjugation and the electrostatic attraction are really part of the same charge delocalization phenomenon. This generalized delocalization is schematically shown by the third of the three Lewis resonance structures shown previously and its seemingly unfavorable formal charges are, in fact, indicative of a significant attractive electrostatic stabilization.

1,4 Overlaps

The 1,4 (O,F) overlap of p orbitals parallel to the NN axis, suggested by Gimarc[59] and Epiotis[60] and evaluated by Schleyer[58] as the most important factor stabilizing D_{2h} N_2O_4, relative to D_{2d}, is approximately the same magnitude (0.004) in the F_2NNO_2 as in N_2O_4. The total O, F overlap is 0.006 showing that the p_z contribution is the largest. There are two problems with this explanation: 1) The overlaps are of similar magnitude in N_2O_4 and F_2NNO_2 but the barrier in the latter is three times that in the former. 2) In F_2NNO_2 the overlap magnitude is one-twentieth that of the FN, NN, or NO bonds yet the conformation barrier is a much larger fraction of these covalent bond energies.

Hyperconjugation

Fluorine or negative hyperconjugation has been discussed many times and a good presentation of the orbital interactions to be expected along with relevant reference has been given.[61] In F_2NNO_2, it is the perpendicular, and not the planar, form that is properly oriented for fluorine hyperconjugation. This effect should be negligible in the planar form. Its negligible contribution is further supported by the normal NF bond lengths obtained, 1.350 for both conformers.

9. F_2NNO

A. Geometry

The 4-31G optimized geometry for F_2NNO is given below:

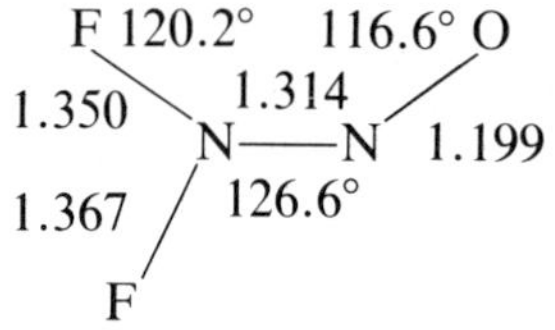

$$
\begin{array}{l}
\text{F } 120.2° \quad 116.6° \text{ O} \\
1.350 \diagdown \quad {}^{1.314} \diagup \\
\qquad \text{N——N } 1.199 \\
1.367 \diagup \quad 126.6° \\
\qquad \text{F}
\end{array}
$$

along with appropriate Lewis resonance structures:

B. Bonding

Molecular Orbital Description and π Conjugation

In this section we give a rather complete description of the valence molecular orbitals of F_2NNO. Our discussion includes both σ and π orbitals, thus going beyond that given for F_2NNO_2 and our motivation is two-fold: 1) We make a detailed comparison with the well-understood HNO molecule. F_2NNO (and F_2NNO_2) and HNO have normal-length bonds even though they are unstable in the laboratory, and establishing close connection with HNO helps support the validity of our calculations. 2) We illustrate the relationship between Lewis resonance structures and molecular orbitals showing not only how the Lewis description provides a succinct and facile summary of the electronic structure for these previously uncharacterized species, but also how the more complete information available from the molecular orbital description leads to greater insight than the Lewis picture.

F_2NNO has C_s point group symmetry, therefore just two symmetry labels, $a'(\sigma)$ and $a''(\pi)$. Figure 10-14 gives the molecular orbital diagram along with that for HNO. There is a large energy gap between the $3a''$ and $11a'$ of

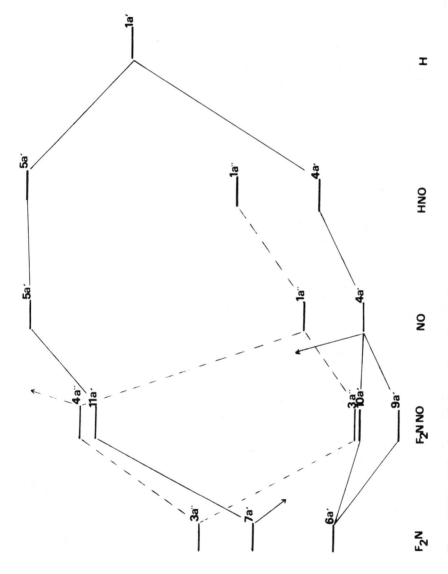

Figure 10-14. Molecular orbital correlation diagram for planar F_2NNO and HNO. Solid lines, σ interactions; dashed lines, π. Orbitals labelled in C_s symmetry.

F_2NNO, comparable to the 1a" to 5a' gap in HNO and the gap between the 1a" and 5a' orbitals of NO; in fact, this is the largest energy gap present in the valence orbitals. The gap in NO is the result of one electron occupying an antibonding orbital. Likewise, the 11a' of F_2NNO and the 5a' of HNO have large antibonding interactions between nitrogen and oxygen lone pairs. On the other hand, the 4a' of HNO and the 10a' of F_2NNO are the in-phase combination of nitrogen and oxygen lone pairs.

The normal NO π bond is accounted for by the 1a" of HNO and the 3a" of F_2NNO. The 1a" and 2a" of F_2NNO are largely the NF_2 π orbitals: that is, the 1a" is bonding FNF and the 2a" consists of two out-of-phase p orbitals on the fluorines. The π conjugation across the NNO fragment of F_2NNO is evident in the small, in-phase π contributions on these atoms in 1a", 2a", and 3a" orbitals. The 4a" orbital is bonding NO, antibonding NF, and antibonding NN. Thus the F_2N π HOMO and the π HOMO of NO give rise to a 4-electron interaction which is mitigated by the 2-electron attractive interaction of the F_2N HOMO and the NO π LUMO. The latter orbital in effect lowers the energy of the F_2NNO 4a" HOMO. The resultant π conjugation is greater in F_2NNO than in F_2NNO_2 because it occurs between the radical HOMOs rather than one level lower as in F_2NNO_2. The sum of the 1a", 2a", 3a", and 4a" orbitals (that is, adding the electron density for the four orbitals together) gives a net π bond over the NO fragment, no net π bonds over the NF_2 fragment, and a partial π bond over the NN fragment.

The σ bonds in HNO are the 1a', the NO bond, and the 2a', the NH bond. The 3a' is the remaining lone pair as required by the Lewis structure. For F_2NNO the four σ orbitals can be identified as follows: the 1a' and 2a' make up the two NF bonds of the NF_2 fragment; the 3a' is the NO bond; and the 4a' is bonding over all five atoms, that is, it is responsible for the NN bond. The remaining σ orbitals, 5a' through 9a', are largely lone pairs, although as usual in many NOs, there is some mixing of bonding and nonbonding features. The 7a' can be singled out as having the largest oxygen contribution, thus it completes the lone pairs required on oxygen (that is, the 10a' and 11a' constitute one nitrogen and one oxygen lone pair) by the dominant Lewis structure (first structure of resonance set in geometry section). The 5a', 6a', 8a', and 9a' complete the fluorine lone pairs along with the two obtained from the a" set. To understand, in molecular orbital terms, how both the first and second Lewis structures come about simultaneously, we recall that many canonical MOs contain both lone-pair and bond-pair contributions. Thus the lone pair on the nitrogen of NF_2 is oriented so that it can delocalize to contribute to the formation of a partial π bond with the nitrogen of the NO group and the electrons of the NO bond can also be part of the lone-pair p orbital on oxygen.

The MO description above also bears on the qualitative orbital theory that Pimentel and Spratley have frequently put forth.[62] Motivated by the unusually long NN bond in $(NO)_2$ and FN bond in FNO, they have postulated the existence of a weak interaction between the antibonding π HOMO of NO with a p orbital of attached atoms or radicals whose electronegativity is high. Thus F_2NNO would be predicted to have an unusually long NN bond, but as noted above, the NO interaction in all of the a″ orbitals of F_2NNO are in-phase, thus in contradiction to the Pimentel and Spratley hypothesis. The origin of this failure is their attempt to attribute the interaction to a single imagined orbital while, in fact, several orbitals, each with somewhat different properties, are responsible for the patterns observed. A similar short-coming of their model in FNO has been pointed out by Grein et al.[26]

Electrostatic Effects and 1,4 Overlaps

The electrostatic effects in F_2NNO are very similar to those in F_2NNO_2 as indicated in Table 10-18. The shorter NN bond length in F_2NNO is reflected in a higher rigid rotation barrier. When the effects of geometry are separated from those of charge redistribution (middle entries in Table 10-18), we again see that the change in atomic charge magnitudes with change in conformation is by far the dominating influence. The information we have available for F_2NNO is less complete than for F_2NNO_2 because we were unable to obtain an optimized solution for the perpendicular conformation. This may well be an artifact of our SCF coverage schemes or a property of the basis set. There is no question, however, that the electrostatic conclusions for F_2NNO and F_2NNO_2 are parallel. As noted in the MO description, there is a generalized delocalization of charge over the NNO framework in the planar form that differentially increases the negative charges on O and decreases it on N and N′ over that of the perpendicular form. Again we estimate that the attractive electrostatic effects are responsible for one-quarter to one-half of the barrier.

TABLE 10-18.   Comparisons Between F_2NNO and F_2NNO_2

Rigid rotation barrier (planar, equilibrium geometry) E_T(kcal/mol)

F_2NNO	42.1
F_2NNO_2	20.7

Electrostatic energy differences, E(kcal/mol)

Rigid rotation, planar equilibrium geometry, planar charges on both conformers. Perpendicular geometry favored in both molecules.

F_2NNO	0.33
F_2NNO_2	0.63

Rigid rotation, planar equilibrium geometry, planar charges for planar conformer, perpendicular charges for perpendicular conformer.

F_2NNO	32.0
F_2NNO_2	23.4

1,4 Overlaps (in planar equilibrium geometry) parallel to NN axis and total, respectively.

F_2NNO	+0.0047	+0.0082
F_2NNO_2	+0.0040	+0.0060

The 1,4 overlaps are also comparable in F_2NNO and F_2NNO_2 and again there are the questions of their small magnitude and the uncertain relationship between them and calculated barrier heights. An additional feature is the rather sensitive dependence of this overlap on the difference in electronegativity between the central atoms (N) and the terminal atoms (F and particularly O) noted by Gimarc.[59] Thus it is not clear whether the small difference in p_z overlaps (those parallel to the NN axis) is related to the barrier heights or to the difference in effective electronegativity of N′ in $F_2NN\overset{-}{O}$ and F_2NNO_2.

Hyperconjugation

Because NNO is a strongly bound linear molecule, one can ask whether this 16-electron fragment might be linear in F_2NNO. While it is immediately apparent that because of the octet rule, attachment of the fluorines prevents stabilization of a triple bond resonance structure between the nitrogens, double bonds between both NN and NO would be possible. This linear structure, however, implies use of p ortibals exclusively and is less stable than a bent configuration with a lone pair on the central nitrogen in place of a NN π bond. The determining factor is the large energy lowering realized when the nitrogen 2s orbital is mixed into the lone-pair hybrid orbital. The stability provided by the nitrogen 2s is also the basic reason why hyperconjugation is not an important factor in this molecule and thus why F_2N retains its free radical geometry when complexed with NO. In terms of Lewis structures the MO-bond resonance form:

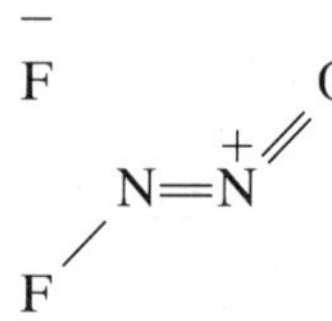

would be a very high energy noncontributing structure because of the large energy required to bend NNO without introducing a lone pair on the central nitrogen.

Triplet State

Given the unusual nature of the bonding between N, O, F radicals, it is conceivable that removal of spin pairing energy by triplet formation could determine the ground state configuration. To investigate this we constructed the lowest lying triplet for planar F_2NNO. This is achieved by the promotion $(4a'')^2 \rightarrow (4a'')(5a'')$ where the 5a″ is NN, NO, and NF antibonding. Optimization of NN (keeping the radical monomers fixed at the singlet geometries) in the triplet elongates this bond from 1.31 to 1.36 Å. The state turns out to lie

30 kcal/mol above the singlet, indicative of the dominating character of the fully antibonding 5a″.

C. Mulliken Populations

Mulliken populations for F_2NNO and F_2NNO_2, as well as relevant N, F and N, O molecules, are collected in Tables 10-15 and 10-16. We have already employed the Mulliken net charges in our analysis of F_2NNO_2 electrostatic effects and discussed these deficiencies and calibration. Here we provide additional data that is applicable to a wide range of N, O, F compounds. Another major deficiency in Mulliken populations is that overlap populations are not separable into lone-pair and bond-pair contributions and this often prevents the establishment of meaningful bond-order relationships between molecules (Table 10-15). For example, the q_F and NF overlaps differ by as much as 47%, but previous research indicates that this is insufficient to establish similarity or differences. Likewise, the NN overlap is greater for N_2F_2 than for N_2F_4, expected for a double versus a single NN bond, but the NN overlaps for both F_2NNO and F_2NNO_2, expected to be intermediate, are significantly smaller than that of N_2F_4.

On the other hand, some results, particularly the N, O Mulliken charges and overlap in Table 10-16, are consistent with other bonding data. For example, the charge on the nitrogen of the NO group in F_2NNO, FNO, and F_3NO, as well as the charge on the NO group itself, increases as the electronegativity of atoms attached, NF_2, F, 3F, increases. Likewise the charge on the nitrogen of the NO_2 group and the charge of the NO_2 group increases in the series F_2NNO_2, $FONO_2$, FNO_2 as does the electronegativity of the atoms attached.

D. Comparison of F_2NNO_2 and F_2NNO with N_2O_2, N_2O_3, N_2O_4, and F_2NNF_2

The NN bond lengths for N_2O_2, N_2O_3, and N_2O_4 are in an opposite category to those in F_2NNO_2 and F_2NNO, the former being very long compared to a normal NN single bond and the latter shorter. They are also opposite in that a correlated wavefunction is required to accurately predict these unusually long bonds, but an SCF solution is adequate for the normal bonds in F_2NNO_2 and F_2NNO. Ha[63] has studied the ground and excited states of N_2O_2 with an extensive configuration interaction treatment and his estimate of the NN bond length is 0.2 Å longer than the current experimental value of 2.18 Å.[64] Dykstra, Boggs et al.[65] and Dykstra et al.[66] have shown that extended s, p basis sets in MO wavefunctions lead to NN bond lengths 0.6 Å too short! Our 4-31G calculation for trans N_2O_2 gives similar results, 1.161 Å. Although this is a very large error, it should be noted that the value predicted is notably longer than a normal NN single bond (1.45–1.49 Å), thereby correctly classifying the bond as unusually long (FNO is a similar case). It is also possible to understand the origin and trends in these long NN bonds even though we cannot predict their absolute magnitudes. NO is the most tightly bound of the four common N, O, F radicals and is best represented by the Linnett structure.[1]

$$\cdot\!\!-\!N\!\!\doteq\!\!O\!-\!\cdot$$

(rather than the Lewis structures $=\!\overset{-}{N}\!\!=\!\!\overset{+}{O}\!-\ \leftrightarrow\ -\!N\!\!=\!\!O\!=$). Thus dimerization to either the cis (or trans) conformation:

$$
\begin{array}{ccc}
O & & O \\
\diagdown & & \diagup \\
& N\!\!-\!\!N &
\end{array}
$$

reduces the number of NO bonds. In contrast to carbon–carbon multiple bonds, those between NN, NO, and OO show a second bond to be stronger than the first and a third stronger than the second. Connecting two NO radicals therefore forces them to give up a stronger bond for a weaker bond and the only NN bonding that results is that which is made possible by the orthogonal lobes of the NO bonding orbitals that point into the NN bond region. NO_2 is the second most tightly bound radical and formally it does not have to give up NO bonds to dimerize, but it clearly wants to retain all possible charge in these bonds. The sequence of long NN bond lengths, 2.18, 1.86, and 1.75 Å for N_2O_2, N_2O_3, N_2O_4, respectively, is therefore as expected. The other two common radicals, F_2N and FO are bound by single bonds without the cooperative multiple bonds found between N and O. Thus when these radicals bind with themselves or with NO and NO_2, the bond linking them is expected to be equal to or shorter than a single bond.

In spite of the wide range of NN bond lengths in F_2NNO_2, F_2NNO, N_2O_2, N_2O_3, and N_2O_4 they share a common preference for planar conformations. Even the 2.18 Å NN separation in N_2O_2 permits a sizable overlap of atomic p orbital tails and the conditions required for the generalized delocalization found in F_2NNO_2 and F_2NNO are present in N_2O_2, N_2O_3, and N_2O_4. In fact, NO and NO_2 both have accessible LUMOs (the other half of the half-filled HOMO), thereby providing an especially favorable energy lowering of the NN antibonding π orbital. On the other hand, neither F_2N nor FO have accessible LUMOs, therefore neither F_2NNF_2 nor $FONF_2$ are planar.

For N_2O_2 there is the further question of trans and cis isomers for which neither F_2NNO_2 nor F_2NNO offers a direct analogy. The cis isomer has the lowest energy, but a significantly longer NN separation (0.73 Å)[62] than trans. Unique among the five molecules, N_2O_2 possess four π electrons, thus permitting a ring with in-phase orbitals between the oxygens that can deferentially stabilize cis. Boggs et al.[61] transformed their MO wavefunctions into localized orbitals and demonstrated such an in-phase bridge between the oxygens. They also displayed oxygen lone pairs with tails which protrude across the NN bond in cis, but not trans, thus suggesting an origin for its long NN bond. As evidenced by the numerous calculations of Boggs et al.[65] and Dykstra,[66] with different basis set levels, including correlation, the balance between trans and cis is very delicate. From our analysis of F_2NNO_2 we expect that generalized π delocalization will tend toward shortening NN and in particular will set its length in trans. In cis the lone pair–lone pair repulsion between nitrogens will be large and cause bond elongation, but the in-phase O O π interaction will still provide enough energy lowering to dominate.

10. NF_4

NF_4 and NF_5 have not yet been detected in the laboratory, but there is a great deal of interest in these molecules as examples of hypervalent nitrogen compounds.

The possible existence of NF_4 as a reaction intermediate in the formation of NF_4^+ salts has been suggested by Christe[67] in the following scheme:

$$F_2 \rightarrow 2F$$

$$F + NF_3 \rightarrow NF_4$$

$$NF_4 + AsF_5 \rightarrow NF_3^+AsF_6^-$$

$$NF_3^+AsF_6^- + F \rightarrow NF_4^+AsF_6^-$$

His thorough and extensive investigation of the reaction conditions and possible alternative reaction mechanisms lends credibility to his conclusion. He has also estimated that the formation of NF_4 from $NF_3 + F$ is approximately thermoneutral, thus leading to prediction of an average bond energy equal to 49.8 kcal/mol.

Another source of supporting evidence is the detection of F_3NO^-, isoelectronic with NF_4, and the first example of a first row 33 valence electron species.[68] Solid solutions of F_3NO in SF_6 were irradiated at $-196°$ and the ESR spectra of F_3NO^- obtained shows three equivalent fluorines on the ESR time scale,[68b] consistent with C_{3v} symmetry.[69]

A. Geometry

A simplified molecular orbital correlation diagram for AB_4 systems in T_d, C_{2v}, and D_{4h} symmetries (Figure 10-15) indicates D_{4h} to be the most likely

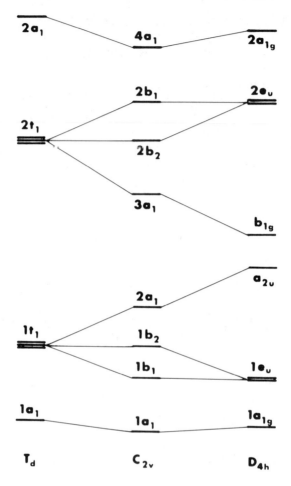

Figure 10-15. AB$_4$ Walsh-Mulliken diagram for T$_d$, C$_{2v}$, and D$_{4h}$, electronegativity of A less than that of B orbitals (from Gimarc, B. M. "Molecular Structure and Bonding". Academic Press: New York, 1979, p. 68).

candidate, with C$_{2v}$ a close competitor. D$_{4h}$ optimized at the 4-31G level gave reasonable bond lengths of 1.487 Å, long when compared to a normal NF bond (1.37 Å, NF$_3$ as reference),[70] but still within the limit of known NF bonds (NF in FNO is 1.512 Å).[71] The D$_{4h}$ structure is a $^2A_{2u}$ electronic state, the singly occupied orbital being of π symmetry and antibonding along the NF bond. Optimization at 6-31G* reduced this length to 1.415 Å, consistent with the shortening of other NF bonds in N, O, F compounds.

Calculations for the T$_d$ configuration at the 4-31G and 6-31G* levels dissociated smoothly to separated atoms. The 4-31G calculation was only carried to 2.95 A at which point the total energy was 1.5 kcal/mol below the separated atoms, but with no evidence of a minimum. This curve was not investigated further because calculations with the higher level basis set, 6-31G*,

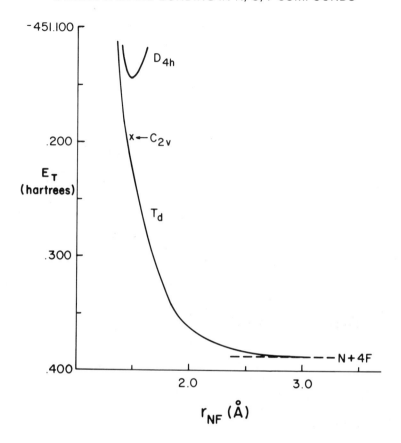

Figure 10-16. Total energy (hartrees) versus NF bond length (Å) for NF_4 in T_d, C_{2v}, and D_{4h}. Single point (x) corresponds to the C_{2v} structure with bond lengths of 1.487 Å and two NF bonds bent 10° out of the plane of the four fluorines.

superseded them. In C_{2v} (4-31G basis) the energy decreases as the two axial bonds elongate smoothly to ∞ and the two equatorial bonds collapse to the NF_2 structure. A summary of these results is given in the potential curves shown in Figure 10-16. The single point on the C_{2v} potential surface shown corresponds to bond lengths of 1.487 Å with two bonds bent 10° above the molecular plane. It is quite clear that no barrier exists between the D_{4h} and the T_d structures, indicating that NF_4 will decompose smoothly.

B. Bonding

Molecular Orbital Diagram

Comparison of the molecular orbitals of NF_4 with those of NF_4^+ can help elucidate the instability of NF_4 and show its relationship to the salt chemistry of NF_4^+. Figure 10-17 gives the 4-31G valence molecular orbitals for NF_4^+ in

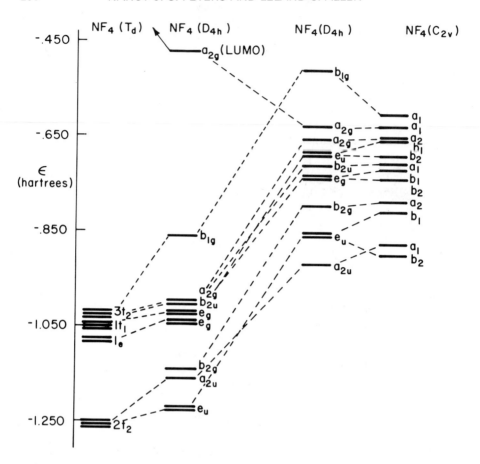

Figure 10-17. Molecular orbital correlation diagram for NF_4^+ in T_d and D_{4h} and NF_4 in D_{4h} and C_{2v} (4-31G basis set).

T_d and D_{4h} and NF_4 in D_{4h} and C_{2v} (the C_{2v} NF_4 does not correspond to an optimized structure, the remaining three do). NF_4^+ is greatly destabilized by forcing its equilibrium T_d structure to become D_{4h} (191.4 kcal/mol higher). Addition of an electron to NF_4^+ (D_{4h}) to give NF_4 (D_{4h}) also causes a large destabilization of the valence orbitals (larger even than that caused by forcing NF_4^+ planar). The preference for a C_{2v} structure for NF_4, as shown in the molecular orbital diagram, is a result of the stabilization of the b_{1g} orbital (see Figure 10-18). Bending two opposite bonds out of the plane of the molecule reduces antibonding interactions without disturbing bonding interactions.

Binding Energy

In spite of the repulsive potential energy surface found above, Christe's kinetics evidence[67] for the existence of NF_4 is sufficiently strong to warrant

Figure 10-18. Atomic orbital contributions to the big molecular orbital of NF_4 in the D_{4h} conformation.

consideration of a binding energy for NF_4. As the proposed kinetic reaction involves the formation of NF_4 from NF_3 and F and the anticipated total binding energy for NF_4 is probably less than or equal to the total in NF_3, we can set the lower limit necessary to permit formation and detection of NF_4. This limit is determined by the endothermicity of formation: an energy input greater than 57.1 kcal/mol would lead to decomposition of NF_3[72] in preference to formation of NF_4. The value thus obtained is 142.1 kcal/mol or 37.2 kcal/mol/bond.

We can provide an independent estimate of what binding might be expected by using reaction 10-7 on the bound D_{4h} structure we have computed.

$$NF_4(^2A_{2u}) + NF_2(^2B_2) \rightarrow 2NF_3(^1A_1) \qquad (10\text{-}7)$$

The UHF/6-31G* (unrestricted Hartree-Fock) ΔE for this reaction is -151.0 kcal/mol (see Table 10-19). Combining this value with the known ΔH_{298} for the formation of NF_3 from NF_2 and F, -57.1 kcal/mol, yields a ΔH_{298} for the formation of NF_4 from NF_3 and F of $+93.9$ kcal/mol. The binding energy for NF_4 is thus 105.3 kcal/mol or 25.4 kcal/mol/bond. As this value falls below the lower bonding limit of 37.2 kcal/mol/bond, we conclude that NF_4 could not be formed from NF_3 and F. This result is independent of, but consistent with, our other calculation of the instability of NF_4.

$NF_4^+AsF_6^-$ Decomposition

The unlikely existence of NF_4 forces reconsideration of Christe's proposed mechanism for the formation and decomposition of $NF_4^+AsF_6^-$. We suggest that some reaction mechanism beyond those considered, perhaps the addi-

TABLE 10-19. E_T for NF_2, NF_3, and NF_4 at the UHF/6-31G* Level

Species	E_T(hartrees)
NF_2	-253.16070
NF_3	-352.55200
NF_4	-451.70266

tion of F_2 to AsF_5 (competing with $F_2 \rightarrow 2F$) to give AsF_6^- and F^+ followed by $NF_3^+F^+$ to give NF_4^+, or the formation of an $NF_3\ AsF_5$ adduct may be responsible.

In conclusion, our calculations indicate that NF_4 is unlikely to be observed even if the specific values we obtain are considerably in error. Nevertheless, it should be clearly stated that the present calculations cannot be taken as definitive. We have not included correlation effects in our wavefunctions and it is absolutely essential to do so to determine whether or not NF_4 will have a detectable existence.

11. NF_5

As previously mentioned, NF_5 is not expected to exist despite the existence of the conjugate acid, NF_4^+. Steric hindrance is also cited as a reason for its nonexistence[73]: in the trigonal bipyramidal structure the axial and equatorial fluorines would be closer than the sum of their van der Waals radii. However, in view of the existence of F_3NO^- and the proposed existence of NF_4, it seems reasonable to consider the quantum mechanical results for NF_5.

All four of the lines of evidence developed below suggest that the existence of NF_5 is unlikely.

A. Geometry

Bond Lengths

The 4-31G optimized structure for NF_5 (Table 10-20) in D_{3h} symmetry has axial bonds of 1.59 Å and equatorial bonds of 1.39 Å. These values are much larger than 1.385 and 1.345 Å, respectively, estimated by Murrell and Scollary[74] from an extensive survey of tri- and penta-halides. The absolute differ-

TABLE 10-20. Geometries (Å) and Energies (Hartrees) for NF_5 and Related Systems

System	Basis set	Bond	Length	Energy
$NF_5(D_{3h})$	4-31G	r_{eq}	1.39	-550.40757
		r_{ax}	1.59	
	6-31G*[a]	r_{eq}	1.39	-551.08868
		r_{ax}	1.59	
		r_{eq}	1.38	-551.09674
		r_{ax}	1.50	
N + 5F	4-31G	—	—	-550.65488
	6-31G*	—	—	-551.18931
$NF_4^+(T_d) + F^-$	4-31G	r_{NF}	1.34	-550.13019

[a] Not optimized at 6-31G*, single point calculations.

ence between axial and equatorial lengths is also much greater than that found for any MX_5 species: eg, PF_5 has lengths of 1.58 and 1.53 Å, respectively.[74,75] Note likewise that the axial bond length in PF_5 is approximately the same as the PF bond length in PF_3, 1.57.[76]

Relation to NF₄

NF_4 is on the margin of stability and, when the electronic charge distribution resulting from sequential addition of fluorine, NF_3, NF_4, NF_5, is examined, there appears little chance for the existence of NF_5. At 4-31G, the optimized bond length in NF_4 is 1.49 Å, again suggesting that the axial bond length found for NF_5 is indicative of decomposition to $NF_3 + F_2$.

B. Bonding

Analysis of Energy versus R Curve

It is well known that molecular orbital wavefunctions separate into ionic fragments at infinite internuclear separation thereby producing energy versus internuclear separation curves with minima that mask a repulsive interaction. However, knowing the separation products and analytically approximating the total energy at large internuclear distances as a Coulomb interaction one can test whether or not the curve minimum is induced by the molecular orbital approximation. Along the axial direction NF_5 separates to $F^- + NF_3 + F^+$ and Figure 10-19 shows that $1/r$, projected inward from r (axial) $= \infty$ matches well to the three 4-31G calculated points on the total energy versus r (axial) curve (r (equatorial) = 1.39 Å). For practical purposes $r = \infty$ was chosen as 10 Å and 16 Å in two separate tests and an equally satisfactory fit was obtained for both. The good match achieved strongly suggests that the 4-31G optimized r (axial) = 1.59 Å is indicative of axial decomposition, $F + NF_3 + F$.

Dissociation Energies

Murrell and Scollary[74] found good agreement between experiment and their nonempirical, valence shell method for the geometry and dissociation energies of the tri- and penta-fluorides and chlorides of N, P, As, and Sb. Their calculations yielded a strong exothermic decomposition process, $NX_5 \rightarrow NX_3 + 2X$ for X = F and Cl. These results are in accord with our finding that at the 6-31G* level, NF_5 lies 63 kcal/mol above the separated atoms and, of course, considerably higher for separation into $NF_3 + 2F$. Although it is always conceivable that a very extensive configuration interaction would show stability, there is as yet no example in the literature which would suggest its likelihood for a system of this type.

Finally, it is interesting to note that the 4-31G optimized D_{3h} structure for NF_5 has a lower total energy than the ionic structure, $NF_4^+F^-$, a two-unit

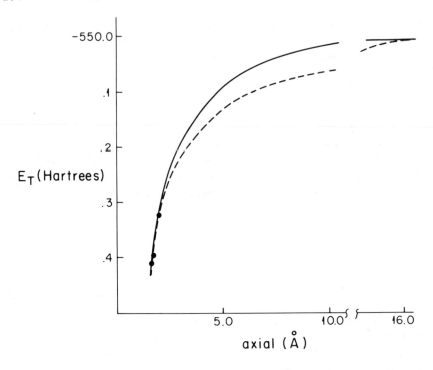

Figure 10-19. Total energy (in hartrees) versus r(axial) (in Å) for NF_5. Three points (4-31G, basis set), R(equatorial) = 1.39 Å. The solid line corresponds to $\frac{1}{4}$ projected inward from $F^+ + NF_3 + F^-$ at 10.0 Å; the dashed line, $\frac{1}{4}$ projected inward from 16.0 Å.

complex in the anticipated crystal structure. It was previously thought that $NF_4{}^+F^-$ might possess a marginal stability,[77] but a recent estimate by Christe[54] suggests that its ΔG of formation (at 298 K) from $NF_3(g)$ and $F_2(g)$ is in excess of 70 kcal/mol. Thus our results confirm the implausibility of detecting the salt.

12. CONCLUSION

Although the behavior of N, O, F compounds is different from those compounds most chemists are most familiar with, ie, carbon compounds, they also comply rigidly with the octet rule. Lewis or Linnett structures can be used successfully to describe the structure and bonding.

Two major differences between N, O, F compounds and carbon compounds exist. First, for N, O, F compounds, multiple bonds are favored over many single bonds. This accounts for the smaller size, usually six atoms or less, for N, O, F compounds. Second, many are unstable, although this is not apparent from Lewis or Linnett structures. The bond energies for X—Y bonds (X, Y = N, O, F) are typically weaker than those for carbon compounds (X, Y = C, H, O, N).

Within the framework of these two major differences, understanding and predictability of N, O, F compounds can be achieved.

REFERENCES

1. Linnett, J. W. "The Electronic Structure of Molecules". Methuen: New York, 1964.
2. Lawless, E. W., Smith, I. C. "Inorganic High Energy Oxidizers". Marcel Dekker: New York, 1968.
3. Siegel, B., Schieler, L. "Energetics of Propellant Chemistry". Wiley: New York, 1964.
4. Advanced Propellant Chemistry, Advances in Chemistry Series #54, ACS, 1966, Holzmann, R. T., Ed.
5. Ditchfield, R.; Hanre, W. H.; Pople, J. A. *J. Chem. Phys.* **1971**, *54*, 724.
6a. Hehre, W. J.; Lathan, W. A.; Ditchfield, R.; Newton, M. D.; Pople, J. A. *Quantum Chemistry Program Exchange* **1973**, *11*, 236.
6b. Binkley, J. S.; Whiteside, R. A.; Krishnan, R.; Seegar, R.; Defrees, D. J.; Schlegel, H. B.; Topiol, S.; Kahn, L. R.; Pople, J. A. *Quantum Chemistry Program Exchange Newsletter* **1980**, *71*, 36.
7. Schlegel, H. B.; Wolfe, S.; Bernardi, F. *J. Chem. Phys* **1975**, *63*, 3632.
8. DeFrees, D. J.; Levi, B. A.; Pollack, S. K.; Hehre, W. J.; Binkley, J. S.; Pople, J. A. *J. Am. Chem. Soc.* **1979**, *101*, 4085.
9. Allen, L. C. *Phys. Rev.* **1960**, *118*, 167.
10. Hurley, A. C. "Introduction to Electronic Theory of Small Molecules". Academic Press: New York, 1976, 231.
11. Bartlett, N.; Passmore, J.; Wells, E. J. *Chem. Comm.* **1966**, *213*.
12. Fox, W. G.; Mackenzie, J. S.; Sukornik, B.; Wamser, C. A.; Holmes, J. R.; Erbech, R. E.; Stewart, B. B. *J. Am. Chem. Soc.* **1966**, *88*, 2604.
13. Muller, A.; Nagarajan, G.; Krebs, B. *Spectroscopic Mol.* **1967**, *16*, 31.
14. Curtiss, E. C.; Pilipovich, D.; Moberly, W. H. *J. Chem. Phys.* **1969**, *46*, 2904.
15. Abramowitz, S.; Levin, I. W. *J. Chem. Phys.* **1969**, *51*, 463.
16. Plato, V.; Hartford, W. D.; Hedberg, K. *J. Chem. Phys.* **1970**, *53*, 3488.
17. Dalby, F. W. *Can. J. Phys.* **1958**, *36*, 1336.
18. Lister, M. W., Sutton, L. E. *Trans. Farad. Soc.* **1939**, *35*, 495.
19. NF$_2$: Bohn, R. K.; Bauer, S. H. *Inorg. Chem.* **1967**, *6*, 304. NF$_3$: Sheridan, J.; Gordy, W. *Phys. Rev.* **1950**, *79*, 513. N$_2$F$_4$: Gilbert, M. M.; Gunderson, G.; Hedbert, K. *J. Chem. Phys.* **1972**, *56*, 1691.
20. (a) Nichikida, K.; Williams, F. *J. Am. Chem. Soc.* **1975**, *97*, 7166. (b) Hasegawa, A.; Hudson, R. L.; Kikuche, O.; Nichikida, K.; Williams, F. *J. Am. Chem. Soc.* **1981**, *103*, 3436.
21. Choplin, F.; Kaufman, G. *J. Mol. Struc.* **1972**, *11*, 381.
22. Olsen, J. F. *J. Fluorine Chem.* **1977**, *9*, 471.
23. Olsen, J. F.; Howell, J. M. ibid., **1977**, *10*, 197.
24. Olsen, J. F.; Howell, J. M. *Theor. Chem. Acta* **1978**, *47*, 39.
25. Olsen, J. F.; Howell, J. M. *J. Fluorine Chem.* **1978**, *12*, 123.
26. Lawlor, L. J.; Vasudevan, K.; Grein, F. *J. Am. Chem. Soc.* **1978**, *100*, 8062.
27. Millen, D. J.; Buckton, K. S.; Legon, A. C. *Trans. Farad. Soc.* **1969**, *65*, 1975.
28. DeFrees, D. J.; Levi, B. A.; Pollack, S. K.; Hehre, W. H.; Binkley, J. S.; Pople, J. A. *J. Am. Chem. Soc.* **1979**, *101*, 4085.
29. The antibonding interaction between oxygen and nitrogen in the Me$_3$NO could be the source of the overestimation of the NO bond length in the geometry optimization study by Radom, L.; Brinkley, J. S.; Pople, J. A. *J. Aust. Chem.* **1977**, *30*, 699.

30. Mulliken, R. S. *J. Chem. Phys.* **1955**, *23*, 1833, 1841, 2338, 2343; **1962**, *36*, 3428.
31. Radom, L.; Hehre, W. J.; Pople, J. A. *J. Am. Chem. Soc.* **1971**, *93*, 289; ibid., **1972**, *94*, 2371.
32. Harmony, M. D.; Laurie, V. W.; Kuczowski, R. L.; Swendeman, R. H.; Ramsay, D. A.; Lovas, F. J.; Lafferty, W. J.; Maki, A. G. *J. Phys. Chem. Ref. Data* **1979**, *8*, 619.
33. Firestone, R. A. *Tetrahedron Lett.* **1968**, 971. *J. Org. Chem.* **1969**, *34*, 2621. *J. Chem. Soc. A* **1970**, 1570. *J. Org. Chem.* **1971**, *36*, 702. ibid., **1980**, *45*, 3604.
33a.The structure of FN_3 has just recently been determined (*Amer. Chem. Soc., J.* **[1988]** *110*, 707). The NF bond is 1.43 Å, and we have published a paper interpreting this bond length in terms of molecular orbital theory (see *Inorg. Chem.* **[1988]**, *27*, 755).
34. Orbital plotting program written by Prof. W. L. Jorgensen, Purdue University. 4-31G and 6-31G* version written by E. Knight at Princeton. Quantum Chemistry Program Exchange, 1977, #340.
35. Pauling, L.; Brockway, L. O. *J. Am. Chem. Soc.* **1937**, *59*, 13.
36. Christe, K. O.; Schack, C. J.; Wilson, R. D. *Inorg. Chem.* **1974**, *13*, 2811.
37. Brandle, K.; Schmeisser, M.; Luttke, W. *Chem. Ber.* **1960**, *93*, 2300.
38. Arvia, A. J.; Cafferata, L. F. R.; Schumacher, H. J. *Chem. Ber.* **1963**, *96*, 1187.
39. Miller, R. H.; Bernitt, D. L.; Hisatsune, J. C. *Spectrochim. Acta, Part A* **1967**, *23*, 223.
40. Shamir, J.; Yellin, D.; Claassen, H. H. *Isr. J. Chem.* **1974**, *12*, 1015.
41. Odeurs, R. L.; van der Veken, B. J.; Herman, M. A. *J. of Molec. Struc.* **1984**, *118*, 81.
42. (a) Christe, K. O.; Guertin, J. P.; Pavlath, A. E. *Inorg. Nucl. Chem. Lett.* **1966**, *2*, 83. (b) *Inorg. Chem.* **1966**, *5*, 1921. (c) United States Patent 3,503,719; 1970.
43. Christe, K. O.; Pilipovich, D. *Inorg. Chem.* **1971**, *10*, 2803.
44. Christe, K. O.; Wilson, R. D.; Axworthy, A. E. *Inorg. Chem.* **1973**, *12*, 2478.
45. Mishra, J. P.; Symons, M. C. R.; Christe, K. O.; Wilson, R. D.; Wagner, R. I. *Inorg. Chem.* **1975**, *14*, 1103.
46. (a) Christe, K. O.; Schack, C. H.; Wilson, R. D. *Inorg. Chem.* **1976**, *15*, 1275. (b) Ibid., **1977**, *16*, 849.
47. Christe, K. O.; Schack, C. J. *Inorg. Chem.* **1977**, *16*, 353.
48. Christe, K. O. *Inorg. Chem.* **1977**, *16*, 2238.
49. Price, W. C.; Passmore, T. K.; Roessler, D. M. *Discuss. Farad. Soc.* **1963**, *35*, 201.
50. Solomon, I. J.; Keith, J. N.; Snilson, A. *J. Fluorine Chem.* **1972**, *2*, 129.
51. Christe, K. O.; Wilson, R. D.; Goldberg, I. B. *Inorg. Chem.* **1979**, *18*, 2572.
52. Christe, K. O. *Spectrochem. Acta* **1980**, *36A*, 921.
53. Charpen, P.; Lance, M.; Bui Huy, T.; Bougon, R. *J. Fluorine Chem.* **1981**, *17*, 479.
54. Bowagon, R.; Bui Huy, T.; Burgess, J.; Christe, K. O.; Peacock, R. D. *J. Fluorine Chem.* **1982**, *19*, 263.
55. Christe, K. O. 8th Winter Fluorene Conference, 26 January 1987, Abstract No. 31.
56. Olsen, J. F.; Howell, J. M. *J. Fluorine Chem.* **1978**, *12*, 123.
57. Gangreli, P. S.; McGee, H. A., Jr. *Inorg. Chem.* **1972**, *11*, 3071.
58. Clark, J.; Schleyer, P. *J. Comp. Chem.* **1981**, *2*, 20.
59. Gimarc, B. M. "Molecular Structure and Bonding". Academic Press: New York, 1979.
60. Epiotis, N. D.; Cherry, W. R.; Shaik, S.; Yates, R. L.; Bernardi, F. *Top. Curr. Chem.* **1977**, *70*, 62.
61. Holtz, D. *Prof. Phys. Org. Chem.* **1971**, *8*, 1. Holtz, D.; Streitwieser, A., Jr.; Jesaitis, R. G. *Tet. Lett.* **1969**, *53*, 4529. Hoffman, R.; Radom, L.; Pople, J. A.; Schleyer, P.; Hehre, W. J.; Salem, L. *J. Am. Chem. Soc.* **1972**, *94*, 6221. DeFrees, D. J.; Bartmess, J. E.; Kim, J. K.; McIver, R. T., Jr.; Hehre, W. J. *J. Am. Chem. Soc.* **1977**, *99*, 6451. Sleigh, J. H.; Stephens, R.; Tatlow, J. C. *J. Fluorine Chem.* **1980**, *15*, 411.
62. Spratley, R. D.; Pimentel, G. C. *J. Am. Chem. Soc.* **1966**, *88*, 2394. Pimentel, G. C.; Spratley, R. D. "Chemical Bonding Clarified Through Quantum Mechanics". Holden-Day: Oakland, 1969, pp. 224–234.
63. Ha, T.-K. *Theor. Chim. Acta* **1981**, *58*, 125.
64. Lipscomb, W. N.; Wang, F. E.; May, W. R.; Lippert, E. L. *Acta Cryst.* **1961**, *14*, 1101.

65. Skaarup, S.; Skancke, P. N.; Boggs, J. E. *J. Am. Chem. Soc.* **1976,** *98,* 6106.
66. Benzel, M. A.; Dykstra, C. E.; Vincent, M. A. *Chem. Phys. Lett.* **1981,** *78,* 139.
67. Christe, K. O.; Wilson, R. D.; Goldberg, I. B. *Inorg. Chem.* **1979,** *18,* 2572.
68. (a) Nishikida, K.; Williams, F. *J. Am. Chem. Soc.* **1975,** *97,* 7166. (b) Hasegawa, A.; Hudson, R. L.; Kikuchi, O.; Nishikida, K.; Williams, F. ibid., **1981,** *103,* 3436.
69. Although F_3PO^- (Shitani, M.; Williams, F., unpublished work) is a trigonal bypyramid, recent results indicate a C_{3v} structure for Ph_3Cl (Symons, M. C. R. *Chem. Phys. Lett.* **1976,** *40,* 226) and $FClO_3^-$ (Hasegawa, A.; Williams, F., submitted for publication).
70. Sheridan, J.; Gordy, W. *Phys. Rev.* **1950,** *79,* 513.
71. Millen, D. J.; Buckton, K. S.; Legon, A. C. *Trans. Farad. Soc.* **1969,** *65,* 1975.
72. Wagman, D. D.; Evans, W. H.; Parker, V. B.; Halow, I.; Bailey, S. M.; Schuman, R. H. *National Bureau of Standards (U.S.) Tech. Note* **1969,** No. 270-3.
73. Wallmeier, H.; Kutzelnigg, W. *J. Am. Soc.* **1979,** *101,* 2804.
74. Murrell, J. N.; Scollary, C. E. *J. Chem. Soc. Dalton* **1975,** 818.
75. Hansen, K. W.; Bartell, L. S. *Inorg. Chem.* **1965,** *4,* 1775.
76. Moreno, Y.; Kuchitsu, K.; Moritani, T. *Inorg. Chem.* **1969,** *8,* 867.
77. Goetschel, C. T.; Campanile, V. A.; Curtis, R. M.; Loos, K. R.; Wagner, C. D.; Wilson, J. N. *Inorg. Chem.* **1972,** *11,* 1969.

CHAPTER 11

Fluorine Substituted Analogs of Nucleic Acid Components

Donald E. Bergstrom and Daniel J. Swartling

Department of Chemistry, University of North Dakota, Grand Forks, North Dakota

CONTENTS

1. INTRODUCTION

Fluorine substituted analogs of the naturally occurring nucleic acid components have become established as antiviral, antitumor, and antifungal agents. A number of potential drugs in which fluorine substitution is a key to biological activity are under intensive study. Moreover, there are a number of new derivatives which are useful as probes for studying biochemical processes, and there may be a future in clinical diagnostics for fluorine substituted analogs. It is not surprising that an exceptionally high percentage of fluorinated nucleoside analogs exhibit significant biological activity, because in many cases researchers have constructed modified structures which differ only slightly in their three-dimensional geometry from the corresponding naturally occurring molecules. Hence, these analogs may bind readily to enzymes for which nucleosides and nucleotides are substrates or allosteres. In these instances, the electronic or polar effects of the fluorine substituent must play a significant role in the expression of biological activity.

How and why does modification of nucleosides lead to analogs with significant biological properties? What predictions can be made about the effect that certain types of substitution patterns will have on structure and chemical properties? Although these questions cannot yet be completely answered, we hope that by this review a greater awareness for the future prospects of research in the area of fluorine modified nucleic acid components will be realized. Recent advances in structure, synthesis, and biological activity are reported for each of the different classes of fluorine substituted nucleosides, nucleotides, and polynucleotides. However, the review is selective rather than comprehensive and numerous worthy examples have been omitted. Where possible, correlations between structure and activity will be treated. In particular, the issue of fluorine substitution of an H versus an OH, and CF_2 as an isopolar-isosteric replacement for O is of interest. The CF_2 group may be regarded as an isosteric replacement for oxygen since it is only slightly larger than an oxygen atom (by way of comparison a P—O bond is typically about 1.57 Å whereas a P—CF bond[1] may be about 1.88 Å). In fact there is no other functional group that can replace an oxygen atom which matches so closely oxygen's electronegativity (this is the sense with which the term isopolar is used in this review) and yet retains a relatively close size match. Furthermore, since biological stability is more often an attribute, rather than a liability in medicinally important nucleoside analogs, the CF_2 group may prove particularly significant as a replacement for oxygen at biochemically labile positions.

Incorporation of fluorine for the purposes of creating a new drug may, in some instances, provide a compound useful as an analytical probe or as a diagnostic. As an example, 5-fluorouracil is an established antitumor agent but has also found use in a study of the structure of tRNA by high resolution NMR spectroscopy because it is readily biosynthetically incorporated into nucleic acids. To be useful drugs, fluorine modified nucleosides or nucleotides must exhibit a differential biological effect between the host and the infectious agent. Currently nucleoside analogs are of most interest as antiviral agents. The nucleoside functions as a prodrug, which must be transformed into the active form; typically a nucleoside 5'-triphosphate which either inhibits DNA synthesis or after incorporation into DNA or RNA effects function of the nucleic acid.

Figure 11-1 illustrates the positions in which fluorine has been incorporated into nucleosides and nucleotides in place of H, OH, or O. These structures are not all inclusive, however, since other derivatives have been created in which modification is more extensive. For example, substitution of the C-5 methyl group of thymidine by a $CH=CHF$, $CH=CHCF_3$ or $(CF_2)_nCF_3$ group gives a series of compounds among which there are several potent antiviral agents, particularly active against HSV-1 (Subsection A, Section 4). This review will include examples of these modified nucleosides where the biological result of modification is significant. Finally, no attempt has been made to review the literature on 5-fluorouracil or 5-trifluoromethyl-2'-deoxythymidine since current research and the resulting literature is primarily of a clinical nature. Information has been incorporated into this review only where new uses or new derivatives of these compounds show either significant new biochemistry or utility as probes.

The rationale for replacing the OH and H groups of biologically significant molecules by F have been reviewed by others,[2-5] and may be summarized as follows:

1. Fluorine closely mimics hydrogen with respect to steric requirements (the van der Waals' radii of F: 1.35 Å; of H: 1.1 Å). Thus in most in-

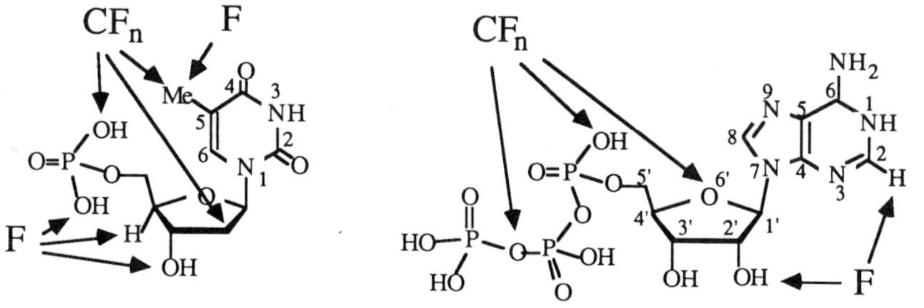

1   **2**

Figure 11-1. Fluorinated Nucleotides Analogs: The positions at which F or CF_n have been substituted for H, O, and CH_n of the naturally occurring nucleotides are illustrated for thymidine 5'-monophosphate (TMP)(**1**) and adenosine 5'-triphosphate (ATP)(**2**).

stances the introduction of fluorine provides a significant change in chemical character without perturbing the molecular geometry. This means that an active substrate and its fluoro-analog would be indistinguishable (sterically) to the active site of an enzyme, and thus the fluoro-analog could act as an anti-metabolite.

2. The replacement of hydrogen by fluorine at or near a reactive site frequently causes inhibition of metabolism because of the high C—F bond energy, which averages 485 kJ/mol (1 kJ = 0.239 kcal). For comparison, the average dissociation energies of other bonds are as follows: C—H, 413; C—O, 358; C—C, 346; C—Cl, 339; C—Br, 284.5, and C—I, 213 kJ/mol.[6] By replacing hydrogen with fluorine, metabolic precursors can be designed so that once metabolized they become irreversibly binding inhibitors; in effect, they become involved in a "lethal synthesis."[4] The carbon fluorine bond is carried through the biosynthetic cycles intact, until at some point a substrate analog is created that blocks a key biochemical process. The classic example of "lethal synthesis" is that of fluoroacetic acid as it undergoes metabolism in the Krebs cycle. When fluoroacetic acid (or a precursor to it, ie, any ω-fluoro even-carbon fatty acid) is ingested, it replaces the natural metabolite and forms fluoroacetyl-coenzyme A, which enters the Krebs cycle and combines with oxalylacetic acid to give α-fluorocitric acid. With the normal substrate (citric acid), dehydration by the enzyme aconitase to give aconitic acid followed by rehydration with reverse orientation yields isocitric acid. However, aconitase is unable to dehydrate α-fluorocitric acid; the substrate binds irreversibly, stopping any further metabolism. Thus the cell, and as a result the organism, dies; a "lethal synthesis" having occurred. Other examples have been described wherein fluorinated analogs function as suicide substrates following enzyme catalyzed elimination of fluoride ion.[3]

3. The replacement of hydrogen by fluorine can increase lipid solubility and thereby increase the rate of transport of biologically active compounds across lipid membranes. The enhanced transport rate of drugs *in vivo* is probably the single most important reason for incorporating a CF_3 group into a drug's design.

4. Due to its high electronegativity, fluorine can be used to mimic or replace oxygen. A fluorine atom replacing an OH group should exert a polar effect of similar magnitude as the oxygen it replaced in a molecule. In some circumstances it may act as a Lewis base in hydrogen bonding, but unlike OH could not function as a donor. CF_2H and CF_2 groups which are both isopolar and isosteric with OH and O, respectively, are discussed in Subsection B of Section 3.

5. The reactivity and stability of functional groups can be influenced by the electron-withdrawing inductive effect of fluorine. The reactivity of neighboring reaction centers can be dramatically altered as well. For example,

1,1,1-trifluoro-2-propanol is not dehydrated by concentrated sulfuric acid, even at 190°C.[4] And unlike acetone, hexafluoroacetone forms stable hydrates, hemiacetals, and hemiaminals. As a further dramatic example, ethylamine has a pK_a of 10.70 while 2,2,2-trifluoroethylamine has a pK_a of 5.7.

It should be emphasized that for studies in which the nucleoside and nucleotide analogs are incorporated into cells, one must normally utilize neutral precursors of the biologically active forms (often the 5'-triphosphate) of the analogs. These modified analogs would have to serve as effective substrates for enzymes necessary to transform them to the biologically active form.

2. CARBOHYDRATE MODIFIED NUCLEIC ACID COMPONENTS

A. C-2'-Fluoronucleosides

Structure

The 2'-fluoronucleoside analogs are of interest first because within this group are analogs that have significant biological activity, and second because these analogs provide useful information about the role of the 2'-OH in nucleic acid structure.

The effect of 2'-fluoro substituents on ribose conformation has been studied extensively. The role of the ribose conformation at the polymer level on binding and biological activity is of particular interest (Subsection B, Section 2). X-ray crystallographic studies on nucleic acid monomers and short oligomers show that the ribose adopts either of two principal forms.[7] One of these is a C3'-endo form and the other a C2'-endo form which are found, respectively, in the A and B families of nucleic acids. Whereas DNA shows substantial flexibility, and forms exist where the deoxyribose assumes either of the two conformational extremes, RNA exits in the A form in which the ribose assumes the 3'-endo conformation. Interconversion between these two puckered forms appears to occur via a pseudorotation pathway in which the pucker moves continuously around the ring.[8,9] These conformational differences have been corroborated by Raman spectroscopy as well.[10] The substituents attached at C1-C4 of the ribose are responsible for the potential energy barrier between the various forms. The two lowest energy forms are referred as the N (3'-endo) and S (2'-endo) forms. At the monomer level (nucleosides), interconversion between these two forms occurs on a timescale of a few nanoseconds.[11] The conformations of the five ring torsion angles τ_j, where $j = 0$–4, are defined with respect to rotations about the bonds $O_{1'}$—$C_{1'}$, $C_{1'}$—$C_{2'}$, $C_{2'}$—$C_{3'}$, $C_{3'}$—$C_{4'}$, and $C_{4'}$—$O_{1'}$. The value of τ_j is given by Equation 11-1.

$$\tau_j = \tau_m \cos [P + 0.8\pi (j - 2)] \qquad (11\text{-}1)$$

where τ_m is the amplitude of puckering, and P is the phase angle of pseudorotation.[8] Since the conformation in solution may be a blend of puckered states it is difficult to predict the conformation on the basis of three-bond proton NMR coupling constants. Nevertheless, some researchers have in fact used a Karplus-like relationship and interpreted the results in terms of only the N and S conformations. These results must be interpreted with some reservation.

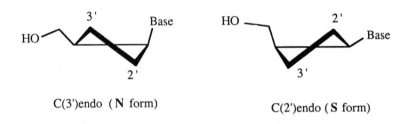

C(3')endo (**N** form)   C(2')endo (**S** form)

NMR studies have demonstrated that the conformation depends largely on the electronegativity of the substituent, rather than on substituent size or hydrogen bonding capacity.[12–17] For example, the deoxyribose of 2'-fluoro-2'-deoxyadenosine (dAfl) **(3)** has an unusually high population of the N (C3'-endo) conformer (67%) while the deoxyribose moieties of 2'-deoxyadenosine **(4)** and 2'-iodo-2'-deoxyadenosine **(5)** have low N conformer populations (19% and 7%, respectively).[17]

Ab initio calculations with the Gaussian 80 program using the split-valence 3-21G basis set give a theoretical average energy barrier of 11.4 kJ/mol between the S and the N states for a model 2-deoxyribofuranose derivative in which the base at C-1 is replaced by an sp^2 hybridized NH_2 and the CH_2OH at C-4 is omitted. The S state 1.2 kJ/mol is more stable than the N state.[16] In comparison the energy barriers for the N $\rightarrow$ S and S $\rightarrow$ N transitions in the 2-fluoro-2-deoxyribofuranose model compound are 20.2 and 7.9 kJ/mol, respectively. The influence of solvent on the N to S interconversion, as well as the possibility of a tunnel transition for the 3'-OH group has yet to be addressed. Analysis of the cooperative movement of the ring atoms and substituents in 2'-fluorodeoxyribose shows that in both the N and S states the most favored orientation is the one that allows the shortest distance between the 3' proton and the fluorine atom.[16] The NMR studies on 2'-fluoro-2'-deoxyribonucleotides in solution indicate that the N conformation is more stable by 1.1 to 2.4 kJ/mol.[17,18] Oligonucleotides constructed from 2'-fluoro-2'-deoxyribonucleotides are like the corresponding natural oligoribonucleotides in physicochemical and biological properties (Subsection B, Section 2).[19]

Synthesis

The synthesis of 2'-fluoronucleosides can be accomplished by at least three different routes: 1) direct fluorination of the appropriately protected nucleoside, 2) fluorination of the appropriately protected carbohydrate, which is then converted to the nucleoside of choice, or 3) construction of a fluorinated carbohydrate from a simple fluorinated precursor. Examples of all three approaches have been published.

A seven-step synthesis of the protected 2-deoxy-2-fluoro-D-arabinofuranose derivative, 1,3-diacetyl-5-O-benzoyl-2-deoxy-2-fluoro-D-arabinofuranose (10) was developed by Reichman, Watanabe, and Fox in 1975.[19] Starting with 1,2:5,6-di-O-isopropylidene-3-O-tosyl-α-D-allofuranose (6), the desired product was obtained in good overall yield through the procedure outlined in Scheme I.

Treatment of 10 with hydrogen bromide-acetic acid in methylene chloride afforded the α-D-glycosyl bromide 11 with only trace amounts of the β anomer. Condensation of 11 with trimethylsilylated N^4-acetylcytosine (12) in methylene chloride followed by deprotection with ammonia afforded nucleoside 13 as the major component. Other nucleic acid analogs can be prepared by condensation of 11 with the appropriate nucleic acid base.[21,22]

A synthesis of two fluorine substituted antiviral agents FMAU and FIAU (39 and 40, cf. Biological Activity, Subsection A, Section 2) reported by C. H. Tann et al. in 1985 adds the fluorine through nucleophilic displacement to the carbohydrate at a later stage in the synthesis. The synthesis starts with commercially available 1-O-acetyl-2,3,5-tri-O-benzoyl-β-D-ribofuranose (14)[23] as shown in Scheme II.

Attempts at direct fluorination of 2-ketofuranose derivatives using DAST to afford the corresponding 2,2-difluoro derivatives have not been successful.[24,25] However, a simple and stereocontrolled synthesis of 2-deoxy-2,2-difluoro-D-ribose has recently been developed by researchers at Lilly Research Laboratories.[26] The R enantiomer of 2,3-(isopropylidenedioxy)propanal, 18, was coupled with ethyl bromodifluoroacetate, using standard Reformatskii conditions, to afford a 3:1 mixture of diastereomers. The major isomer was subjected to hydrolytic removal of the isopropylidene group with closure to the γ lactone. The lactone was silylated to give the 3,5-bis(*tert*-butyldimethylsilyl) γ lactone, 20. Reduction of this prod-

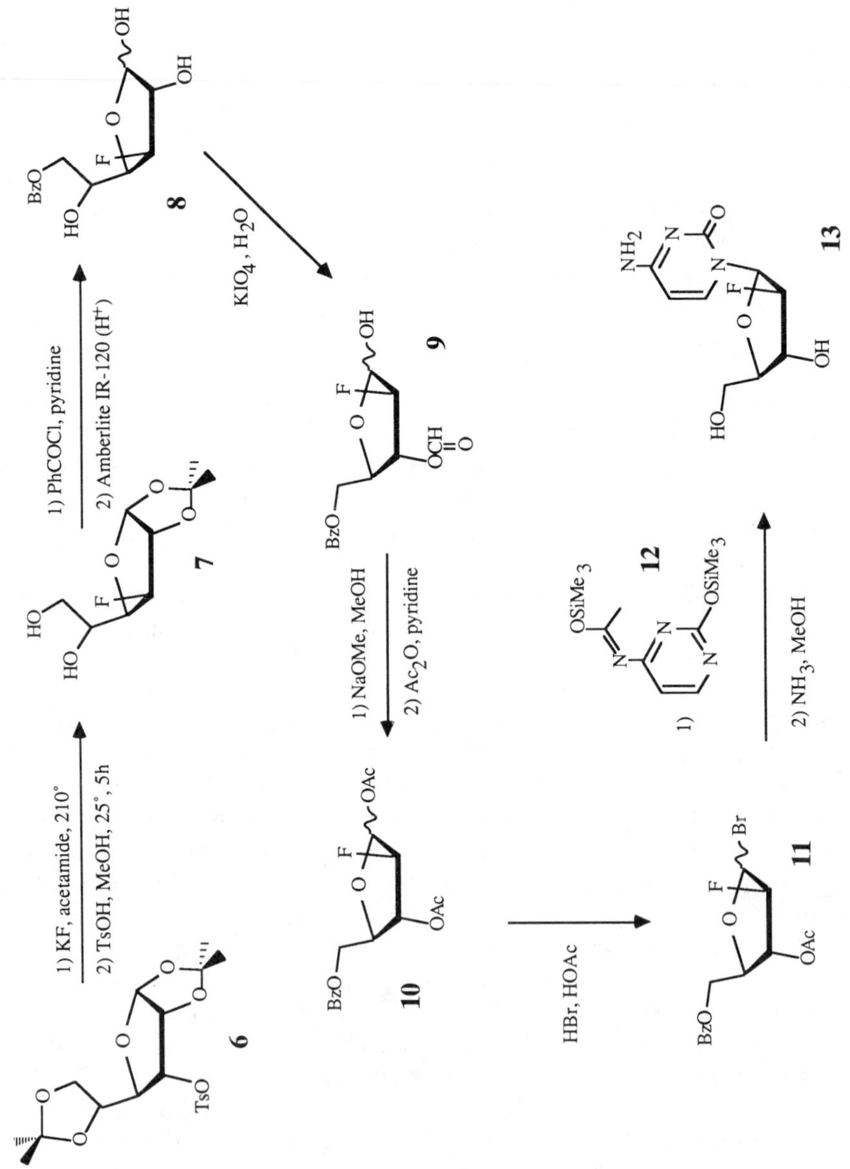
1) KF, acetamide, 210°
2) TsOH, MeOH, 25°, 5h
1) PhCOCl, pyridine
2) Amberlite IR-120 (H+)
KIO4, H2O
1) NaOMe, MeOH
2) Ac2O, pyridine
HBr, HOAc
2) NH3, MeOH
6
7
8
9
10
11
12
13

Scheme I

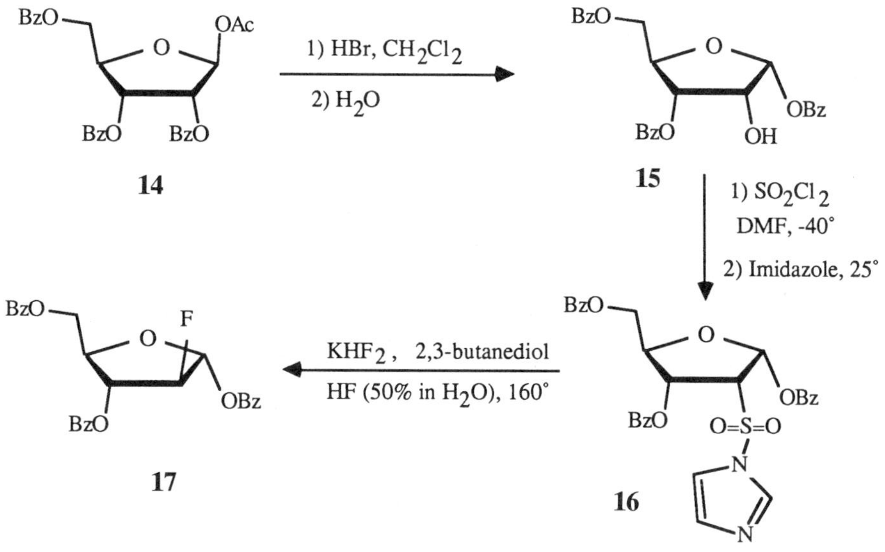

Scheme II

uct with DIBAL gave the disilyl lactol **21.** Removal of the silyl groups afforded the 2-deoxy-2,2-difluoro-D-ribose. Nucleosides were constructed by activating the protected ribose derivative with mesyl chloride and condensing with silylated nucleosides as shown for uracil in Scheme III.

Bobek and An recently came upon an interesting new class of nucleoside analog when they attempted to convert 3,4-di-O-acetyl-2-deoxy-2,2-difluoro-D-erythro-pentopyranoside **(25)** to a 1-bromo derivative in preparation for coupling to a trimethylsilylated heterocyclic base (Scheme IV).[27] Hydrogen bromide resulted in the formation of **26** which with trimethylsilylated uracil and $HgO-HgBr_2$ in methylene chloride gave nucleoside **27** in 23% yield as a mixture of α and β anomers in a 2:1 ratio.

Biggadike and co-workers have been successful in preparing carbocyclic analogs of FMAU **(33)** and FIAU **(40)** (Figure 11-3) by a procedure in which the nucleoside is constructed from protected 1-amino-2-fluoro-2-deoxyarabinofuranose **(30)** (Scheme V).[28]

In order to investigate the structure–function relationship of DNA and RNA, Ikehara and coworkers have prepared a number of 2'-substituted-2'-deoxypurinenucleosides, of which 2'-fluoro-2'-deoxyadenosine **(39)** is an example (Scheme VI).[29] Starting from adenosine, 8,2'-anhydro-8-oxy-9-β-D-arabinofuranosyladenine, **36** is obtained in approximately 40% yield. Compound **36** is converted, through the 3',5'-protected 9-β-D-arabinofuranosyladenine, **38,** to the 2'-fluoro compound in an overall yield of about 8%. This route can be used to attach other groups at the 2' position as well.

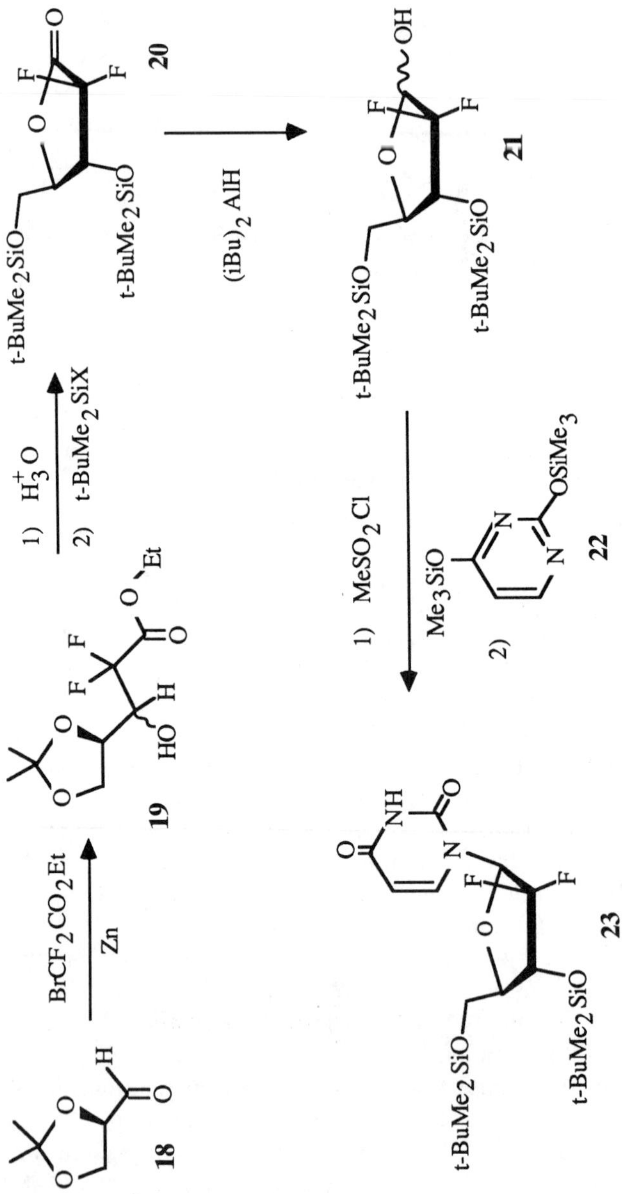
18
BrCF2CO2Et
Zn
19
1) H3+O
2) t-BuMe2SiX
t-BuMe2SiO
t-BuMe2SiO
20
(iBu)2AlH
21
1) MeSO2Cl
2)
Me3SiO
OSiMe3
22
23

Scheme III

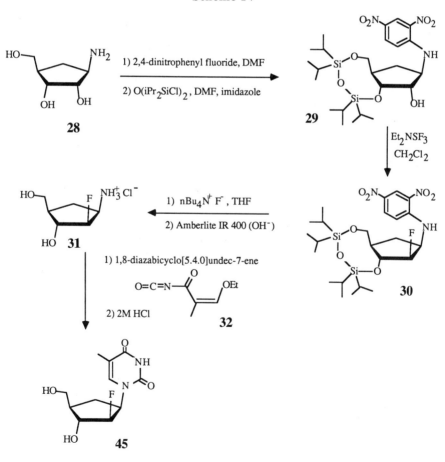

Scheme IV

Scheme V

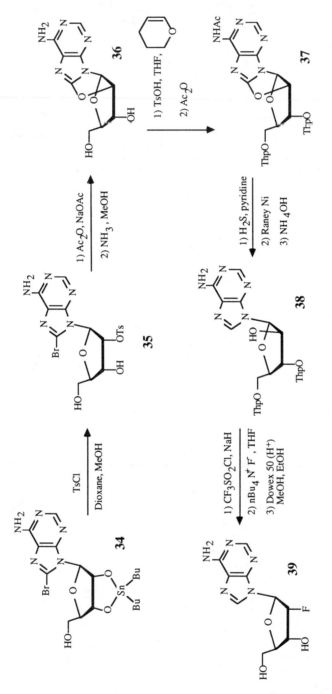

Scheme VI

Biological Activity

The most significant compounds of this class may be the 2' fluoro modified thymidine and cytidine derivatives which are active against herpes viruses. 2'-Fluoro-5-iodo-1-β-D-arabinofuranosylcytosine (FIAC) **(43)**, 2'-fluoro-5-methyl-1-β-D-arabinofuranosyluracil (FMAU) **(33)**, and 2'-fluoro-5-iodo-1-β-D-arabinofuranosyluracil (FIAU) **(40)** are potent and selective inhibitors of herpes simplex virus type 1 and type 2, varicella zoster virus, and cytomegalovirus.[30-32] Current evidence suggests that these derivatives act as antiviral agents via their 5'-triphosphates which are substrates for incorporation into viral DNA via viral DNA polymerases.[33]

Typically, once a modification, such as the 2'-fluoro substituent in pyrimidine nucleosides is discovered to give biologically active compounds, medicinal chemists combine other types of modifications that also alone give active compounds. This empirical approach has been only partially fruitful in the 2'-F arabinofuranosyl series. Although a number of nucleoside analogs are known in which replacement of O-4' with a CH_2 group gives biologically active compounds (referred to as carbocyclic nucleosides), carbocyclic FMAU, **45**, is 10^3 times less active than FMAU. Neither carbocyclic 2',2'-difluoro-2'-deoxythymidine, **46,** nor carbocyclic 2'-fluoro-2'-deoxycytidine, **47,** are active as antiviral agents.[28] The parent ribosyl compounds are active.[34] On the other hand, incorporating various larger C-5 substituents into analogs of the 2'-arabino series appears to have yielded some potent new antiviral compounds. 2'-Fluoro-5-ethyl-1-β-D-arabinofuranosyluracil **(41)** is only 2 to 10 times less active as an antiviral agent than FMAU, but unlike FMAU does not block the incorporation of [^{3}H]dThd or [^{3}H]dUrd into rat bone marrow cells and L1210 cells, unless given at extremely high dose. 2'-Fluoro-5-(2-fluoroethyl)-1-β-D-arabinofuranosyluracil (FEFAU, **42**) has been recently reported as a potent inhibitor of HSV-1, HSV-2, and VZV replication in the concentration range 0.03 to 0.3 μm and also like the 5-ethyl analog is not nearly as toxic towards mammalian cells.[35]

Enzyme studies on 2'- and 3'-fluoronucleoside analogs have been pursued parallel to biological studies with some interesting results. As an example, 2'-fluoro-1-β-D-arabinofuranosylcytosine **(13)** is deaminated by cytidine deaminase at only 10% of the rate of cytidine, whereas 2'-fluoro-1-β-D-arabinofuranosyladenine **(48)** is deaminated more rapidly by adenosine deaminase than is adenosine.[36]

2'-Halo-2'deoxyribonucleoside 5'-di- and tri-phosphates have been instrumental in elucidating the mechanism of action of the enzyme ribonucleotide reductase obtained from *E. coli* and *Lactobacillus leichmannii*.[37] The 5'-diphosphates of 2'-deoxy-2'-fluoroadenosine, 2'-deoxy-2'-fluorocytidine, and 2'-deoxy-2'-fluorouridine **(49–51)** inactivate ribonucleotide diphosphate reductase through a complex radical mechanism with the loss of fluoride ion and the generation of an unstable 3'-ketonucleotide.[38] Qualitatively the results are similar to that obtained for the 2'-chloro derivatives which have been more extensively studied.

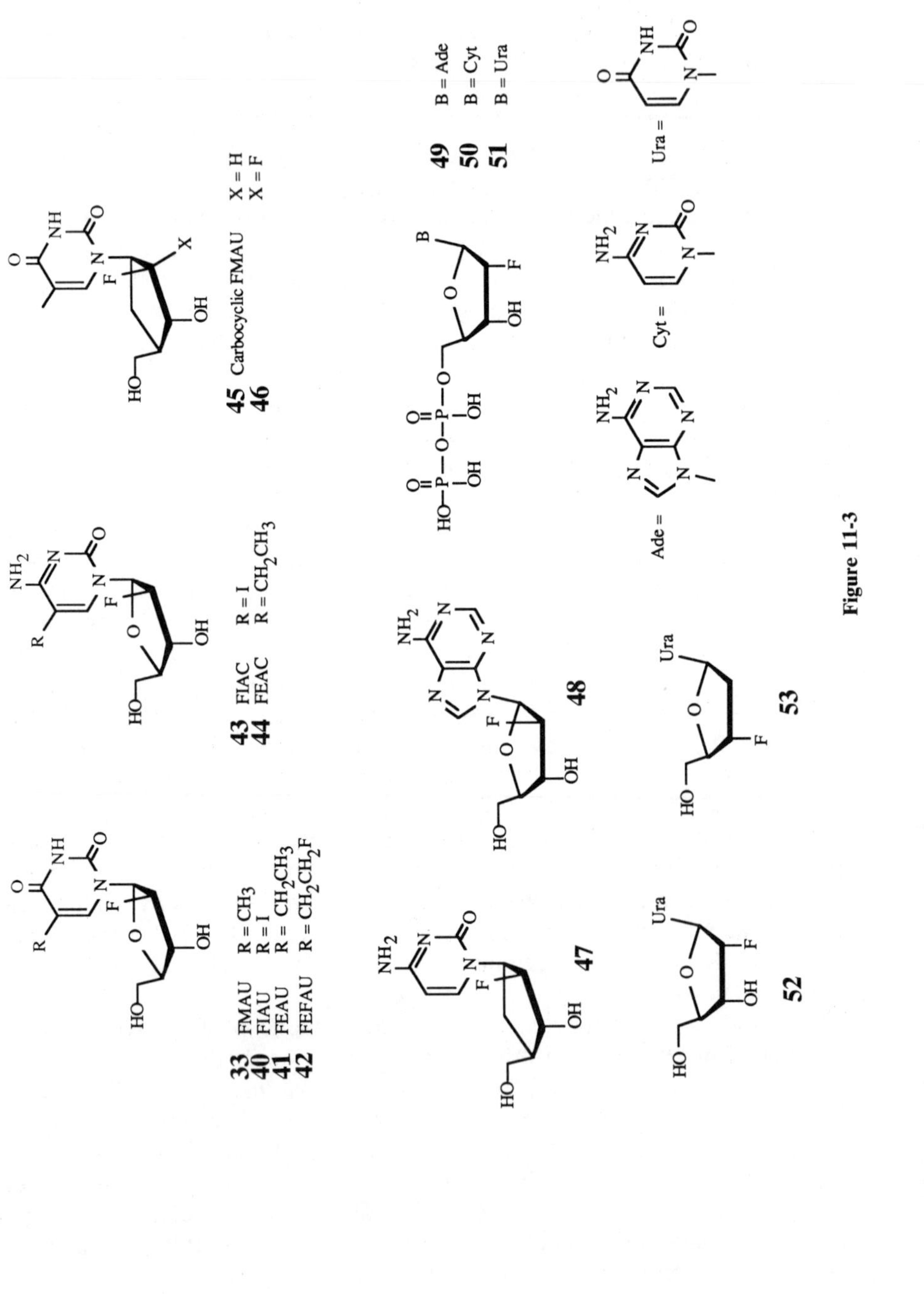

45 Carbocyclic FMAU X = H
46   X = F

43 FIAC R = I
44 FEAC R = CH₂CH₃

33 FMAU   R = CH₃
40 FIAU   R = I
41 FEAU   R = CH₂CH₃
42 FEFAU R = CH₂CH₂F

47

48

49 B = Ade
50 B = Cyt
51 B = Ura

52

53

Ade =

Cyt =

Ura =

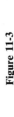

Figure 11-3

B. C-2′ Fluorinated Analogs of Polynucleotides

Oligomers and polymers of 2′-fluoronucleosides have been prepared in order to study the role the 2′ substituent plays in the physical and biological properties of both RNA and DNA. The most sustained effort in this area has come from Ikehara and co-workers.[12,17,39–42]

After synthesizing 2′-fluoro-2′-deoxyadenosine (dAfl) **(39)**, NMR studies were done and it was determined that the compound had a strongly N-type favored conformation, whereas normal ribonucleosides usually favor the S-type configuration.[9,17] Three dinucleoside monophosphates containing dAfl and adenosine (A) were prepared (A-dAfl, dAfl-A, and dAfl-dAfl) and compared to the A-A dimer. The results of UV, CD, and NMR studies showed that the three dimers exist in a stacked conformation with a geometry similar to A-A but with a greater extent of base-base overlapping.[39] The magnitude of the circular dichroism bands is in the order dAfl-dAfl > dAfl-dA > dA-dAfl, which also parallels the decrease of the magnitudes of the Tm's of the complexes of these dimers with poly(U).[12] The three 2′-fluoro-containing dimers formed complexes with poly(uridylic acid) at 0.8°C. Melting temperature (T_m) studies showed that the dAfl-dAfl·poly(U) complex was more thermally stable than the A-A·2poly(U) complex by 7°C. It could not be determined if the increased thermal stability was due to the fluorine substituent only or if it was due to the N-type conformation preference.

RNA can exist only in the A-form structure; the 2′ position of each ribosyl group bearing an OH group, which intrinsically prefers the N-type conformation. DNA can exist in either the A form or the B form; the 2′ positions bearing only hydrogen. Fluorine is only slightly larger in size than hydrogen and much smaller than OH, yet the dAfl-dAfl dimer prefers the N-type conformation. Thus, it was postulated that the reason RNA does not take up the B form structure may be due to the steric and electrostatic repulsions of the 2′-OH groups in adjacent residues in the B-form structure.

Poly(2′-fluoro-2′-deoxyadenylic acid) [poly(dAfl) and other [poly(2′-halo)adenylic acid]s] have been tested in vitro as messenger RNAs in protein synthesizing systems.[41] All of these modified polynucleotides showed activity as messengers and lead to the incorporation of [^{14}C]lysine into polypeptides. The initial velocity of polylysine formation was greater using poly-(dAfl) as messenger than in the case of poly (rA), and all of the synthetic messengers were longer-lived in the protein synthetic system.

NMR, UV, and CD studies of 2′-deoxy-2′-fluoroguanylyl-(3′-5′)-uridine (Gfl-U) showed that it preferred to take up a right-handed, weakly stacked, N-type conformation.[43]

Poly(2′-deoxy-2′-fluoroinosinic acid) [poly (IA)] has been synthesized by polymerization of 2′-deoxy-2′-fluoroinosine 5′-diphosphate catalyzed by *E. coli* polynucleotide phosphorylase, and its properties compared to poly(inosinic acid).[40] Poly (Ifl) forms a 1 : 1 complex with poly(C) that has a melting curve similar to that of poly(I)·poly(C) complex, which is a well-known interferon inducer. The poly(Ifl)·poly(C) complex melts 10°–12° higher than poly(I)·poly(C). In contrast, poly(Cfl) forms a 1 : 1 complex with poly(I) which is less stable than poly(I)·poly(C). A fiber X-ray pattern of poly(Ifl)·poly(C) demonstrated that the overall conformation was similar to that of poly(I)·poly(C), but apparently in the former, C-4' assumes an exo form which is slightly different than the 3'-endo form assumed by poly(I) in poly(I)·poly(C).

Poly(dIfl) has been shown to be an effective template for the RNA-directed DNA polymerase (reverse transcriptase) from Moloney murine leukemia (MuLV) virus.[42] Poly(inosinic acid) [poly(I)] showed little, if any, template activity when assayed under the same conditions. Poly(dIfl) assumed the role of primer for the reverse transcriptase reaction of poly(2'-methylcytidylic acid) and other templates, whereas poly(I) failed to do so here as well. From kinetic data Lineweaver-Burk plots were constructed and the apparent K_m value for the template activity of poly(dIfl) was found to be almost $\frac{1}{10}$ that of the apparent K_m value for poly(I). From these studies it is apparent that the conformational differences induced by the electronegative fluorine atom at C-2' (Subsection A, Section 2) affect the polymer structure and hence binding properties which in turn affect biological activity.

C. C-3' Fluoronucleosides

Structure

The 3'-fluoro substituent has an effect on the 3' position of ribose that is quantitatively similar to the effect that a 2'-fluoro substituent has on the 2' position of ribose. NMR studies suggest that 2'-fluoro-2'-deoxyuridine (52) exists 78% in the N form while 3'-fluoro-2',3'-dideoxyuridine (53) exists 88% in the S conformation.[44] The effect of the fluorine on configuration is similar at the dinucleotide level.[45]

Synthesis

In comparison to the C-2' fluoronucleosides, there are far fewer nucleosides reported that have been modified in the C-3' position by a fluorine substituent. The synthesis of 3'-fluoro-2',3'-dideoxythymidine (56) from 2'-deoxythymidine (54) in an overall yield of 22–25% via 2,3'-anhydro-1-(2-deoxy-5-O-mesyl-β-D-xylofuranosyl)thymine (55) has been achieved as shown in Scheme VII.[46,47]

A similar route was followed to obtain 3'-deoxy-3'-fluorouridine (58) via 2,3'-anhydro-1-(β-D-xylofuranosyl)uracil (57).[48] Although the 3-isomer was obtained in 31% yield, the major product from the treatment of 57 with HF/

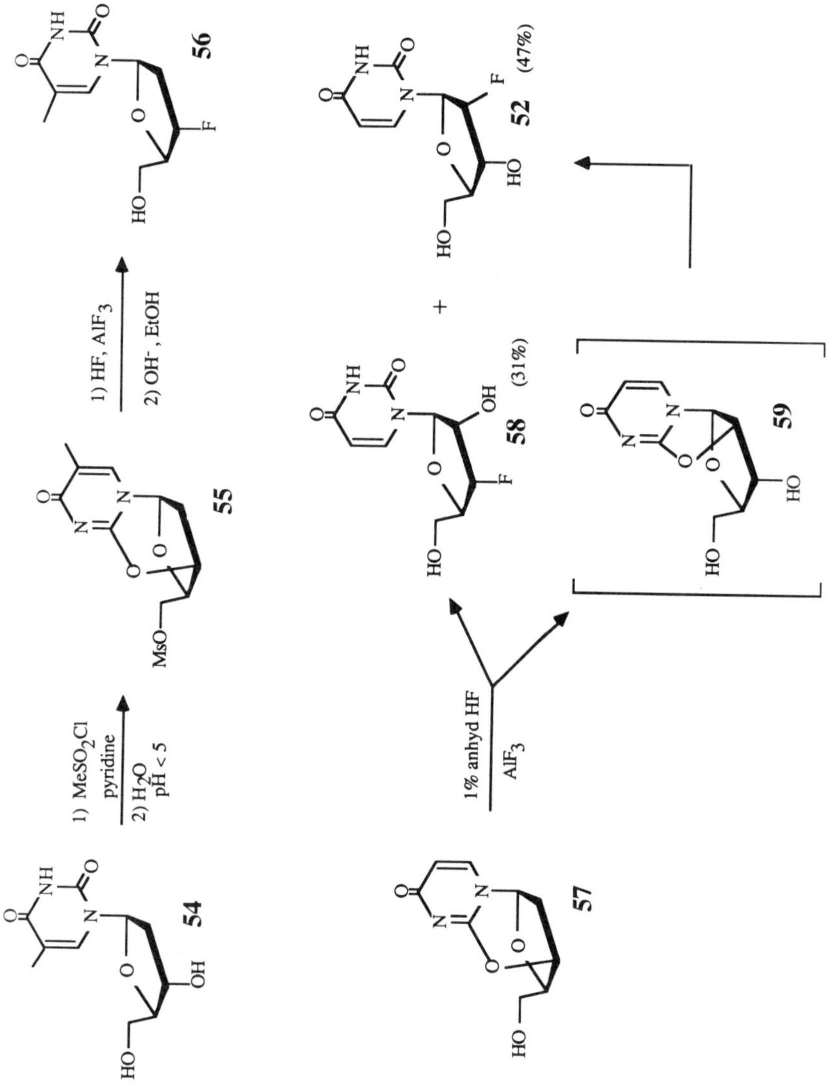

56

52 (47%)

1) HF, AlF₃

2) OH⁻, EtOH

55

+

58 (31%)

59

1) MeSO₂Cl
pyridine

2) H₂O
pH < 5

1% anhyd HF

AlF₃

54

57

Scheme VII

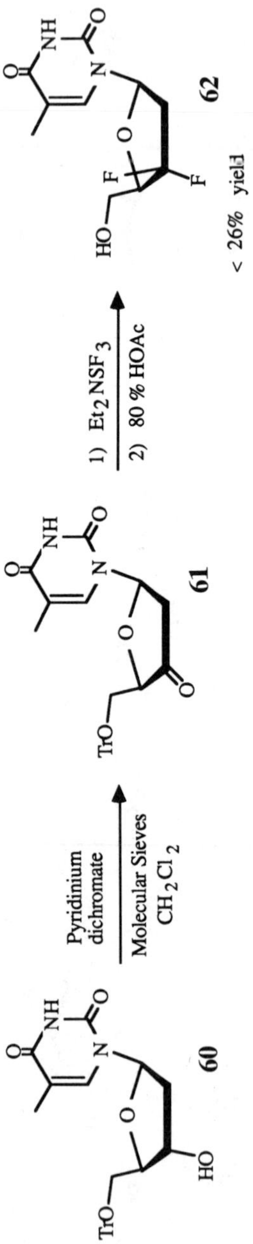
TrO
HO
NH
O
O
N
60
Pyridinium
dichromate
Molecular Sieves
CH_2Cl_2
61
1) Et_2NSF_3
2) 80 % HOAc
F
62
< 26% yield

Scheme VIII

AlF$_3$ was 2'-deoxy-2'-fluorouridine (52), which was obtained in 47% yield. Under the strongly acidic reaction conditions 2,3'-anhydro-1-(β-D-xylofuranosyl)uracil may rearrange to a 2,2'-anhydro intermediate (59) before ring opening to give nucleoside (52). 2',3'-Dideoxy-3',3'-difluorothymidine (62) can be obtained from 5'-O-tritylthymidine (60) in three steps via the 3'-keto derivative, 61 (Scheme VIII).[50]

Biological Activity

Following the reported synthesis in 1969,[51] the biological and biochemical studies of 2',3'-dideoxy-3'-fluorothymidine (56) have appeared in a number of papers. It was determined that the nucleoside can be phosphorylated in vivo up to the triphosphate level and can be incorporated into DNA. This might be expected, because of the close structural resemblance to 2'-deoxythymidine.

2',3'-Dideoxy-3'-fluorothymidine 5'-triphosphate (65) strongly inhibits DNA polymerase from *Micrococcus lutens* and *Streptomyces hygroscopicus*.[52] The analog is incorporated into DNA by both of these microbial polymerases, and hence blocks further chain elongation since there is no 3'-OH available for chain extension from the 3' terminus of the DNA. In contrast, 2',3'-dideoxy-3'-fluorothymidine 5'-triphosphate is a competitive inhibitor of phage T4 wild-type and mutant DNA polymerases, but does not appear to be incorporated as a chain terminator.[53] The sensitivity of the assay was only sufficient to determine incorporation levels greater than one fluorinated nucleoside per 1000 thymidine molecules. The dideoxy analog, 2',3'-dideoxythymidine 5'-triphosphate (66) is incorporated into DNA by the T4 polymerase, albeit at only 10^{-3} the rate of the natural substrate.

2',3'-Dideoxy-3'-fluororibonucleoside 5'-triphosphates (67) of the four common DNA bases, adenine, cytosine, guanine, and thymine have been compared as substrates for *E. coli* DNA polymerase I.[54] All were found to serve as terminators for DNA synthesis and were proposed as alternative substrates for DNA sequencing since they gave clearer sequence patterns and the optimal concentration was an order of magnitude lower than that required for 2',3'-dideoxyNTPs which are the current standard. The 3'-fluoronucleoside 5'-triphosphates are also substrates for calf thymus terminal deoxyribonucleotidyl transferase and avian myeloblastosis virus reverse transcriptase, but not for calf thymus DNA polymerase α.[55,56]

In an attempt to find a substrate analog that would behave similar to 5-fluoro-2'-deoxyuridine (68) with respect to inhibition of dTMP synthetase, but which would not be degraded by thymidine phosphorylase, Ajmera et al. investigated 2',3'-dideoxy-3',5-difluorouridine (3'-FFdUrd) (69).[57] As hoped, 3'-FFdUrd was not a substrate for dThd phosphorylase from Lewis lung carcinoma, but unfortunately it was also not a very good inhibitor of L1210 leukemia cell proliferation. 3'-FFdUrd gave an IC$_{50}$ of only 3.4 $\times$

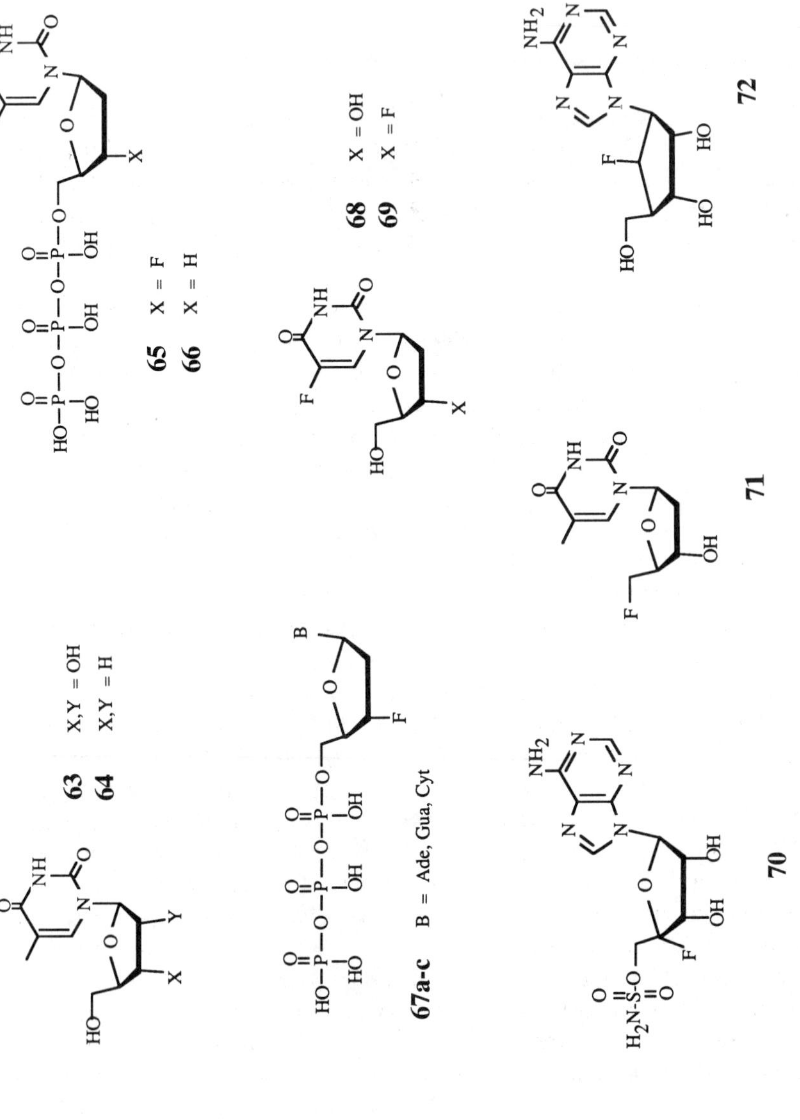

65   X = F
66   X = H

63   X,Y = OH
64   X,Y = H

68   X = OH
69   X = F

67a-c   B = Ade, Gua, Cyt

71

70

72

Figure 11-4

10^{-6} μM in comparison to FdUrd at 5×10^{-10} M. The K_i value for 3'-FFdUrd for inhibition of dTMP synthetase was 0.13 mM compared to a K_i for FdUMP of about 0.4 μM. The 3' hydroxyl group is clearly a very important structural feature for binding to dTMP synthetase. From the results of kinetic studies it was suggested that the weaker initial interaction was tied, via an "induced fit" phenomenon, to a much lower rate of conversion to the tight-binding complex.

D. C-4′ and C-5′ Fluoronucleosides

Very little work has been done with these two classes of compounds despite the occurrence in nature of one of the few fluorine-containing natural products, the nucleoside antibiotic nucleocidin (70), which possesses a 4′-fluorine.[58] The synthesis of nucleocidin was reported in 1976.[59]

The synthesis of 5′-fluoro-2′,5′-dideoxyuridine (71) has been reported and a number of NMR studies have been done with this nucleoside and related derivatives.[60]

E. C-6′ Fluoronucleosides

Carbocyclic nucleosides, analogs in which the ribosyl ring oxygen is replaced by a CH_2 group, are of interest because the analogs often retain biological properties while gaining resistance to enzymes that catabolize the normal nucleoside substrates. Unfortunately, many nucleoside analogs that act as antiviral agents show lower activity when the O-4′ oxygen is replaced by a methylene group (for purposes of common consistent nomenclature, O-4′ becomes C-6′ when replaced by a methylene group; C-4′ and C-5′ are already present in the ribose molecule).[61,62] If the O-4′ oxygen is required for some polar interaction at an active site, then potentially a CFH or CF_2 group would be a more suitable replacement than CH_2. Currently the only analogs of this type reported in the literature are the β-6′-fluoro analog of ($\pm$) aristeromycin (72),[63] and the two diastereomers, carbocyclic 1-(2′-deoxy-6′-β-fluorofuranosyl)-5-iodouracil (76) and carbocyclic 1-(2′-deoxy-6′-α-fluorofuranosyl)-5-iodouracil (79).[64] The latter two compounds have been synthesized via carbocyclic 1-aminoribosyl analogs as shown in Scheme IX.

This study established that the configuration at C-6′ significantly influenced biological activity. Carbocyclic 1-(2′-deoxy-6′-α-fluorofuranosyl)-5-iodouracil (79) was almost as active against herpes simplex type 1 as the antiviral agent acyclovir, while carbocyclic 1-(2′-deoxy-6′-β-fluorofuranosyl)-5-iodouracil (76) was at least two orders of magnitude less active.

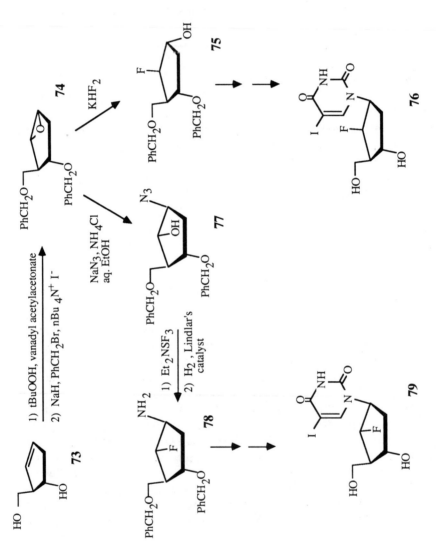

Scheme IX

3. Phosphate Modified Analogs of Nucleotides

A. Fluorophosphates

Structure and Properties

Fluorophosphate analogs of nucleotides have been used in enzyme studies to examine the role of the phosphate group. When fluorine is substituted for a phosphate hydroxyl the resulting nucleotide analog carries only a single negative charge at physiological pH. Consequently, the importance of the phosphate charge in enzyme-substrate binding can be assessed. Other types of phosphate modified nucleotides, such as the methylphosphonate substituted nucleosides, are less suitable because the increased steric bulk may hinder binding. The fluorine atom occupies slightly less space than the hydroxyl group it replaces (average P—X bond lengths are approximately 1.52 and 1.57 Å, respectively, for X = F and O).[65] Electronically the fluorine substituent has a more profound effect than other substituents because of its high electronegativity. The pK_as of a number of phosphoric acid derivatives that are commonly incorporated into nucleotide analogs are shown in Table 11-1.[66]

The pK_a for ionization to the monoanions range from 0.55 for fluorophosphoric acid to 3.0 for phosphoramidic acid. Nucleoside monoesters derived from each of these derivatives would be expected to exist as the monoanion at physiologic pH. Of course, the natural phosphate esters which have two ionizable OH groups are primarily in the dianion form at physiologic pH. The enhanced stability of the fluorophosphate anions of nucleotide analogs **80–86** (Figure 11-5) will weaken binding to metal cations relative to the natural phosphate. This has been demonstrated in Mg^{2+} titration studies with the ADP and ATP analogs **81** and **82.** Both bind to Mg^{2+} relatively poorly compared to ADP, ATP, or analogs in which the oxygen linking the two end phosphorus atoms (β,γ) is replaced by a methylene group.[67]

TABLE 11-1.

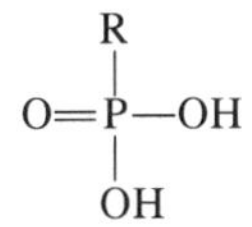

$$O=\overset{\overset{\displaystyle R}{|}}{\underset{\underset{\displaystyle OH}{|}}{P}}-OH$$

R	F	H	OH	CH₃	NH₂
pKa 1	0.55	1.3	2.0	2.3	3.0
pKa 2	4.8	6.7	6.8	7.9	8.15

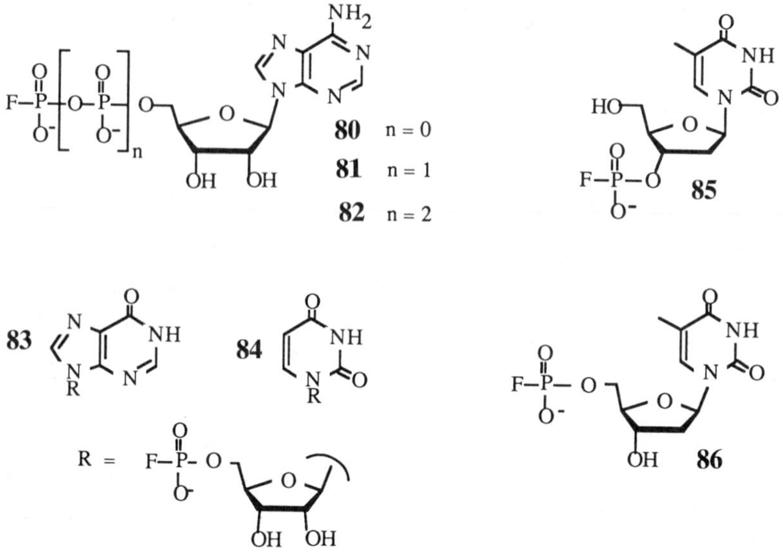

Figure 11-5

Synthesis

The majority of nucleoside monofluorophosphates (84–86) have been synthesized by the method of Wittman, wherein a nucleoside 3'- or 5'-phosphate is allowed to react with excess 2,4-dinitrofluorobenzene and tri-n-butylamine in N,N-dimethylformamide.[68] The 3'-phosphorofluoridate of a ribonucleoside could not be obtained by this method because of cyclization of the activated intermediate to the 2',3'-cyclic phosphate.

Inosine 5'-phosphorofluoridate (83) was obtained by the reaction between inosine 2',3'-carbonate, fluorophosphoric acid, and dicyclohexylcarbodiimide in pyridine. The carbonate protecting group is rapidly removed by ammonia.[69]

Biochemistry

Inosine 5'-monophosphate analogs in which one of the phosphate oxygens was replaced by CH_3, CH_2Cl, H, or F were found to be only weak noncompetitive inhibitors of inosine 5'-phosphate dehydrogenase of *Aerobacter aerogenes* and adenylosuccinate synthetase of *E. coli*.[69] The failure of inosine 5'-phosphorofluoridate (83) to serve as a substrate for either of these enzymes precludes a mechanism which involves nucleophilic attack to generate enzyme-phosphorus bonds.

Kucerova and Skoda determined that uridine 5'-fluorophosphate (84) was not a substrate for the alkaline phosphatase [EC 3.1.1.1] and acid phosphatase [EC3.1.3.2] under conditions that cleaved uridine 5'-monophosphate completely in one hour.[70] 5'-Nucleotidase [EC 3.1.3.5] cleaves uridine 5'-fluorophosphate to uridine, while snake venom phosphodiesterase yielded uridine 5'-phosphate at a rate five times greater than that of uridylyl-3'-5'-uridine under the same conditions.

Nuclear exoribonucleases isolated from Ehrlich ascites tumors and HeLa cells are irreversibly inhibited by thymidine 3'-fluorophosphate (85), but not by thymidine 5'-fluorophosphate (86).[71]

Because of the small steric size of the fluorine atom, nucleoside fluorophosphates can be expected to be useful probes for determining whether the mono- or di-anionic form of a nucleotide is necessarily involved in certain enzyme activity. Withers and Madsen studied the effect of adenosine 5'-fluorophosphate on a number of enzymes that are normally activated by 5'-AMP.[72] Adenosine 5'-fluorophosphate did not activate glycogen phosphorylase b even though it did inhibit the activation of this enzyme by AMP with a K_i of 3 mM. This was taken as proof that this enzyme has a strict requirement for the nucleoside diphosphate in a dianionic form.

B. Fluoroalkylphosphonates

Properties

Phosphate esters and anhydrides constitute two of the most significant structural entities in living organisms. The transfer and exchange of phosphate groups is an essential occurrence in an almost endless array of biochemical reactions.[73] Detailed studies of the chemical transformations that occur during these biochemical processes have been greatly aided by the use of phosphonate analogs in which a methylene group has been incorporated into a substrate or cofactor in place of an oxygen.[74] Although the CH_2 group of alkylphosphonate is a suitable steric replacement for the oxygen of biological phosphonates, electronically it is quite different since carbon is much less electronegative than oxygen. G. Michael Backburn set out a number of years ago to demonstrate that α-fluoroalkylphosphonates would be more effective analogs because they can satisfy both the steric and electronic criteria.[75] The electronic (or polar) effect is best illustrated by comparison of the pka's of pyrophosphate and the three isosteric analogs shown in Table 11-2.[76,77] The data show that the CF_2 analog of 87 is a close electronic match to pyrophosphate. Whereas the CF_2 analog 90 will exist almost entirely as the trianion at physiological pH (~7.2), the CH_2 analog 88 will be principally in the dianion form. Similar effects on polarity would be expected for other fluoroalkylphosphonate analogs.

Blackburn has noted three physical parameters for comparing the electronic effect of the substituent R, in $RP(O)(OH)_2$, to assess the electronic

TABLE 11-2. Effect of the Bridging Atom on Pyrophosphate Analog pK_a.[76] (The numbers in parentheses were determined from ^{31}P NMR data.[77])

	Pyrophosphate analog	pKa_2	pKa_3	pKa_4
87	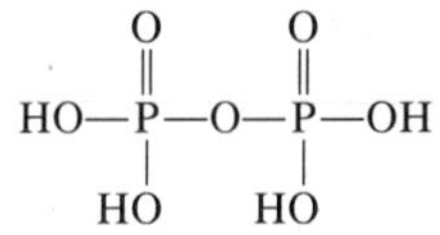 HO—P(=O)(OH)—O—P(=O)(OH)—OH	2.36	5.77	8.22
88	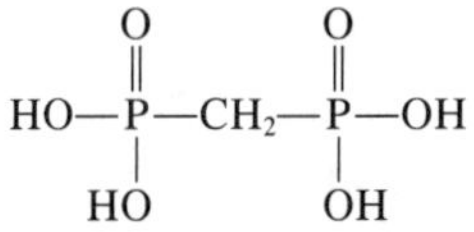 HO—P(=O)(OH)—CH$_2$—P(=O)(OH)—OH	2.87	7.45	10.96
89	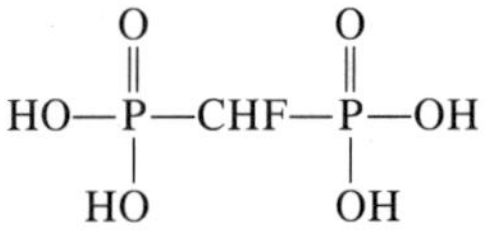 HO—P(=O)(OH)—CHF—P(=O)(OH)—OH	(2.78)	(6.62)	(10.1)
90	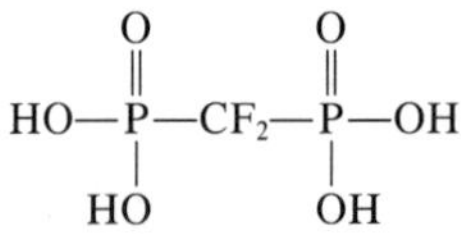 HO—P(=O)(OH)—CF$_2$—P(=O)(OH)—OH	2.60 (2.57)	5.80 (6.08)	8.00 (9.07)

suitability of a replacement group for the oxygen of phosphomonoesters.[78] These include the second acid dissociation constant, pKa_2, the stretching frequency of the P=O bond, and the ^{31}P NMR chemical shift. Data is compared for phenyl phosphate and two benzylphosphonate analogs in Table 11-3.

Data from Table 11-3 support the hypothesis that the CHF group is a good isopolar replacement for oxygen, and establishes a trend that suggests that the CF$_2$ group may possibly be an electronically closer match to oxygen. One can envision replacing oxygen and incorporating the CF$_2$ unit into nucleoside monophosphates or into the phosphodiester linkage between nucleosides as depicted in Figure 11-6. Of these three types of analogs only the synthesis of derivatives of type 94 has so far been realized.

Synthesis

α-Fluoroalkylphosphonates have been synthesized by a number of different methods. α-Fluorobenzylphosphonate esters can be obtained in good yield from α-hydroxybenzylphosphonate esters by reaction with diethylaminosulfur trifluoride.[78] But the reaction is not general since secondary alkylphosphonates possessing β hydrogens undergo dehydration rather than substitution. An alternate route employs the carbanion generated from diisopropyl

TABLE 11-3. Phosphonate Group Parameters Effected by Substituents Attached at Phosphorus

	Compound	pKa_2	$v_{P=O}(cm^{-1})$ of diethyl ester	$\delta\ ^{31}P\ (ppm)$ of diethyl ester
91	PhO—P(=O)(OH)—OH	6.20	1275	6.8
92	PhCHF—P(=O)(OH)—OH	6.50	1266	14.71
93	PhCH₂—P(=O)(OH)—OH	7.60	1254	27.06

fluoromethylphosphonate[79] or tetraisopropyl fluoromethylenebisphosphonate,[80] which react with aldehydes and ketones. This approach was used to introduce a CF group in place of O-5 of ribose (Scheme X). Application to the synthesis of α-fluoroalkylphosphonate analogs of nucleoside 5'-phosphates appears to be forthcoming.[81]

Difluoromethylphosphonate esters are most easily prepared via the reaction of difluoromethyl phosphonochloridate (CHF_2POCl_2) with alcohols. The chloridate is prepared in two steps from diethyl phosphite and chlorodifluoromethane.[82] Difluoromethyl phosphonochloridate reacts with thymidine

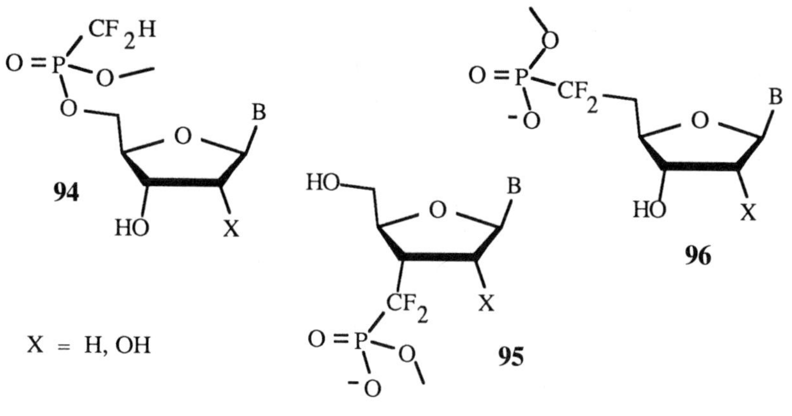

X = H, OH

Figure 11-6

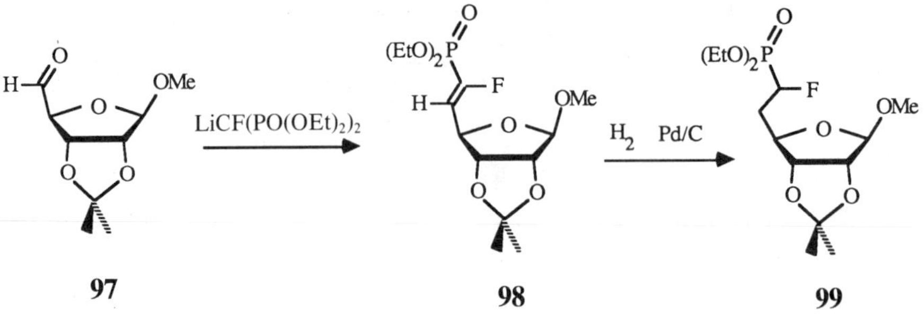

Scheme X

in trimethylphosphate (modified Yoshikawa procedure[83]) to give, after aqueous workup and chromatography on DEAE-Sephadex with an ammonium bicarbonate gradient, a mixture of the 3'- and 5'-phosphonylated analogs **100–102** (Scheme XI). These have been separated by HPLC on a C-18 reverse phase column. The pure 3'-phosphonylated derivative, **101,** can also be obtained by reaction of difluoromethyl phosphonochloridate with 5'-tritylthymidine followed by aqueous workup and detritylation in 80% aqueous acetic acid, while the 5' analog, **100,** has been obtained by phosphonylation of 3'-O-acetylthymidine.[84]

Preparation and characterization of the protected TpT dimer analog, **107,** was achieved with somewhat greater effort. Following a procedure established by van Boom and co-workers for the preparation of oligonucleotides using 1-hydroxybenzotriazole activated phosphotriester intermediates,[85,86] it was determined that the difluoromethylphosphonate linkage could be introduced as a 3'-5' linkage.[87] Sequential reaction of difluoromethyl-O,O-*bis* (1-benzotriazolyl) phosphonate **(104)** with a 5'-O-dimethoxytrityl-2'-deoxythymidine **(103)** and 3'-O-levulinyl-2'-deoxythymidine **(106)** gave two dimeric products (Scheme XII).

The principal product was the symmetrical 3'-3' dimer, **108,** which could be separated from the unsymmetrical 3'-5' dimer **107** (10% yield) by chromatography on silica gel (gradient elution with MeOH/CHCl₃). In the ¹⁹F NMR, **108** showed two sets of four peaks (dd) separated by 6 Hz. Each set represents one of the fluorines, which are diastereotopic and hence show different chemical shifts in the spectra. The 3'-5' dimer was a mixture of two diastereomers (1 : 1) that could be separated by careful chromatography on silica gel eluting with ethyl acetate/methanol. The ¹⁹F NMR spectra of the separated diastereomers are very different and must reflect the distinct environments of the CF₂H groups. One of the isomers shows only four narrow peaks (dd, J_{HF} = 48 Hz and J_{FP} = 94 Hz) in the proton undecoupled spectrum. The two fluorine atoms have exactly the same chemical shift despite their stereochemical uniqueness. The second isomer shows sixteen peaks as a result of the chemical shift difference between the two diastereotopic fluorines (two ddd, J_{HF} = 48 Hz, J_{FP} = 93 Hz, and J_{FF} = 462 Hz).

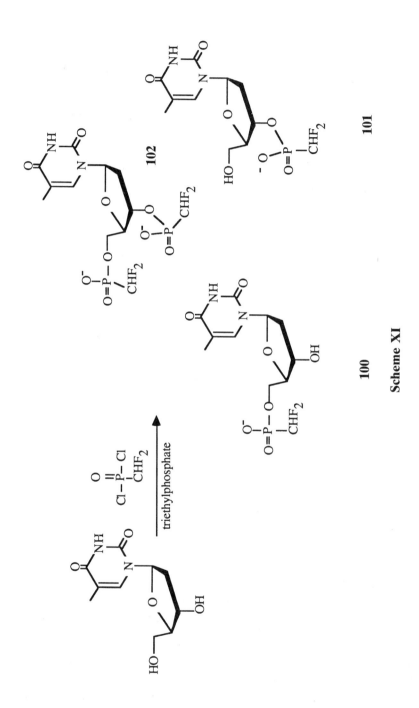

Scheme XI

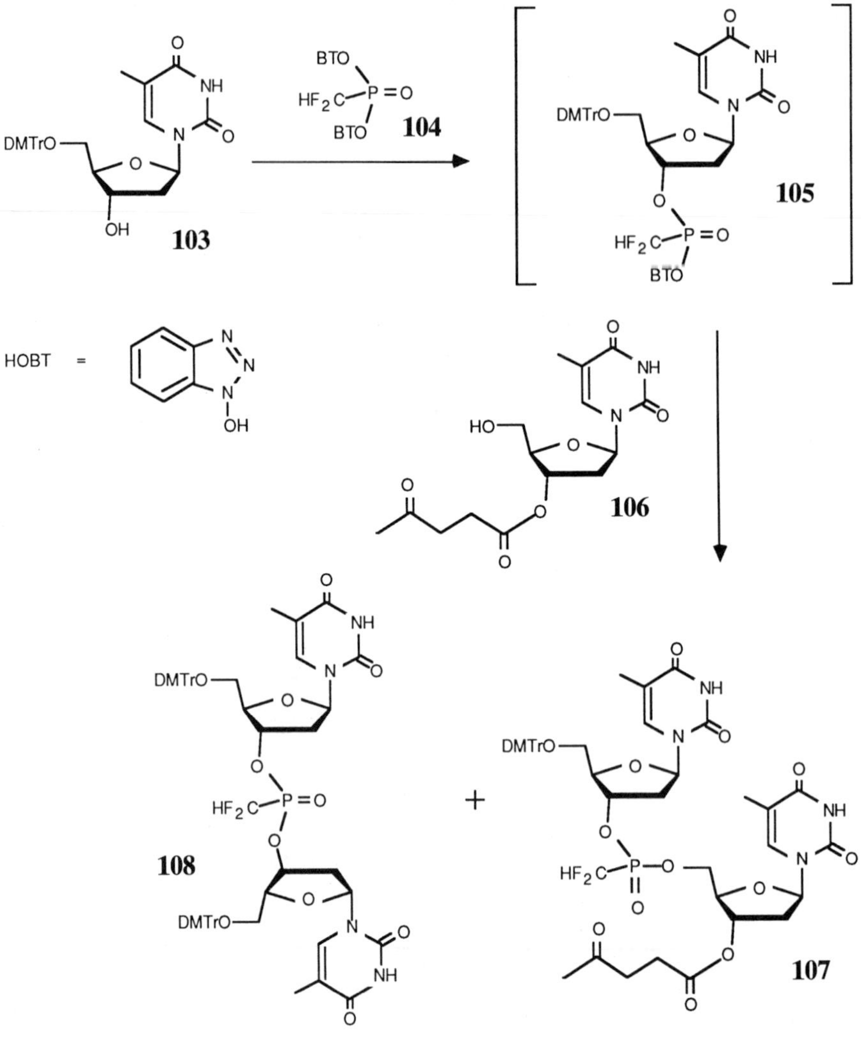

Scheme XII

To compare the biochemical effects of various isosteric linkages between phosphates, Blackburn and co-workers synthesized a series of ATP analogs in which the β,γ linkage in the triphosphate group was O, CF_2, NH, CCl_2, CH_2, and $C{\equiv}C$ (Figure 11-7).[88] The electronic influence of the CF_2 group shows up in the fourth ionization constant of the triphosphate group. The pKa's of ATP, $p_{CH_2}ppA$, $p_{CFH}ppA$, and $p_{CF_2}ppA$ were determined to be, respectively, 7.1, 8.4, 7.4, and 6.7.

Synthesis of **109** can be accomplished via difluoromethylenebisphos-

Compound	X	Abbreviation
109	CF_2	$p_{CF_2}ppA$
110	CH_2	$p_{CH_2}ppA$
111	CHF	$p_{CHF}ppA$
112	O	pppA (ATP)

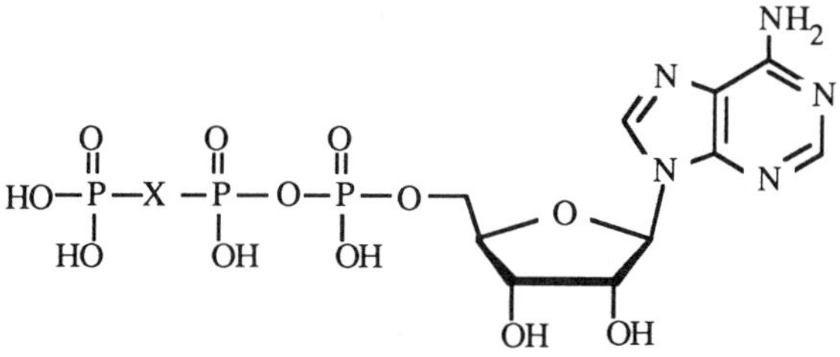

Figure 11-7

phonic acid, obtained from a tetraalkyl difluoromethylenebisphosphonate, which in turn can be prepared by the reaction of sodium dialkyl phosphite with dialkyl bromodifluoromethylphosphonate.[89–91] Nucleoside triphosphate analogs can then be obtained by conventional condensation between a 5'-nucleotide monophosphoromorpholidate and the pyrophosphate analog.[92]

Biochemical Studies

Although only preliminary results have been published, the β,γ-difluoromethylene analog, $p_{CF_2}ppA$ **(109)**, appears to affect biochemical processes requiring ATP to a greater extent than the β,γ-methylene analog, $p_{CH_2}ppA$ **(110)**.[88] $p_{CF_2}ppA$ was found to be a more effective inhibitor ($K_i = 0.90$ mM) of hexokinase than $p_{CH_2}ppA$ ($K_i = 2.5$ mM), but did not bind quite as tightly as the natural substrate, ATP ($K_m = 0.25$ mM). $p_{CF_2}ppA$ is also a competitive inhibitor ($K_i = 2.5$ mM) of rabbit muscle pyruvate kinase ($K_m = 1.4$ mM for ATP), and a good substrate for bovine heart adenylate kinase.[81]

(2'-5')Oligoadenylate synthetase from HeLa cells transforms $p_{CF_2}ppA$ into a 2-5 A oligomer composed of 3 to 8 adenylate units terminated at the 5' end by the $p_{CF_2}pp$ group. These analogs appear to behave in many respects like the natural ppp(A2'p)nA oligomers. They are degraded by snake venom phosphodiesterase and resistant to T2 RNase. Since the chemical stability is enhanced and they show little cytotoxicity at submicromolar concentration, it has been speculated that such analogs may be useful antagonists of the 2-5A system.[81]

4. HETEROCYCLIC MODIFIED NUCLEOSIDES AND NUCLEOTIDES

A. C-5 Substituted Pyrimidine Nucleosides

Synthesis

The fluorine substituted pyrimidine nucleosides, 5-fluoro-2'-deoxyuridine (68), 5-trifluoromethyl-2'-deoxyuridine (113), are well established as thera peutic agents. Most current studies, particularly with 5-fluoro-2'-deoxyuridine and 5-fluorouracil, are concerned with details of the pharmacology and biochemistry with little emphasis on new synthetic methods or on the effects of the fluorine atom on structure. Only a few studies are referenced in this review.

Direct fluorination (F_2 or CF_3OF) of uracil in aqueous solvent gives 5-fluorouracil and 5-fluoro-6-hydroxy-5,6-dihydrouracil which can be dehy-drated to 5-fluorouracil. This reaction is the basis for commercial production of 5-fluorouracil.[93] In alcohol solvents uracil and F_2 give 6-alkoxy-5-fluorouracil in good yield.[94] This reaction has opened the door for the con-struction of some unusual dicyclouracil nucleosides as shown in Scheme XIII. 6,2'-cyclouracil arabinoside (114) reacts with fluorine gas diluted with nitrogen in acetic acid to give 6,2':6,5'-dianhydro-5,5-difluoro-6,6-dihy-droxy-5,6-dihydrouracil arabinoside (115).[95]

Placing fluorine atoms on carbon side chains attached to C-5 calls for different methodologies depending on the location and bonding of the fluo-rine. Whereas CF_3 can be attached directly by coupling 3',5'-di-O-acetyl-5-iodo-2'-deoxyuridine (116) with trifluoromethyl iodide and copper powder,[96] attachment of CHF_2 and CH_2F requires quite a different strategy.[97] Light-catalyzed free radical bromination of the C-5 methyl group of suitably pro-tected thymidine analogs can be used to attach either one or two bromine atoms.[98] Since the bromine atoms in these derivatives are benzylic-like they are easily displaced. The preparation of 5-difluoromethyl-2-deoxyuridine (119, X = H) is illustrated in Scheme XIV. The *tert*-butyldiphenylsilyl pro-

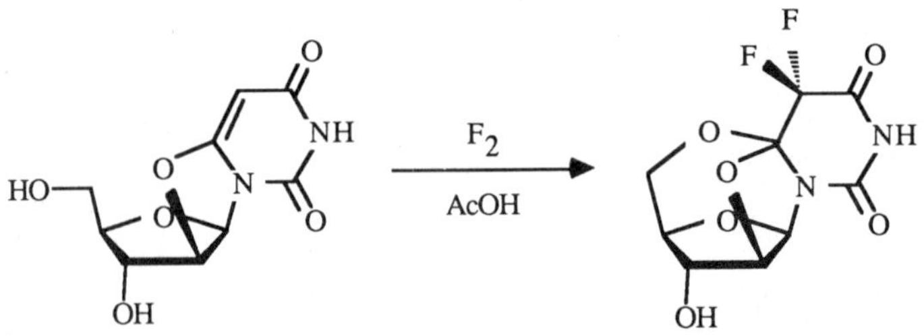

Scheme XIII

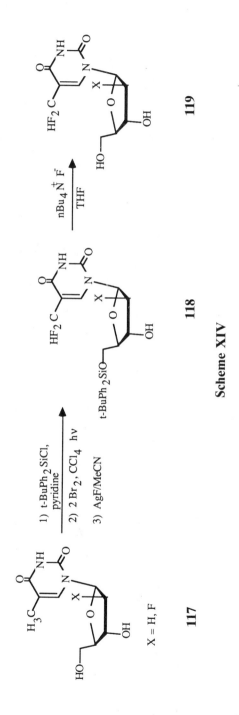

Scheme XIV

tecting group was a key to the success of the procedure since it can be removed under the same conditions ($nBu_4N^+F^-$ in THF) necessary to displace the bromine atoms.

The bromomethyl group was also used as an intermediate in the synthesis of (E)-5-(2-fluoroethenyl)-2'-deoxyuridine (122). Displacement of the α-bromo by the sodium enolate of ethyl fluoroacetate gave nucleoside 121 after the ester groups were saponified by NaOH, and the sugar hydroxyls re-esterified with acetic anhydride. The carboxyl group on the side chain was replaced by a bromine, which was subsequently eliminated with triethylamine to give 3',5'-O-diacetyl-5-(2-fluoroethenyl)-2'-deoxyuridine which was transesterified with alkoxide to give a mixture of the E and Z isomers, 122 and 123.

Methodology for attaching longer fluorine substituted side chains is actually simpler. The trifluoropropenyl side chain can be added in two steps from 2'-deoxyuridine by the sequence shown in Scheme XVI.[100]

A short procedure for attaching perfluoroalkyl groups at C-5 is the reaction of bis(perfluoroalkyl)mercury with unprotected nucleosides. Yields are low, but for the purpose of obtaining new compounds rapidly for biological studies, the method is useful. For example, 5-pentafluoroethyl-2'-deoxyuridine (126) was obtained by reaction of bis(pentafluoroethyl)mercury with 2'-deoxyuridine using the free radical initiator 2,2'-azobisisobutyronitrile in 6.44% yield.[101]

Antiviral Activity

Recently, there has been an interest in substituents larger than methyl attached at C-5 of pyrimidine nucleosides, stemming from the observation that 2'-deoxyuridine substituted by larger side chains in this position have potent activity against herpes viruses.[102] Particularly important, these analogs show a selectivity towards inhibition of viral replication not shown by two classic antiherpes drugs, 5-iodo-2'-deoxyuridine and 5-(3,3,3-trifluoromethyl)-2'-deoxyuridine (F_3CdU, 113, Table 11-4). For example, when the trifluoromethyl group is attached via a CH=CH spacer the toxicity towards mammalian cells drops precipitously while the activity against herpes simplex type 1 (HSV-1) increases. Research in this area has been largely empirical. The possible variations of C-5 modified deoxyuridines is virtually unlimited, but a substantial number of compounds have been synthesized solely to explore the relationship of structure to antiviral activity.[103,104]

In primary rabbit kidney cells, F_3CdU (113) shows antimetabolic activity at a concentration of 0.05 μg/mL. However, the concentration required to inhibit cytopathogenicity of herpes simplex type 1 by 50% is 0.2 μg/mL.[105] In contrast, the antimetabolic activity of F_3CCH=CHdU, 125, is less than 10^{-4} that of F_3CdU. Data in Table 11-4 shows also how the HSV-2 activity of F_3CdU is similar in magnitude to its HSV-1 activity, whereas F_3CCH=CHdU is nearly inactive against HSV-2.[100] It is not the fluorine substitu-

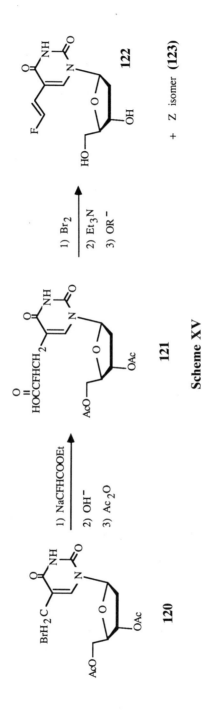

Scheme XV

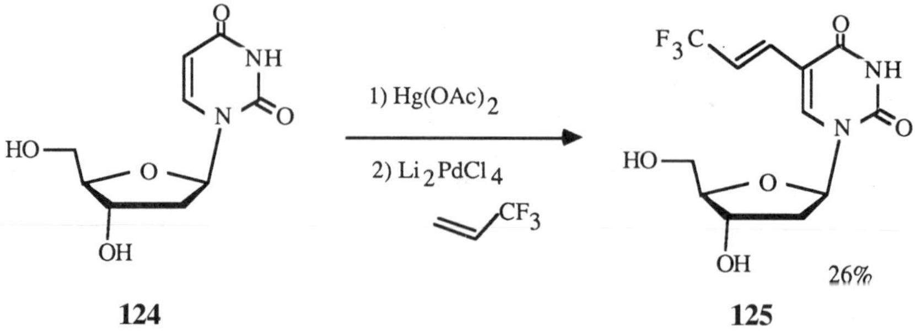

Scheme XVI

tion per se that is responsible for the activity and selectivity shown by this class of compounds. The size and geometry of the C-5 side arm may be more important.[106] If the CF_3 group of (E) $F_3CCH{=}CHdU$ is replaced by a bromine atom, the resulting analog, (E) $BrCH{=}CHdU$, **131,** still has relatively little antimetabolic activity, but is three to four times more active against HSV-1. (*E*)-5-(2-fluorovinyl)-2′-deoxyuridine **(122)** is slightly less active than either (E) $F_3CCH{=}CHdU$ or (E) $BrCH{=}CHdU$.[107] Insufficient knowledge about the structures of the enzymes, HSV thymidine kinase and DNA polymerase preclude more detailed predictions relating analog structure to activity. That the corresponding Z isomers of many of the C-5 substituted alkenyls are much less active demonstrates that there are geometric restrictions.[104]

There is no obvious reason for the lack of antiviral activity of $CF_3CF{=}CFdU$, **129,**[108] in comparison to $F_3CCH{=}CHdU$. Other derivatives containing perfluorinated hydrocarbon substituents in the C-5 position (C_2F_5dU and C_4F_9dU) also appear to lack significant antiviral activity.[101] The CF_2H side chain behaves, as expected, like a CF_3 group. The antiviral activity of CF_2HdU is only slightly less than that of CF_3dU.[97] On the other hand, FCH_2dU did not show antiviral activity. This may be a result of the reactivity of the CH_2F group towards nucleophiles which could prevent significant amounts of the compound from completing the journey from nucleoside to nucleoside 5′-triphosphate.

B. Fluorinated Purine Nucleoside Analogs

The best method for induction of a fluorine into the purine ring appears to be nonaqueous diazotization. Robins and Uznanski have reported on the synthesis of 2-fluoroadenosine triacetate in 48% yield via the reaction of 2,6-diamino-9-(2′,3′,5′-tri-O-acetyl-β-D-ribofuranosyl)purine **(132)** with *t*-butyl nitrite in HF/pyridine at −20°C.[109] Deprotection was achieved by NH_3 in methanol to give 2-fluoroadenosine **(134)** (Scheme XVII).

TABLE 11-4

Compound	R	Anti HSV-1 Activity	Anti HSV-2 Activity
113	CF_3 [b]	0.7	1.2
119	CHF_2 [a]	0.39	0.46
122	(E) CH=CHF [c]	3.8	9.2
123	(Z) CH=CHF [c]	Inactive	Inactive
125	(E) CH=CHCF$_3$ [b]	0.03	100
126	C_2F_5 [c]	93	166
127	CH_2F [a]	>100	>100
128	CH=CH$_2$ [d]	0.02	0.07
129	CF=CFCF$_3$ [d]	Inactive	Inactive
130	(E) CH=CHCH$_3$ [b]	0.15	18
131	(E) CH=CHBr [b]	0.008	3

a - Inhibitory concentration in human foreskin fibroblasts (% inhibition unspecified).

b - Concentration required to inhibit virus yield by 99% in PRK cells.

c - Concentration in µg/mL, required to reduce virus plaque formation by 50% in HELF cell cultures.

d - Minimal inhibitory concentration (concentration in µg/mL, required to reduce virus-induced cytopathic effects by 50% in PRK cells).

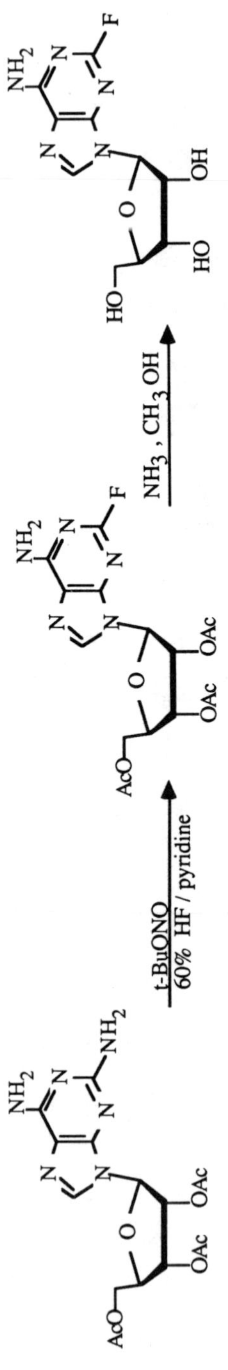

Scheme XVII

Another route, aromatic nucleophilic displacement of trimethylamine by fluoride ion, works well for the conversion of trimethylammoniopurines to fluoropurines.[110] 6-Fluoro-9-β-D-ribofuranosylpurine (137) has been obtained in 23% yield by the reaction of 9-β-D-ribofuranosylpurin-6-yl trimethylammonium chloride[111] with KF in DMF at 50°C.

C. Other Base Modified Analogs

Synthesis

Some interesting modified nucleosides have been obtained by the reaction of pyrimidine nucleosides with fluorocarbenes. Phenyl(dichlorofluoromethyl)-mercury on heating gives chlorofluorocarbene which adds to the 5,6 double bond of protected pyrimidine nucleosides. The reaction is not particularly stereoselective, since for example, 3-methyl-2′,3′-isopropylidine-5′-acetyl-uridine (138) gave four stereoisomers in about equal amounts (Scheme XIX).[112] As illustrated in the scheme the carbene adducts undergo base catalyzed rearrangement to ring expanded nucleosides.[113] Unfortunately, the compounds apparently do not have significant biological activity.

More recently Pein and Cech investigated the reaction of difluorocarbene with 4-O-trimethylsilylnucleosides.[114] In this case the trimethylsilyl group is replaced by a CF_2H group as shown in Scheme XX. The difluorocarbene was generated by thermolysis of bistrifluoromethylmercury. In addition to the thymidine derivative 142 shown in Scheme XX, Pein and Cech synthesized a series of 4-difluoromethoxypyrimidine nucleosides with different C-5 substituents and with the sugar component ribo-, 2′-deoxyribo-, or arabino-furanosyl.

A somewhat more unusual pyrimidine nucleoside analog is 3-fluoropyridinone nucleoside 147, in which a CF group replaces the 3 position nitrogen of cytidine.[115] This is a more interesting compound because rather than simply serving as another substituent with no special role, here the CF may be considered an isopolar–isosteric nitrogen analog. Unlike most of the other analogs described in this section, nucleoside 147 required de novo synthesis as outlined in Scheme XXI.

Biological Activity

All the derivatives discussed in the synthesis section above were designed as either antiviral or antitumor agents. Generally the fluorine substituted nucleosides were synthesized as part of a series of compounds in which the fluorine substitution simply represented one more member of the series; that is, no particular emphasis was placed on the fluorine as a substituent. The exception is the fluoropyridine derivative designed by McNamara and Cook. This derivative was tested against L1210 lymphoid leukemia in mice and

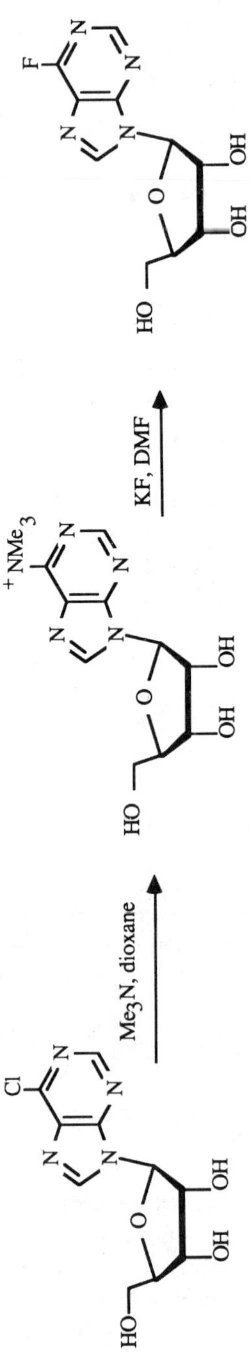

Scheme XVIII

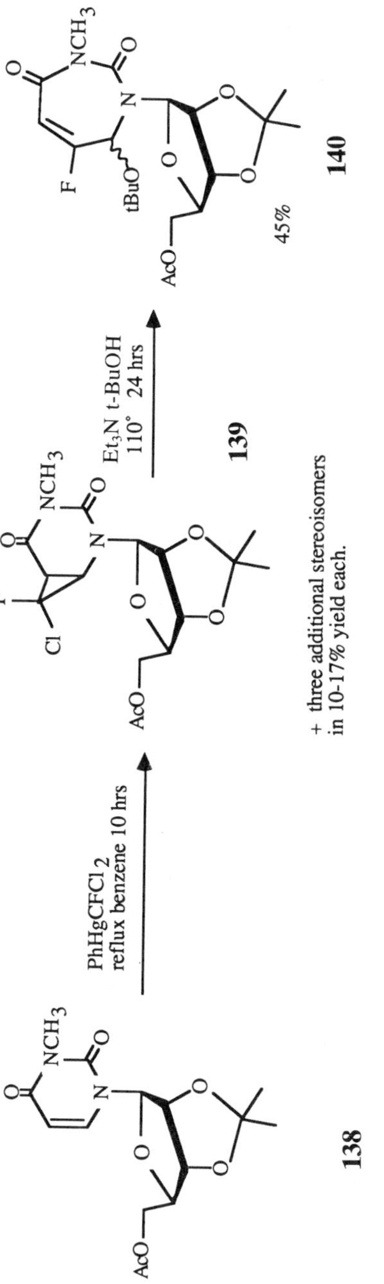

138

PhHgCFCl$_2$
reflux benzene 10 hrs

139

+ three additional stereoisomers
in 10-17% yield each.

Et$_3$N t-BuOH
110° 24 hrs

140

45%

Scheme XIX

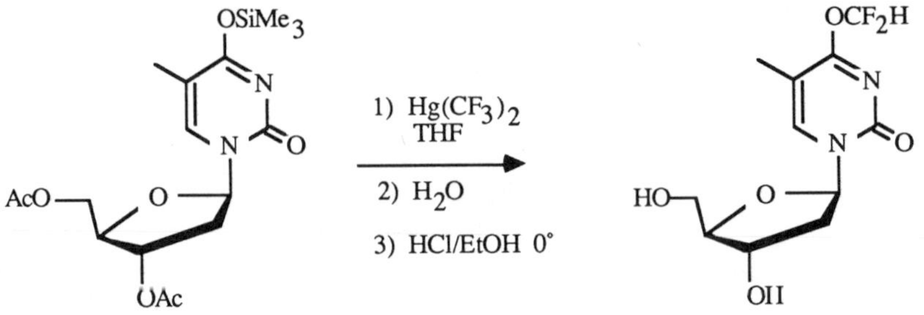

141 **142**

Scheme XX

found to be highly active (net cell kill of 5.54 at 100 mg/kg). Biochemical studies are currently underway to investigate the mode of action of **147**.[115]

The 4-difluoromethoxy derivative of arabinofuranosylthymine synthesized by Pein and Cech proved to be the most potent of any of the 4-difluoromethoxy pyrimidine nucleoside analogs tested against herpes simplex virus type 1 (active at 1.5 μM).[111] However, since there is no information about the metabolism of this compound it cannot be certain whether this is a result of hydrolysis in vivo to arabinosylthymine. The chlorofluorocarbene adducts from Pandit's laboratory were not reported to have biological activity.

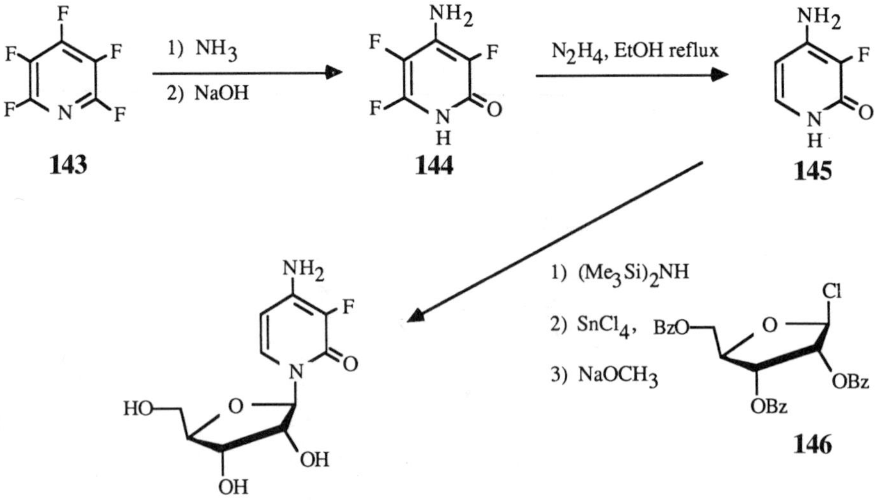

143 **144** **145** **146** **147**

Scheme XXI

5. FLUORINE SUBSTITUTED NUCLEOSIDES, NUCLEOTIDES, AND NUCLEIC ACIDS AS BIOLOGICAL AND BIOCHEMICAL PROBES

A. Positron Emission Transaxial Tomography

[18]F labeled analogs of biologically significant molecules can be used for in vivo imaging studies. Positron emission transaxial tomography (PETT)[116] requires that a compound be labeled with a positron-emitting nuclide that, after administration to a patient, will be transported to, and accumulate in, a site for which tissue physiological and biochemical function need to be measured. Positron-emitting nuclides, which include ^{11}C, ^{18}F, ^{15}O, and ^{13}N, typically have short half-lives for emission of a positron. Of these nuclides, ^{18}F has the longest half-life ($t_\frac{1}{2}$ = 110 min versus 20.4 min for ^{11}C). A longer half-life facilitates ease of preparation and less rushed use in PETT scanning devices in the clinic.

[18]F is prepared by irradiation of [20]Ne with deuterium in a cyclotron or accelerator.

$$^{20}Ne + {}^{2}H \rightarrow {}^{18}F + {}^{4}He$$

The ^{18}F produced can be exchanged with [19]F_2 to give [18]F–[19]F which is then transformed into an appropriate fluorinating agent (XeF$_2$, HF, etc) for incorporation into an organic compound.

When the ^{18}F emits a positron it is converted into [18]O. In tissue a positron will encounter an electron within a range of a few millimeters. The resulting annihilation generates two gamma rays which travel in opposite directions and are detected by coincidence detectors which scan simultaneously 180° apart. As the detector moves around the object being scanned simultaneous gamma rays are detected and the spatial distribution of these events recorded from which an image is computed.

[18]F substituted nucleoside analogs represent potentially useful imaging substances. It has been established that 5-fluorouracil and 5-fluoro-2'-deoxyuridine are taken up selectively by animal tumor cells.[117,118] 5-[^{18}F]fluoro-2'-deoxyuridine has been successfully used to obtain PETT images of a rabbit tumor.[119] However, it has been suggested that biochemically more stable analogs would be advantageous. By incorporating a second fluorine atom at C-2', an analog is created (2',5-difluoro-2'-deoxyuridine, **151**) that is easily transported and relatively resistant to degradation by pyrimidine phosphorylase.[118] Studies with the ^{14}C labeled compound in vivo show that 2',5-difluoro-2'-deoxyuridine has a favorable tumor-to-blood distribution ratio (10.3) and is resistant to phosphorylytic cleavage and dehalogenation. The ^{18}F label could be introduced at either the C-5 or C-2' position by the reactions shown in Scheme XXII.

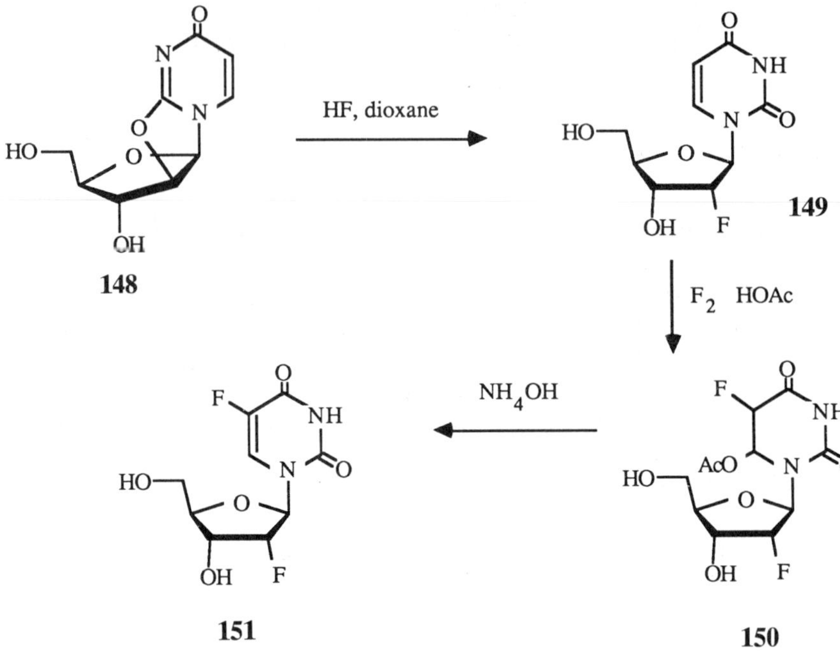

Scheme XXII

The limitations imposed by the need for specialized equipment (cyclotron; PETT scanner) and rapic synthetic methods for incorporating ^{18}F into biologically significant molecules will continue to slow development in this area. There are examples of fluorine substituted analogs that would be useful as imaging agents if ^{18}F could be incorporated into the molecule. 2'-Fluoro-5-methyl-1-β-D-arabinosyluracil (FMAU, **39**) labeled with carbon 14 has been used for autoradiographic mapping of rat brain infected by herpes simplex virus encephalitis.[120] Unfortunately the synthesis of this compound appears too complex to garner any optimism about eventual ^{18}F labeling. Nevertheless, more generally the synthesis of new ^{18}F labeled compounds is an active area of research as evidenced by the publication of thirty-one abstracts on ^{18}F labeled biologically significant compounds presented at the Sixth International Symposium on Radiopharmaceutical Chemistry in 1986.[121]

B. [19]F Nuclear Magnetic Resonance Studies

[19]F Nuclear magnetic resonance of nucleic acid components remains a relatively untapped area of research. But with the sophisticated high field NMR instrumentation now readily available, one can expect to see many more experiments using ^{19}F labeled probes. In the study of the macromolecular

structure and dynamics in solution and as a method for probing interactions between biological molecules, NMR spectroscopy of ^{19}F labeled nucleoside, nucleotide, and polynucleotide analogs should find many uses. Unlike ^{1}H, ^{13}C, or ^{31}P, which occur throughout living organisms, ^{19}F is found at significant levels only in calcified material. Hence, a ^{19}F labeled probe offers the advantage of little interference from other nuclei. Furthermore, the chemical shift of ^{19}F is quite sensitive to environment.[122]

A good example is *E. coli* tRNA$_1^{Val}$, wherein fourteen uracil residues have been replaced by 5-fluorouracil.[123] The labeling is accomplished by growing *E. coli* B in medium containing 5-fluorouracil. The modified tRNA behaves biochemically like the native unmodified tRNA. In the ^{19}F NMR spectrum at 47° all of the resonances are resolved and fall in the range 1.8 to 7.7 ppm downfield from the parent base, 5-fluorouracil. The large variation in chemical shift for the fluorine resonances appears to be related to the secondary and tertiary structure of the tRNA since all of the resonances coalesce into one at a temperature at which the tRNA undergoes complete denaturation (80°). When the ^{19}F NMR signals were correlated with other data the peaks could be tentatively assigned as being due to non-base paired fluorouracils located in loops, base paired fluorouracils located in helical regions, and fluorouracils involved in tertiary structure.

Replacement of uracil residues in *E. coli* 5S RNA with 5-fluorouracil gave a labeled RNA in which only 25–35% of the 5-fluorouracil residues were estimated to be exposed to the solution.[124] Measurement of the spin-relaxation time for 5-fluorouracil labeled *E. coli* 5S RNA gave a value of less than 0.4 second, which indicates a relatively rigid solution structure.

Kremer, Mikita, and Beardsley prepared self-complementary oligodeoxyribonucleotide 12mers in which a single fluorouracil was incorporated in place of thymine.[125] In buffered solution, the N-3 hydrogen bond between the 5-fluorouracil and adenine was not as stable as the hydrogen bond in an A-T base pair. However, the presence of the 5-fluorouracil did not destabilize the corresponding region of the double helix.

A somewhat different type of study involves the binding of fluorine labeled analogs of intercalating drugs in order to follow the process of binding to DNA. When the fluoroquinacrine, 3-fluoro-7-chloro-9-(diethylamino-1-methylbutylamino)acridine **(152)**, interacts with DNA, the changes in the uv-vis absorption spectrum parallels the changes in ^{19}F chemical shift.[126]

Each of the fluorinated acridine derivatives shown in Figure 11-8 gives only one signal when bound to DNA. Compounds **153** and **154** show singlets at 12.1 and 12.5 ppm downfield from external aqueous fluoride. In the absence of DNA both show peaks near 8 ppm downfield from fluoride. On the other hand **153** which cannot undergo bis intercalation, and must exist, in the presence of DNA, with one ring intercalated and the other in solution, shows a peak at 10.6 ppm in the presence of DNA and at 9.1 ppm in the absence of DNA. This result was interpreted as a rapid exchange between intercalated and non-intercalated rings.[127]

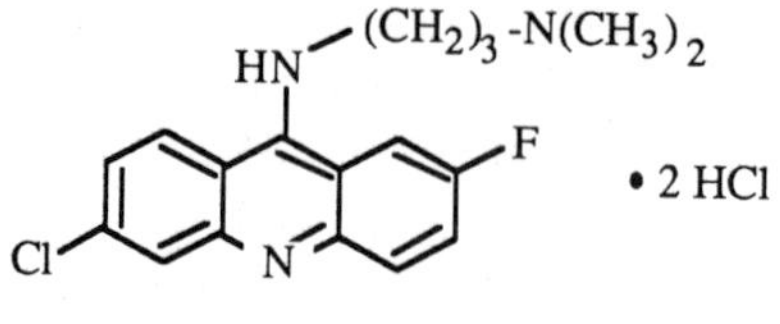

Figure 11-8

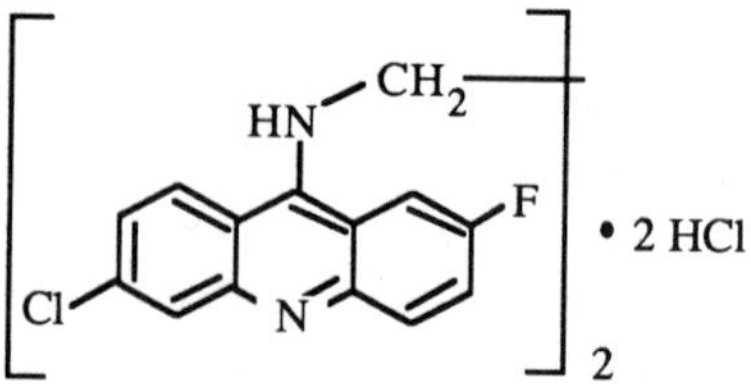

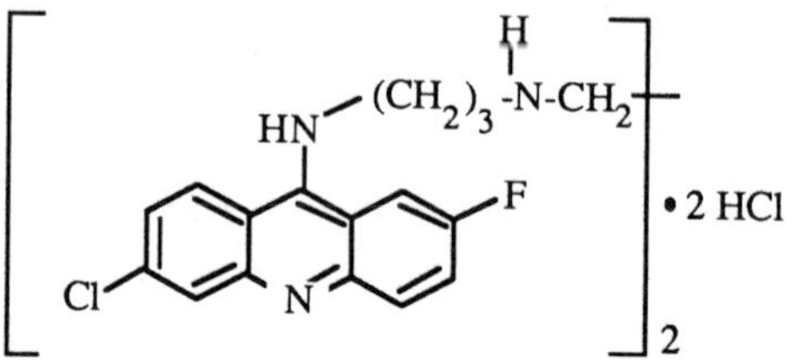

^{19}F NMR has been used to monitor the in vivo metabolism of 5-fluorouracil and 5-fluorouridine in novikoff hepatoma cells (N1S1) and in a murine sarcoma cell line (S-180).[128] The incorporation of the base into nucleotides and into low molecular weight RNA was observed by following the development and peak broadening of a signal at 4.9 ppm downfield from the 5-fluorouracil signal. Other similar studies with fluorine labeled molecules demonstrate that in vivo NMR is a promising and useful technique.[129-131]

^{19}F NMR imaging of biological fluids and tissue is in its infancy. Although very few experiments to localize fluorine-containing molecules in biological organisms have been done, some preliminary results have been reported.[132] Especially now that NMR has been used to image single cells,[133] one can imagine many opportunities for studying the fate of nucleic acid components labeled by fluorine.

REFERENCES

1. Dixon, D. A.; Smart, B. E. *J. Am. Chem. Soc.* **1986**, *108*, 7172.
2. Filler, R. *J. Fluorine Chem.* **1986**, *33*, 361.
3. Walsh, C. "In Methods in Enzymology"; Meister, A., Ed.; Wiley: New York, 1983, Vol. 55, p. 1997.
4. Schlosser, M. *Tetrahedron* **1978**, *34*, 3.
5. Kollonitsch, J. *Israel J. Chem.* **1978**, *17*, 53.
6. Banks, R. E.; Tatlow, J. C. *J. Fluorine Chem.* **1986**, *33*, 227.
7. Sundaralingam, M., In "The Jerusalem Symposia on Quantum Chemistry and Biochemistry"; Bergmann, E. D.; and Pullman B., Eds., The Israel Academy of Sciences and Humanities, Jerusalem, 1973, Vol. 5, p. 417.

8. Olson, W. K.; Sussman, J. L. *J. Am. Chem. Soc.* **1982**, *104*, 270.
9. Altona, C.; Sundaralingam, M. *J. Am. Chem. Soc.* **1972**, *94*, 8205.
10. Manfait, M.; Theophanides, T. In "Spectroscopy of Biological Systems"; Clark, R. J. H.; and Hester, R. E., Eds.; Wiley: New York, 1986, p. 327.
11. Keepers, J. W.; James, T. L. *J. Am. Chem. Soc.* **1982**, *104*, 929.
12. Ikehara, M. *Heterocycles* **1984**, *21*, 75.
13. Uesugi, S.; Kaneyasu, T.; Imura, J.; Ikehara, M.; Cheng, D. M.; Kan, L. S.; Ts'o, P. O. P. *Biopolymers* **1983**, *22*, 1189.
14. Cheng, D. M.; Kan, L. S.; Ts'o, P. O. P.; Uesugi, S.; Takatsuka, Y.; Ikehara, M. *Biopolymers* **1983**, *22*, 1427.
15. Joecks, A.; Koppel, H.; Schleintiz, K. D.; Cech, D. *J. Prakt. Chem.* **1983**, *325*, 881.
16. Lesying, B.; Marck, C.; Guschlbauer, W. *Int. J. Quantum Chem.* **1985**, *28*, 517.
17. Uesugi, S.; Miki, H.; Ikehara, M.; Iwahashi, H.; Kyogoku, Y. *Tetrahedron Letters* **1979**, *42*, 4073.
18. Guschlbauer, W.; Jankowski, K. *Nucleic Acids Research* **1980**, *8*, 141.
19. Guschlbauer, W. Nucleic Acids Research, Symposium Series No. 11, 1982, 113.
20. Reichman, U.; Watanabe, K. A.; Fox, J. J. *Carbohydrate Res.* **1975**, *42*, 233.
21. Su., T.-L.; Watanabe, K. A.; Schinazi, R. F.; Fox, J. J. *J. Med. Chem.* **1986**, *29*, 151.
22. Watanabe, K. A.; Reichman, U.; Hirota, K.; Lopez, C.; Fox, J. J. *J. Med. Chem.* **1979**, *22*, 21.
23. Tann, C. H.; Brodfuehrer, P. R.; Brundidge, S. P.; Sapino, C., Jr.; Howell, H. G. *J. Org. Chem.* **1985**, *50*, 3644.
24. Sharma, R. A.; Kavai, I.; Fu, Y. L.; Bobek, M. *Tetrahedron Letters* **1977**, *39*, 3433.
25. Bergstrom, D. E.; Mott, A. W. Unpublished results.
26. Hertel, L. W.; Kroin, J. S.; Misner, J. W.; Tustin, J. M. *J. Org. Chem.* **1988**, *53*, 2406–2409.
27. An, S.-H.; Bobek, M. *Tetrahedron Letters* **1986**, *27*, 3219.
28. Biggadike, K.; Borthwick, A. D.; Evans, D.; Exall, A. M.; Kith, B. E.; Roberts, S. M.; Stephenson, L.; Youds, P.; Slawin, A. M. Z.; Williams, D. J. *J. Chem. Soc. Chem. Commun.* **1987**, 251.
29. Uesugi, S.; Kaneyasu, T.; Matsugi, J.; Ikehana, M. *Nucleosides and Nucleotides* **1983**, *2*, 373.
30. Watanabe, K. A.; Su., T.-L.; Klein, R. S.; Chu, C. K.; Matsuda, A.; Chun, M.-W.; Lopez, C.; Fox, J. J. *J. Med. Chem.* **1983**, *26*, 152.
31. Watanabe, K. A.; Su., T.-L.; Reichman, U.; Greenberg, N.; Lopez, C.; Fox, J. J. *J. Med. Chem.* **1984**, *27*, 91.
32. Fox, J. J.; Lopez, C.; Watanabe, K. A. In "Medicinal Chemistry Advances"; De Las Heras, F. G., Ed.; Pergamon: New York, **1984**, *27*, 91.
33. Herada, K.; Matulic-Adamic, J.; Price, R. W.; Schinazi, R. F.; Watanabe, K. A. *J. Med. Chem.* **1987**, *30*, 226.
34. Eli Lilly and Co., Pat, U.K., Appl., G.B. 2,136,425 A., 1984.
35. Griengl, H.; Wanek, E.; Schwarz, W.; Streicher, W.; Rosenwirth, B.; De Clercq, E. *J. Med. Chem.* **1987**, *30*, 1199.
36. Wright, J. A.; Wilson, D. P.; Fox, J. J. *J. Med. Chem.* **1970**, *13*, 269.
37. Ashley, G. W.; Stubbe, J. *Pharmac. Ther.* **1987**, *30*, 301.
38. Stubbe, J.; Kozarich, J. W. *J. Biol. Chem.* **1980**, *255*, 5511.
39. Uesugi, S.; Takasuka, Y.; Ikehara, M.; Cheng, D. M.; Kan, L. S.; Ts'o, P. O. P. *Biochemistry* **1981**, *20*, 3056.
40. Ikehara, M.; Kakiuchi, N.; Fukui, T. *Nucleic Acids Res.* **1978**, *5*, 3315.
41. Fukui, T.; Kakiuchi, N.; Ikehara, M. *Biochem. Biophys. Acta* **1982**, *697*, 174.
42. Fukai, T.; De Clercq, E.; Kakiuchi, N.; Ikehara, M. *Cancer Letters* **1982**, *16*, 129.
43. Ikehara, M.; Imura, J. *Chem Pharm. Bull.* **1981**, *29*, 2408.
44. Joecks, A.; Koppel, H.; Schleinitz, K. D.; Cech, D. *J. Prak. Chem.* **1983**, *325*, 881.
45. Rosenthal, A.; Cech, D.; Joecks, A. *Z. Chem.* **1985**, *25*, 26.

46. Von Janta-Lipinski, M.; Langen, P.; Cech, D. *Z. Chem.* **1983,** *23,* 335.
47. Kowollik, G.; Etzold, G.; Von Janta-Lipinski, M.; Gaertner, K.; Langen, P. *J. Prakt. Chem.* **1973,** *315,* 895.
48. Kowollik, G.; Gaertner, K.; Langen, P. *J. Carbohydrates Nucleosides Nucleotides* **1975,** *2,* 191.
49. Etzold, G.; Hintsche, R.; Kowollik, G.; Langen, P. *Tetrahedron* **1971,** *27,* 2463.
50. Bergstrom, D. E.; Romo, E. H.; Mott, A. W. Abstr. CARB 38, 185th ACS National Meeting, Seattle, Washington, March 20–25, 1983.
51. Langen, P.; Etzold, G.; Hintsche, R.; Kowollik, G. *Acta Biol. Med. German.* **1969,** *23,* 759.
52. Waehnert, U.; Langen, P. *Proc. Hung. Annu. Meet. Biochem.* **1979,** *19,* 27.
53. Schroeder, C.; Jantschak, J. *Z. Allg. Mikrobiol.* **1980,** *20,* 657.
54. Chidgeavadze, Z. G.; Scamrov, A. V.; Beabealashvilli, R. Sh.; Kvasyuk, E. I.; Zaitseva, G. V.; Mikhailopulo, I. A.; Kowollik, G.; Langen, P. *FEBS Lett.* **1985,** *183,* 275.
55. Chidgeavadze, Z. G.; Beabealashvilli, R. Sh.; Krayevsky, A. A.; Kukhanova, M. K. *Biochim. Biophys. Acta* **1986,** *868,* 145.
56. Beabealashvilli, R. Sh.; Scamrov, A. V.; Kutateladze, T. V.; Mazo, A. M.; Krayevsky, A. A.; Kukhanova, M. K. *Biochim. Biophys. Acta* **1986,** *868,* 136.
57. Ajmera, S.; Bapat, A. R.; Danenberg, K.; Danenberg, P. V. *J. Med. Chem.* **1984,** *27,* 11.
58. Suhadolnik, R. J. "Nucleosides as Biological Probes". Wiley: New York, **1979,** p. 102.
59. Jenkins, I. D.; Verheyden, J. P. H.; Moffatt, J. G. *J. Am. Chem. Soc.* **1976,** *98,* 3346.
60. Gottfried, H.; Staske, R.; Cech, D. *Z. Chem.* **1978,** *7,* 258.
61. Goodchild, J.; Wadsworth, H. J.; Sim, I. S. *Nucleosides and Nucleotides* **1986,** *5,* 571.
62. Shealy, Y. F.; O'Dell, C. A.; Shannon, W. M.; Arnett, G. *J. Med. Chem.* **1983,** *26,* 156.
63. Madhavan, G. V. B.; Prisbe, E. J.; Verheyden, J. P. H.; Martin, J. C. Abstract CARB 11, 193rd ACS Meeting, Denver, Colorado, April 5–10, 1987.
64. Biggadike, K.; Borthwick, A. D.; Exall, A. M.; Kirk, B. E.; Roberts, S. M.; Youds, P.; Slawin, A. M. Z.; Williams, D. J. *J. Chem. Soc., Chem. Commun.* **1987,** 255.
65. Corbridge, D. E. C. "The Structural Chemistry of Phosphorus". Elsevier: Amsterdam, 1974.
66. Corbridge, D. E. C. "Phosphorus". Elsevier: Amsterdam, 1978.
67. Vogel, H. J.; Bridger, W. A. *Biochemistry* **1982,** *21,* 394.
68. Wittman, R. *Chem. Ber.* **1963,** *96,* 771.
69. Nichol, A. W.; Nomura, A.; Hampton, A. *Biochemistry* **1967,** *6,* 1008.
70. Kucerova, Z.; Skoda, J. *Biochim. Biophysica Acta* **1971,** *247,* 194.
71. Sporn, M. B.; Berkowitz, D. M.; Glinski, R. P.; Ash, A. B.; Stevens, C. L. *Science* **1969,** *164,* 1408.
72. Withers, S. G.; Madsen, N. B. *Biochem. Biophys, Res. Commun.* **1980,** *97,* 513.
73. Westheimer, F. H. *Science* **1987,** *235,* 1173.
74. Engel, R. *Chem. Revs.* **1977,** *77,* 349.
75. Blackburn, G. M. *Chem. Ind. (London)* **1981,** 135.
76. Blackburn, G. M.; England, D. A.; Kolkmann, F. *J. Chem. Soc., Chem. Commun.* **1981,** 930.
77. McKenna, C. E.; Shen, P. *J. Org. Chem.* **1981,** *46,* 4573.
78. Blackburn, G. M.; Kent, D. E. *J. Chem. Soc., Perkin Trans. I.* **1986,** 913.
79. Blackburn, G. M.; Parratt, M. J. *J. Chem. Soc., Perkin Trans. I.* **1986,** 1425.
80. Blackburn, G. M.; Parratt, M. J. *J. Chem. Soc., Perkin Trans. I.* **1986,** 1417.
81. Blackburn, G. M.; Perrée, T. D.; Rashid, A.; Bisbal, C.; Lebleu, B. *Chemica Scripta* **1986,** *26,* 21.
82. Soborovskii, L. Z.; Baina, N. F. *J. Gen. Chem. USSR* **1959,** *29,* 1115.
83. Yoshikawa, M.; Kato, T.; Takenishi, T. *Bull. Chem. Soc. Japan* **1969,** *42,* 3505.
84. Bergstrom, D. E.; Romo, E. H. Unpublished results.
85. Marugg, J. E.; McLaughlin, L. W.; Piel, N.; Tromp, M.; van der Marel, G. A.; van Boom, J. H. *Tetrahedron Letters* **1983,** *37,* 3989.

86. Marugg, J. E.; de Vroom, E.; Dreef, C. E.; Tromp, M.; van der Marel, G. A.; van Boom, J. H. *Nucleic Acids Research* **1986**, *14*, 2171.
87. Bergstrom, D.; Romo,E.; Shum, P. *Nucleosides and Nucleotides* **1987**, *6*, 53.
88. Blackburn, G. M.; Eckstein, F.; Kent, D. E.; Perrée, T. D. *Nucleosides and Nucleotides* **1985**, *4*, 165.
89. Burton, D. J.; Flynn, R. M. *J. Fluorine Chem.* **1980**, *15*, 263.
90. Blackburn, G. M.; Kent, D. E. *J. Chem. Soc. Chem. Commun.* **1981**, 511.
91. Burton, D. J.; Pietrzyk, D. J.; Ishihara, T.; Fonong, T.; Flynn, R. M. *J. Fluorine Chem.* **1982**, *20*, 617.
92. Blackburn, G. M.; Kent, D. E.; Kolkmann, F. *J. Chem. Soc., Chem. Commun.* **1981**, 1188.
93. Schuman, P. D.; Tarrant, P.; Warner, D. A.; Westmoreland, G.; U.S. Pat. 3 954 758 to PCR, Inc., 1976.
94. Miyashita, O.; Matsumura, K.; Shinadzu, H.; Hashimoto, N. *Chem. Pharm. Bull.* **1981**, *29*, 3181.
95. Kumadaki, I.; Nakazawa, M.; Kobayshi, Y. *Tetrahedron Letters* **1983**, *24*, 1055.
96. Kobayashi, Y.; Kumadaki, I.; Yamamoto, K. *J. Chem. Soc., Chem. Comm.* **1977**, 536.
97. Matulic-Adamic, J.; Watanabe, K. A.; Price, R. W. *Chemica Scripta* **1986**, *26*, 127.
98. Baerwolff, D.; Langen, P. In "Nucleic Acid Chemistry—Part 1"; Townsend, L. B.; and Tipson, R. S., Eds.; Wiley: New York, **1978**, 359.
99. Baerwolff, D.; Reefschlaeger, J.; Langen, P. *Nucleic Acids Symp. Ser.* **1981**, *9*, 45.
100. Bergstrom, D. E.; Ruth, J. L.; Reddy, P. A.; De Clercq, E. *J. Med. Chem.* **1984**, *27*, 279.
101. Schwarz, B.; Cech, D.; Reefschlaeger, J. *J. Prakt. Chemie* **1984**, *326*, 985.
102. De Clercq, E.; Walker, R. T. In "Progress in Medicinal Chemistry"; Ellis, G. P.; and West, G. B., Eds.; Elsevier: Amsterdam, **1986**, *23*, 187.
103. De Clercq, E.; Walker, R. T. *Pharmac. Ther.* **1984**, *26*, 1.
104. Goodchild, J.; Porter, R. A.; Raper, R. H.; Sim, I. S.; Upton, R. M.; Viney, J.; Wadsworth, H. J. *J. Med. Chem.* **1983**, *26*, 1252.
105. De Clercq, E.; Descamps, J.; Barr, P. J.; Jones, A. S.; Serafinowski, P.; Walker, R. T.; Huang, G. F.; Torrence, P. F.; Schmidt, C. L.; Mertes, M. P.; Kukilowski, T.; Shugar, D. In "Antimetabolites in Biochemistry, Biology and Medicine"; Skoda, J.; and Langen, P., Eds.; Pergamon: New York, **1979**, 275.
106. Sim, I. S.; Raper, R. H. *Antiviral Research* **1984**, *4*, 159.
107. Reefschlaeger, J.; Baerwolff, D.; Langen, P. *Acta Virol* **1984**, *28*, 282.
108. Coe, P. L.; Harnden, M. R.; Jones, A. S.; Noble, S. A.; Walker, R. T. *J. Med. Chem.* **1982**, *25*, 1329.
109. Robbins, M. J.; Uznanski, B. *Can. J. Chem.* **1981**, *59*, 2608.
110. Kiburis, J.; Lister, J. H. *J. Chem. Soc.* **1971**, 3942.
111. Gerster, J. F.; Jones, J. W.; Robins, R. K. *J. Org. Chem.* **1963**, *28*, 945.
112. Thieller, H. P. M.; Koomen, G. J.; Pandit, U. K. *Tetrahedron* **1977**, *33*, 1493.
113. Thieller, H. P. M.; Koomen, G. J.; Pandit, U. K. *Tetrahedron* **1977**, *33*, 2609.
114. Pein, C. D.; Cech, D. *Tetrahedron Letters* **1985**, *26*, 4915.
115. McNamara, D. J.; Cook, P. D. *J. Med. Chem.* **1987**, *30*, 340.
116. Dagani, R. *Chem. Engineering News*, Nov. 9, 1981, 30.
117. Major, P.; Egan, E.; Herrick, D.; Kufe, D. W. *Cancer Res.* **1982**, *42*, 3005.
118. Mercer, J. R.; Knaus, E. E.; Wiebe, L. I. *J. Med. Chem.* **1987**, *30*, 670.
119. Abe, Y.; Fukuda, H.; Ishiwata, K.; Yoshioka, S.; Yamada, K.; Endo, S.; Kubota, T.; Sato, T.; Matsuzawa, T.; Takahashi, T.; Ido, T. *Eur. J. Nucl. Med.* **1983**, *8*, 258.
120. Saito, Y.; Price, R. W.; Rottenberg, D. A.; Fox, J. J.; Su, T.-L.; Watanabe, K. A.; Philips, F. S. *Science* **1982**, *217*, 1151.
121. Presented at the Sixth International Symposium on Radiopharmaceutical Chemistry, Boston, June 29–July 3, 1986 and published in *J. Labeled Compounds and Radiopharmaceuticals* **1986**, *23* (10–12), 1035.

122. Gerig, J. T. In "Biological Magnetic Resonance, Vol. 1"; Berliner, L. G-.; and Reubens, J., Eds.; Plenum: New York, 1978, Chapter 4.
123. Hardin, C. C.; Gollnick, P.; Kallenback, N. R.; Cohn, M.; Horowitz, J. *Biochemistry* **1986,** *25,* 5699.
124. Marshall, A. G.; Smith, J. L. *J. Am. Chem. Soc.* **1977,** *99,* 635.
125. Kremer, A. B.; Mikita, T.; Beardsley, G. P. *Biochemistry* **1987,** *26,* 391.
126. Mirau, P. A.; Shafer, R. H.; James, T. L.; Bolton, P. H. *Biopolymers* **1982,** *21,* 909.
127. Delbarre, A.; Shafer, R. H.; James, T. L. *Biopolymers* **1983,** *22,* 2497.
128. Keniry, M. A.; Benz, C.; James, T. L.; Shafer, R. *Proc. Annu. Meet. Am. Assoc. Cancer Res.* **1985,** *26,* 18.
129. Selinsky, B. S.; Thompson, M.; London, R. E. *Biochem. Pharm.* **1987,** *36,* 413.
130. Stevens, A. N.; Moris, P. G.; Iles, R. A.; Sheldon, P. W.; Griffiths, J. R. *Br. J. Cancer* **1984,** *50,* 113.
131. Wyrwicz, A. M.; Pszenny, M. H.; Schofield, J. C.; Tillman, P. C.; Gordon, R. E.; Martin, P. A. *Science* **1983,** *222,* 428.
132. Digenis, G. A.; Hawi, A. A.; Yip, H.; Layton, W. J. *Life Sciences* **1986,** *38,* 2307.
133. Aguayo, J. B.; Blackband, S. J.; Schoeniger, J.; Mattingly, M. A.; Hintermann, M. *Nature* **1986,** *322,* 190.

CHAPTER 12

Fluorine Chemistry without Fluorine: Substituent Effects and Empirical Mimicry

Joel F. Liebman

Department of Chemistry University of Maryland, Baltimore County Campus Baltimore, Maryland

CONTENTS

1. INTRODUCTION

The chemistry of fluorine and its compounds is unquestionably unique. Indeed, how many other elements have entire journals and book series devoted to them, as well as its own specialized division of the American Chemical Society? Monographs such as that which contains this chapter are just part of the research literature. Except for carbon, fluorine is the only element receiving such singular attention. Many of the chapters in this monograph address features of the seemingly unique substituent effects of groups containing fluorine and generally atypical behavior of fluorinated species which sustains this research interest. This chapter, however, is devoted to a rather different type of study of fluorine and its compounds—how can one quantitatively mimic the effects of fluorine-containing substituents by another group or groups that lack this element? While in fact various individual idiosyncracies can be duplicated, it will be seen that no single group comes close to reproducing the numerous, yet general, changes on molecular structure and energetics that fluorine imparts to a wide variety of species. Fluorine, its compounds, and their mimics offer the chemist a unique constellation of facets and facts that are found nowhere else in the study of chemistry.

2. MIMICKING THE PROTON AFFINITIES OF TRIFLUOROMETHYL SUBSTITUTED BASES

In this section, we discuss "the proton affinity," which is to be recognized as the gas phase energy quantity corresponding to basicity. The proton affinity of an arbitrary base is an experimentally measurable quantity and is defined simply as the negative of the heat of the following reaction:

$$B + H^+ \rightarrow BH^+ \tag{12-1}$$

Indeed, the proton affinity was intentionally chosen here instead of any other property that relates to basicity in order to eliminate any condensed phase (eg, solvent, lattice) effects on the basicity. Its use also avoids all entropic contributions that may differentially stabilize the neutral or cation. Furthermore, this choice is a convenient one since there are recently evaluated data compendia[1] for proton affinity which provided all the numbers desired for this chapter.

Substituents are well known to have large effects in modulating proton affinity.[2] For example, the trifluoromethyl, —CF_3, group is a powerful electron withdrawing substituent and CF_3 substitution decreases the basicity of whatever species it is attached to relative to the parent unsubstituted compound. Conversely, substitution by the hydrogen analog to CF_3, methyl, —CH_3, should increase the basicity since methyl is established to be a rather respectable electron donor. The cyano, —CN, group with its carbon bound to the electronegative nitrogen is powerfully electron withdrawing and thus the cyano substituted species should have a weakened basicity relative to the parent species. By contrast, monofluoro substitution, the conceptual replacement of one hydrogen by one fluorine, is expected to result in rather unpredictable changes in basicity because while fluorine is a powerful σ withdrawing substituent, it is also mildly π donating. While this might suggest that the basicity of the monofluoro derivative of an arbitrary species should be less than that of the unsubstituted compound, the basicity of a compound is modified by both σ and π effects in generally a seemingly nonadditive, inseparable, and unpredictable way.

What follows are the proton affinities of trifluoromethyl substituted bases, their cyano, methyl, and fluoro analogs, and the unsubstituted species. To simplify reading, two tables are presented, one for bases that protonate on nitrogen and the other for bases that protonate on either oxygen or carbon.

TABLE 12-1. Proton Affinities of Trifluoromethyl, Cyano, Methyl, and Fluoro Substituted Compounds and the Corresponding Unsubstituted Bases that Protonate on Nitrogen. (All values are taken from Reference 1 and are given in kcal/mol.)

R=	CF_3	CN	H	CH_3	F
a) Nitriles					
RCN	164.3	162	171.4	188.4	
RCF$_2$CN	167.1				164.3
RCF$_2$CF$_2$CN	167.4				167.1
b) Pyridines and other sp² nitrogen					
2-RC$_5$H$_4$N	211.5	208.1	220.8	225.0	210.6
3-RC$_5$H$_4$N	212.6	209.3	220.8	224.1	214.3
4-RC$_5$H$_4$N	212.8	210.3	220.8	225.2	216.6
RNO		169			
c) Aliphatic and alicyclic amines					
RN(CH$_3$)$_2$	193.8	205[a]	220.6	225.1	
RCH$_2$NH$_2$	202.5	197.4	214.2	217.0	
RCH$_2$NHCH$_3$	209.8	206.0	220.6	222.8	
RCH$_2$N(CH$_3$)$_2$	215.1	211.1	225.1	227.5	
RCH$_2$CH$_2$NH$_2$	210.6	207.6	217.0	217.9	212.3
R(CH$_2$)$_3$NH$_2$	214.3		217.9	218.4	217.8
RCH(CH$_2$CH$_2$)$_2$NH (4-R-piperidine)	219.4	216.6	226.4		
RC(CH$_2$CH$_2$)$_3$N (4-R-quinuclidine)	224.9	221.8	232.1		
d) Aromatic amines					
3-RC$_6$H$_4$NH$_2$	204.2	200.7	209.5	213.4	207.0
3-RC$_6$H$_4$N(CH$_3$)$_2$	217.4	214.4	223.4	224.5	
4-RC$_6$H$_4$N(CH$_3$)$_2$	216.3	213.2	223.4	225.6	
3,5-R$_2$C$_6$H$_3$N(CH$_3$)$_2$	212.3		223.4	227.0	

[a] It is highly probable that the protonation is on the nitrogen of the cyano group and not on the basic site of the remainder of the molecule, the amino group.

TABLE 12-2. Proton Affinities of Trifluoromethyl, Cyano, Methyl, and Fluoro Substituted Compounds and the Corresponding Unsubstituted Bases that Protonate on Oxygen or on Carbon (All values are taken from Reference 1 and are given in kcal/mol.)

R=	CF_3	CN	H	CH_3	F
Bases that Protonate on oxygen					
a) Aldehydes and ketones					
RCHO	165.1		171.7	186.6	
RCOCH₃	174.2	179.1[a]	186.6	196.7	
RCOCF₃	161.5		165.1	174.2	160.2
R₂CO	161.5		171.7	196.7	160.5
4-RC₆H₄CHO	191.0	187.0	202.0	203.7	199.2
b) Carboxylic acid derivatives					
RCOOH	169.0		178.8	190.2	
RCOF	160.2				160.5
HCOOCH₂R	179.4		188.9	197.8	
RCOOCH₃	178.8		188.9	197.8	
RCOOC₂H₅	184.6	179.5	193.1	200.7	
RCOOCH₂CH₂F	178.6				
RCOO-n-C₄H₉	185.8		194.8		
RCONH-n-C₄H₉	203.6				
c) Alcohols and ethers					
RCH₂OH	169.0		181.9	188.3	
R₂CHOH	165.0		181.9	191.2	
R₃COH	163.1		181.9	193.7	
R₂(CH₃)COH	167.0		188.3	193.7	
RCH₂OC₂H₅	186.4		196.4	200.2	
d) Other					
RSO₃H	169				
Bases that protonate on carbon					
3-RC₆H₄C≡CH	192.8		200.2		195.4
3-RC₆H₄CH=CH₂	194.6		202.0		
4-RC₆H₄C(CH₃)=CH₂	199.6		207.0	211.0	206.7

[a] It is highly probable that the protonation is on the nitrogen of the cyano group and not on the basic site of the remainder of the molecule, ie, ketonic oxygen.

It is seen that our expectations about substituent effects on basicity are fulfilled. As long as the site of protonation in the base is the same, the values of proton affinities for the —CF₃ and —CN substituted species seem to parallel and are numerically considerably closer to each other than to either the parent or the methyl substituted compound. Indeed, we find the following statistical results for all of the bases for which data exists by which trifluoromethyl and cyano substituted bases can be compared:

$$PA(XCF_3) = 0.989PA(XCN) + 5.697 \ (n = 15, r = 0.9986) \quad (12\text{-}2)$$

$$PA(XCF_3) = PA(XCN) + 3.47 \pm 0.82 \quad (12\text{-}3)$$

wherein Equation 12–2 allowed the slope and intercept to vary while Equation 12–3 forced the slope to equal unity. (Quite obviously, we limited our attention to those species that are protonated on the nominally basic part of the molecule, ie, we ignore dimethylcyanamide, $NCN(CH_3)_2$, because this species is expected to be protonated on the cyano, and not dimethylamino, nitrogen.) An ever so slightly better statistical correlation was likewise found using only those bases that protonate on the non-cyano nitrogen for the twelve nitrogen bases:

$$PA(XCF_3) = 1.006 \, PA(XCN) + 2.112 \, (n = 13, r = 0.9988) \quad (12\text{-}4)$$

$$PA(XCF_3) = PA(XCN) + 3.30 \pm 0.72 \quad (12\text{-}5)$$

In Equations 12–2 and 12–4 just presented, the slope is close to unity, the intercepts small, and the correlation coefficient close to unity as well. We thus deduce that while the trifluoromethyl and cyano substitution are clearly not identical, the resulting proton affinities show significant similarities. Equivalently, the cyano group qualifies as a successful mimic for the trifluoromethyl group in the study of proton affinities.

Since this is a chapter in a volume on fluorine chemistry, let us now use the data in the above tables to further understand fluorine compounds. In particular, how well do trifluoromethyl and fluoro substituents track? Consider the following derived equations:

$$PA(XCF_3) = 0.920PA(XF) + 13.597 \, (n = 14, r = 0.9934) \quad (12\text{-}6)$$

$$PA(XCF_3) = PA(XF) - 1.81 \, (\pm 3.16) \quad (12\text{-}7)$$

It would appear that the proton affinities of trifluoromethyl and fluoro substituted species are also comparable and likewise statistically are rather well correlated although not quite as well as trifluoromethyl and cyano containing species. In Equation 12-6 above, the slope is rather far from unity and the intercept is far from small. Likewise, in Equation 12-7, the standard deviation is large enough to "betray" the fact that we cannot even predict whether the proton affinities of monofluorinated and trifluoromethylated bases is larger.

For our comparative study of proton affinities, other mimics for the trifluoromethyl group may also seem plausible. These include fluorocarbonyl, carbethoxy, and other carboxylic acid derivatives, nitro, alkylsulfinyl and alkylsulfonyl, and even other poly and perfluoroalkyl groups. However, to date, meaningful comparisons (ie, not isolated pairs of compounds) have been thwarted for the following interrelated reasons:

a. difference in protonation sites, eg, while trifluoromethyl-ethylene (3,3,3-trifluoropropene) no doubt protonates on carbon, carbethoxy-ethylene (ethyl acrylate) no doubt protonates on oxygen. No direct comparison of the two proton affinities can thus be made.
b. general lack of data, eg, while there is considerable data on numerous mono and polyalkylpyridines, the data on any of the fluorinated deriva-

tives of these bases is seemingly limited to the three isomeric trifluoro-methylpyridines given in Table 12-1. It would thus appear that there are more data than one thinks and less than one needs.

3. THE HEATS OF FORMATION OF FLUORINATED SPECIES AND MIMICRY BY SUITABLE OXYGEN-CONTAINING COMPOUNDS

Fluorine and oxygen are the two most electronegative elements that are generally considered in a chemical context. Both elements have small atomic radii and combine with almost all other elements. It is thus not surprising that many features of molecular structure and energetics are shared by compounds of the two elements. At the risk of seeming egotistic, let me chronicle three of the many "informal" similarities, all discussed earlier in papers authored or coauthored by the author of this chapter:

1. the relative ease of formation of radical anions of the formally saturated perfluoroalkanes and perfluorocycloalkanes that parallels the radical anion formation by numerous species with adjacent carbonyl groups. More precisely, paralleling the ease of formation of semidiones via single electron reduction of α-diketones (and the formation of numerous radical anions, dianions, and even polyanions from oxocarbons), octafluorocyclobutane (and numerous other perfluorinated alkanes and cycloalkanes) readily add an electron without concommitant loss of either atomic fluorine or fluoride ion.[3]
2. the comparative instability of species that have adjacent CF_2 groups and so parallels the instability of α-diketones. Such lack of stability is exemplified by the facile extrusion of CF_2 itself from hexafluorocyclopropane under mild thermolytic conditions.[4]
3. the high degree of stabilization in the unsaturated and seemingly more strained 3,3-difluorocyclopropene and cyclopropenones, but marked destabilization in the related saturated, but seemingly less strained, 1,1-difluorocyclopropanes and cyclopropanones.[5]

However useful these observations are as qualitative conclusions, one may still aspire for quantitation. Of these earlier studies cited above, only 2. above was quantitated in that the group increment $C(F)_2(CF_2)_2$ defined for CF_2 bonded to two other CF_2 groups was shown to have a considerably higher heat of formation, and hence thermodynamically less stable, than $C(F)_2(C)_2$ in which the central CF_2 is bonded to two CH_2 groups.[4,5] This is a breakdown of the validity of the usual assumption that next-nearest neighbor effects can be neglected. Can one be more quantitative? Recall the brief comment by Benson[7] that fluoro and hydroxo groups constitute a "homothermal'' pair—the heats of formation of solid fluorosulfates and bisulfates (hydrogen sulfates) are nearly identical. Woolf[8] extended this observation by contrasting the heats of formation of numerous fluorinated species and their formally related oxygen-containing analogs. In this section we investigate this Benson/Woolf quantitatable analogy where for brevity we consider solely organic compounds. Furthermore, so as not to be troubled by questions of intermolecular interactions, we consider only gas phase species. From Woolf's analysis, whenever the fluorinated compound has a sole fluorine on a given carbon, that is an isolated fluorinated substituent, this compound will be contrasted with the corresponding alcohol compound with an —OH replacing the —F. When there are two geminal fluorines in a compound, the fluorinated compound will be contrasted with the corresponding carbonyl compound and water to compensate for the general absence of thermochemical data on gem-diols. Finally, should there be a trifluoromethyl group with its three geminal fluorines, the fluorinated compound will be contrasted with the corresponding carboxylic acid and water because gem-triols are seemingly always nonisolable. Admittedly, the above ''recipes'' may strike the reader as odd—they did the current author. Why is it valid to assume that a carbonyl compound and water is equivalent to one containing two geminal —OH groups?

Nonetheless, the above procedure defines a unique method for obtaining isoelectronic heats of formation. It also offers a possible, and certainly well-defined and so unique, mimic for any fluorinated organic compound of interest. What follows is a brief table that gives the experimentally determined[9] gas phase heats of formation of most of Woolf's corresponding pairs of fluorinated and oxygen-containing organic compounds along with annotations that include what assumptions Woolf opted to make, as well as other commentary by the current author.

It is to be seen that the oxygen-containing analog has always a higher heat of formation that the fluorinated species of interest. Furthermore, it is seen that the difference between the two species per fluorine is almost always ca. 6 kcal/mol wherein admittedly the scatter may hide some interesting chemistry. As such, so long as data solely for the gas phase is used, the oxygen-containing analog is a better mimic for the fluorinated species than Woolf himself concluded from his analysis: ''We do not expect the same accuracy in matching heat data as obtained by various methods of group additivity. . . . [The] ideal values serve as a reference from which deviations can be rationalized in terms of specific interactions and hence trends can be assessed.''[12]

We close this section with an affirmation of this philosophy: it is to be noted that the two examples involving CO_2 have less stabilization per fluorine than almost any other. It is well established that CF_4 enjoys considerable stabilization[13] as indeed, it is the consummate example of gem-fluorination.[5]

The current numbers and accompanying text of Table 12-3 document CF_4,

TABLE 12-3. Hearts of Formation of Gas Phase Isoelectronic Pairs of Fluorinated and Oxygen-containing Species (Unless said, all data in this table are from Ref. 9, and are given here in kcal/mol.)

F-compound	ΔH_f	O-compound	ΔH_f	$\delta\Delta H_f/n_F{}^a$
n-C$_3$H$_7$F[b]	−68	n-C$_3$H$_7$OH	−61	7
CH$_3$COF[c]	−106	CH$_3$COOH	−103	3
C$_6$H$_5$F[c]	−28	C$_6$H$_5$OH	−23	5
p-C$_6$H$_4$F$_2$[d]	−73	p-C$_6$H$_4$(OH)$_2$	−63	5
COF$_2$	−153	CO$_2$ + H$_2$O	−152	1
		CO(OH)$_2$[e]	−146	3.5
CCl$_2$F$_2$	−114	COCl$_2$ + H$_2$O	−110	2
CF$_2$ClCF$_2$Cl[b]	−219	ClCOCOCl + 2H$_2$O	−196	6
CHF$_3$[b]	−166	HCOOH + H$_2$O	−148[f]	6
CF$_3$COOH[g]	−246	HOOCCOOH + H$_2$O	−230	5
C$_6$H$_5$CF$_3$[b]	−143	C$_6$H$_5$COOH + H$_2$O	−128	5
C$_2$F$_6$[d]	−321	HOOCCOOH + 2H$_2$O	−288	5
CF$_4$	−223	COF$_2$ + H$_2$O	−211	6
		CO$_2$ + 2H$_2$O	−210[h]	2

[a] This quantity is the difference in the heats of formation of the fluorinated and oxygen-containing species divided by the number of fluorines.
[b] Woolf chose data for the liquid phase.
[c] Woolf chose data for the mean of the liquid and gas phase.
[d] Woolf chose data for the solid phase.
[e] Interestingly, gas phase H$_2$CO$_3$ has been recently synthesized (Ref. 10) with a heat of formation suggested by these authors of −146 kcal/mol. Encouragingly, nearly the identical value, −147 kcal/mol, can be obtained by assuming thermoneutrality for the following macroincrementation[11] reaction:

$$CO(OH)_2 = CO(OC_2H_5)_2 + 2(CH_3COOH - CH_3COOC_2H_5).$$

[f] A value of −144 kcal/mol for the heat of formation of HC(OH)$_3$ can be estimated by use of assuming thermoneutrality for the following macroincrementation[11] reaction:

$$HC(OH)_3 = HC(OCH_3)_3 + 3(C_2H_5OH - C_2H_5OCH_3).$$

It may be argued that methyl orthoformate is more strained than orthoformic acid. An approximate correction for this using the following macroincrementation reaction to estimate the strain energy:

$$HC(CH_3)_3 = HC(CH_2CH_3)_3 + 3(C_2H_5CH_3 - C_2H_5CH_2CH_3).$$

As the heat of reaction is 3 kcal/mol, it suggests that the "correct" heat of formation of HC(OH)$_3$ is −47 kcal/mol, encouragingly close to the value for "HCOOH + H$_2$O" as suggested by Woolf's analysis.
[g] Woolf chose data for the mean of the solid and gas phases.
[h] Woolf did not explicitly include this entry perhaps because it is, of course, obtainable from the numbers for COF$_2$. We present an estimate for the heat of formation of C(OH)$_4$ making use of macroincrementation[11] reactions using the orthocarbonate ester C(OCH$_3$)$_4$ and steric correction terms analogous to that for orthoformic acid above. In particular, consider the following reaction:

TABLE 12-5. Heats of Vaporization of Variously Fluorinated Halogenated Hydrocarbons and of their Hydrogenated Analogs (All data in kcal/mol.)

Entry	Compound	ΔH_v
1.	$ClCF_2CF_2Cl^a$	5.6
	$ClCH_2CH_2Cl^a$	8.5
2.	$BrCF_2CF_2Br$	6.8
	$BrCH_2CH_2Br$	10.0
3.	CF_3CCl_3	6.8
	CH_3CCl_3	7.8
	CF_2ClCCl_2F	6.8
	$CH_2ClCHCl_2$	9.6
4.	$CF_2BrCHFCl$	7.2
	CH_2BrCH_2Cl	9.1
5.	$CF_3CH_2CF_2Cl^a$	8.1
	$CH_3CH_2CH_2Cl$	9.2
6.	$CF_3CFClCF_2Cl$	6.5
	$CH_3CHClCH_2Cl$	8.6
7.	$C_6F_5Cl^a$	9.0
	C_6H_5Cl	9.8
8.	$C_6F_5Br^a$	10.3
	C_6H_5Br	10.7
9.	$C_6F_5I^a$	13.8
	$C_6H_5I^a$	11.4

^{a} From Ref. 9 "deriving" ΔH_v as the average of their recommended heats of vaporization (and negative of the heats of condensation) used by these authors to obtain the heats of formation of the liquid and gaseous species.

le is fluorinated as the carbons are less able to donate electron
these halogens. The first factor would not change the heat of
n, while the second would decrease it because there is less inter-
dipolar attraction. The reason we suggest why the heat of vapor-
he iodine case increases upon fluorination is that the iodine has
essentially neutral in the hydrocarbon case to rather positive (cf.
n chemistry of another perfluoro iodide, CF_3I^{16}) and so the dipolar
s in fact increased in this case.
turn to the effect of fluorination of the heats of vaporization of a
of compounds with a rather varying set of oxygen-containing
ties. Table 12-6 shows that the effects are again generally rather
seemingly unpredictable either in sign or absolute magnitude.
y, it appears that fluorination has generally but a small effect on
vaporization of an arbitrary organic compound.

$COF_2(+H_2O)$ and $C(OH)_4$ have the expected differences in heats of formation. It is thus CO_2 that is the anomaly. Of course, we could have omitted this discussion altogether because CO_2 is such a "small" and almost "inorganic" molecule. It was included, however, because all the needed data are extant, there is precedent for its stability,[14] and carbon dioxide is much more important than any orthocarbonate (and a fortiori the parent acid) in the general study of chemical phenomena.

4. HYDROGEN AND FLUORINE: ESTIMATION OF THE HEATS OF VAPORIZATION OF ORGANIC COMPOUNDS

The equivalence of fluorine and hydrogen that are suggested below may seem to constitute a most unlikely mimic. After all, it is the uniqueness of fluorinated species usually taken by comparison to their hydrogen-containing parent from which we derive the fascination of so many of the compounds of interest. Nonetheless, both hydrogen and fluorine are small, univalent, and contribute rather little to total molecular polarizabilities. As such, numerous physical properties may be less affected by equating fluorine and hydrogen than most chemical properties. Such properties include the interrelated heat of vaporization and boiling point. We opt to consider the former because it is a single thermodynamic quantity, while the latter depends on both enthalpic and entropic components. In addition, the former is essentially independent of such external variables as pressure, although one must stipulate the temperature of the measurement. Thermochemical convention suggests the uniform conditions of 1 atm and 25°C and these conditions will be assumed implicitly below for all heats of vaporization given in this chapter.

It is more than interest in thermochemistry per se, however, that accounts for our interest in heats of vaporization. Most organic compounds of inter-

$$C(OH)_4 = C(OCH_3)_4 + 4(C_2H_5OH - C_2H_5OCH_3)$$

and the resulting uncorrected heat of formation of −193 kcal/mol, and the strain energy correction of 4 kcal/mol from the reaction,

$$C(CH_3)_4 = C(CH_2CH_3)_4 + 4(C_2H_5CH_3 - C_2H_5CH_2CH_3)$$

resulting in a predicted heat of formation of $C(OH)_4$ of −197 kcal/mol. This corresponds to a difference of ca. 6 kcal/mol per F stabilizing the fluorinated species over the hydroxo analog. The consistency of this result with our earlier findings again documents that CO_2 is anomalously stable.

est, whether they contain fluorine or not, are liquids or solids under most laboratory conditions such as the above standard ones. However, most conceptual models of the energetics of organic chemistry, eg, strain, resonance energy, aromaticity, refer to isolated, and hence gaseous, molecules. It is this quantity discussed here, the heat of vaporization, that interrelates the liquid and gaseous phases of experimental practice and conceptual understanding, respectively.

Indeed, it is part of the folklore of organofluorine chemistry that perfluorination of an arbitrary hydrocarbon has but a small effect on the boiling point and so one might surmise that the heat of vaporization is likewise largely unaffected. We note that this assumption was recently used in deriving the heat of formation of a fluorine-containing group, that of CF_2 bonded to two other CF_2 groups. What follows is a somewhat more careful test of this conjecture as well as investigation of its validity for partial fluorination and for fluorinated derivatives of other classes of organic compounds. It is to be emphasized that we are referring to organic compounds: H_2 and its "perfluorinated analog" F_2 are low boiling gases, but the partially fluorinated HF is almost a liquid under ambient conditions. The explanation is rather obvious—HF is extensively hydrogen bonded while the other two species are not. This alerts us to another convention of fluorine chemistry: perfluorination is not taken to mean replacement of "essential" hydrogens in a substituent, eg, one does not change the —OH group in alcohols or in carboxylic acids to —OF along with changing all of the C—H bonds into C—F bonds.

Table 12-4 presents the desired heats of vaporization for hydrocarbons and their fluorinated derivatives (ie, fluorocarbons and their hydrogenated derivatives) are given. Implicitly, all data comes from the recent critically evaluated compilation Reference 15 unless otherwise explicitly stated. It is seen that for hydrocarbons and their fluorinated derivatives that the heat of vaporization is generally affected by only a kcal/mol or so as one varies the number of fluorines. However, it is also seen that there is apparently no way of predicting whether perfluorination will increase or decrease the heat of vaporization, nor any way of predicting the effects of partial fluorine replacement of hydrogens.

In Table 12-5 we present the related data (and sources) for halogenated hydrocarbons (ie, those containing the "heavier" halogens, chlorine, bromine, and iodine) and their fluorinated derivatives (ie, those replacing some or all the hydrogens by fluorines).

It would appear that a general rule is that fluorination decreases the heat of vaporization and halogenated hydrocarbons, although even this is seemingly reversed for the sole iodine compound. It is tempting to explain these findings by asserting the following: a) the intrinsic "framework" heat of vaporization is unaffected upon fluorination as is found for the hydrocarbons and b) the net negative charge on the chlorines or bromines is decreased when

TABLE 12-4. Heats ⸻ Variously Fluorinated ⸻ their Hydrogenated An⸻ kcal/m⸻

Entry	Compo⸻
1.	C_6F_6
	C_6HF_5
	$1,2\text{-}C_6H_4F_2$
	$1,3\text{-}C_6H_4F_2$
	$1,4\text{-}C_6H_4F_2$
	C_6H_5F
	C_6H_6
2.	$cyclo\text{-}C_6F_{10}{}^a$
	$cyclo\text{-}C_6H_{10}$
3.	$cyclo\text{-}C_6F_{12}{}^a$
	$cyclo\text{-}C_6H_{12}$
4.	$C_6F_5CF_3{}^a$
	$C_6F_5CH_3$
	$3\text{-}FC_6H_4CF_3$
	$C_6H_5CF_3$
	$4\text{-}FC_6H_4CH_3$
	$C_6H_5CH_3$
5.	$cyclo\text{-}C_6F_{11}CF_3{}^{a,b}$
	$cyclo\text{-}C_6H_{11}CH_3$
6.	$n\text{-}C_7F_{16}{}^a$
	$n\text{-}C_7H_{16}$
7.	$cyclo\text{-}C_6F_{11}C_2F_5{}^a$
	$cyclo\text{-}C_6H_{11}C_2H_5$
8.	$n\text{-}C_8F_{18}$
	$n\text{-}C_8H_{17}F$
	$n\text{-}C_8H_{18}$
9.	$cyclo\text{-}C_6F_{11}C_3F_7{}^a$
	$cyclo\text{-}C_6H_{11}C_3H_7$
10.	bicyclo[4.4.0]dec⸻
	trans, perfluoro-
	cis, perfluoro-
	trans, perhydro-^{a}
	cis, perhydro-^{a}
11.	$cyclo\text{-}C_6F_{11}C(CF_3⸻$
	$cyclo\text{-}C_6H_{11}C(CH⸻$

^{a} From Ref. 9 "deriving" Δ⸻ their recommended heats ⸻ negative of the heats of co⸻ these authors to obtain th⸻ of the liquid and gaseous ⸻
^{b} It seems quite inexplicabl⸻ the heat of vaporization ⸻ less than $cyclo\text{-}C_6F_{12}$.

the molec⸻
density to⸻
vaporizati⸻
molecular ⸻
ization is ⸻
gone from ⸻
the reactic⸻
attraction ⸻
We now⸻
collection ⸻
functional⸻
small and ⸻
In summa⸻
the heat o⸻

TABLE 12-6. Heats of Vaporization of Variously Fluorinated Hydrocarbons with Assorted Oxygen Functionalities and of their Hydrogenated Analogs (All data in kcal/mol.)

Entry	Compound	ΔH_v
1.	ClCHFCF$_2$OCHF$_2$	7.8
	ClCHFCF$_2$OCH$_3$	8.3
2.	ClCHFCF$_2$OCHFCl	9.0
	ClCHFCF$_2$OCH$_2$Cl	10.1
3.	CF$_3$CH$_2$CH$_2$OCH$_2$CH$_2$CF$_3$[a]	9.8
	CHF$_2$CF$_2$CH$_2$OCF$_2$CHFCF$_3$	10.3
	CH$_3$CH$_2$CH$_2$OCH$_2$CH$_2$CH$_3$	8.6
4.	CHF$_2$CF$_2$CF$_2$CF$_2$CF$_2$OCF$_2$CHFCF$_3$	12.7
	CH$_3$CH$_2$CH$_2$CH$_2$CH$_2$OCH$_2$CH$_2$CH$_3$	10.3
5.	n-C$_4$F$_9$O-n-C$_4$F$_9$	9.7
	n-C$_4$H$_9$O-n-C$_4$H$_9$	10.8
6.	n-C$_5$F$_{11}$O-n-C$_5$F$_{11}$[a]	11.3
	n-C$_5$H$_{11}$O-n-C$_5$H$_{11}$	13.0
7.	CF$_3$COCH$_2$COCF$_3$ (enol)	7.3
	CF$_3$COCH$_2$COCH$_3$ (enol)	8.9
	CH$_3$COCH$_2$COCH$_3$ (enol)[b]	10.3
8.	CF$_3$CH$_2$OH	10.5
	CH$_3$CH$_2$OH	10.1
9.	CF$_3$CF$_2$CH$_2$OH[a]	12.8
	CH$_3$CH$_2$CH$_2$OH	11.3
10.	CF$_3$COOH	9.2
	CH$_3$COOH	12.3
11.	CF$_3$CF$_2$COOCH$_3$[a]	8.3
	CH$_3$CH$_2$COOCH$_3$	8.6
12.	CH$_3$COF[c]	6.0
	CH$_3$CHO	6.2
13.	CF(NO$_2$)$_3$[a]	8.3
	CH(NO$_2$)$_3$[a]	7.8
14.	CF$_2$NO$_2$)$_2$[a]	10.4
	CH$_2$(NO$_2$)$_2$[a]	11.0

[a] From Ref. 9 "deriving" ΔH_v as the average of their recommended heats of vaporization (and the numerical negative of the heats of condensation) used by these authors to obtain the heats of formation of the liquid and gaseous species.

[b] Note that "acetylacetone" is in fact a mixture of the enol and the diketone in the gas phase. The value given is the difference of the heats of formation of liquid and gaseous enol, ie, "pure" (Z)-3-penten-2-on-4-ol, as given by Ref. 17.

[c] Strictly speaking, this entry violates our rule of fluorination in that the —CHO functionality is normally assumed not to be changed to —COF by definition since even perfluoroaldehydes still have the —CHO moiety that defines the aldehyde functionality.

TABLE 12-7. Heats of Sublimation of Variously Fluorinated Species and of their Hydrogenated Analogs (All data in kcal/mol.) All inexplicitly referenced data are taken from the recent compendium Ref. 18. The mean temperature of the measurement of heat of sublimation is tacitly assumed to be 298 K = 25°C unless otherwise given in degrees Kelvin in parentheses.

Entry	Compound $(T)^a$	ΔH_s
1.	FCN (166)	5.8
	HCN^b (228)	9.0
2.	CF_4 (76)	4.1
	CH_4 (77)	2.3
3.	C_2F_6 (103)	6.2
	C_2H_6 (90)	4.9
4.	CF_3CONH_2	18.6
	$CH_3CONH_2{}^a$	18.8
5.	$cyclo\text{-}C_5F_{10}$ (115)	9.1
	$cyclo\text{-}C_5H_{10}$ (120)	10.1
6.	$n\text{-}C_5F_{12}$ (145)	10.4
	$n\text{-}C_5H_{12}$ (143)	10.0
7.	C_6F_5OH	16.1
	C_6H_5OH	16.4
8.	$cyclo\text{-}C_6F_{12}$ (313)	8.7
	$cyclo\text{-}C_6H_{12}$ (273)	8.9
9.	$HOCH_2CF_2CF_2CF_2CF_2CH_2OH^a$	21.3
	$HOCH_2CH_2CH_2CH_2CH_2CH_2OH^a$	19.9
10.	C_6F_5COOH	21.9
	$p\text{-}FC_6H_4COOH$	21.8
	C_6H_5COOH	21.7
11.	$C_{10}F_8{}^a$	18.9
	$C_{10}H_8{}^c$	17.4
12.	$(C_6F_5)_2$	20.4
	$2,2'\text{-}(C_6H_4F)_2$	22.7
	$4,4'\text{-}(C_6H_4F)_2$	21.8
	$(C_6H_5)_2{}^c$	19.5
13.	$n\text{-}C_{16}F_{34}$ (295)	25.0
	$n\text{-}C_{16}H_{34}{}^d$ (289)	19.9

^{a} A superscript "a" after the value given means that no temperature was reported in Ref. 18 or no value at all was given in Ref. 9.

^{b} This is no doubt an "unfair" example to cite because HCN is extensively hydrogen-bonded in the condensed phase while FCN is not. As such, this comparison is the "organic" counterpart of that of the heats of vaporization of H_2, F_2, and HF.

^{c} When Ref. 18 gave several conflicting values, the value was taken as the average of recommended values of heats of sublimation from Ref. 9.

^{d} The value of the heat of sublimation of $n\text{-}C_{16}H_{34}$ given here was chosen because it was done by a common investigator and technique to that of n-

5. HYDROGEN AND FLUORINE: ESTIMATION OF THE HEATS OF SUBLIMATION OF ORGANIC COMPOUNDS

Experience has shown us that estimation of the heat of sublimation of a compound of interest, ie, the difference of the heats of formation of a compound in its solid and in its gaseous state, is much less numerically reliable than estimation of the heat of vaporization.[18,19] Yet the heat of sublimation is as much a necessary energetics quantity as heat of vaporization in order to interrelate experimental data and theoretical constructs. For fluorinated species, however, two approaches suggest themselves. The first recalls the success of our recent semiempirical technique[19] in which the heat of sublimation was estimated by summing the experimentally measured heat of fusion (s → l) and a heat of vaporization (l → g) estimated using an earlier empirical approach of ours that made use solely of the number of (quaternary and nonquaternary) carbons,[20]

$$\Delta H_S = \Delta H_{fus} (T_M) + \Delta H_V(\text{est}) \tag{12-8}$$

$$\Delta H_S = \Delta H_{fus} (T_M) + 1.1 \, \bar{n}_C + 0.3 \, n_Q + 0.7 \tag{12-9}$$

As such, for fluorinated species we suggest a formally related approach in which the heat of sublimation of a compound is estimated by summing its experimental heat of fusion and the heat of vaporization of the corresponding hydrocarbon. Owing to the lack of thermochemical data on organic compounds in general and in all three phases, solid, liquid, and gas, it is surprising that unavoidable that there is but one piece of data that we know to test this conjecture. This is for C_6F_5Cl in which the heat of fusion is 2.0 kcal/mol. Combining this number with the heat of vaporization of C_6H_5Cl, 9.8 kcal/mol, results in a predicted heat of sublimation of 11.8 kcal/mol for C_6F_5Cl. The experimental value is 11.0 kcal/mol in satisfactory agreement.

The second approach recalls that fluorine and hydrogen are both small and rather unpolarizable. This suggests that fluorination should have a small effect on heats of sublimation. Table 12-7 gives the admittedly few cases we can find to test this conjecture. It is seen that our conjecture is encouragingly valid. The table also suggests that fluorination has but a small effect on heat of fusion. Unfortunately, there is little independent evidence.

$C_{16}F_{34}$. However, the heat of vaporization for n-$C_{16}H_{34}$ reported in Ref. 9 is 19.5 kcal/mol from which we derive a value of 0.4 kcal/mol for the heat of fusion. This last value seems low. There are two more recent reports on the heat of sublimation of n-$C_{16}H_{34}$, both giving a value of 32.3 kcal/mol. The derived heat of fusion is 12.8 kcal/mol but somehow this last value seems rather high.

6. POLYVINYLIDENE FLUORIDE AND CONCEPTUAL SELF-CONSISTENCY

In the penultimate section of this chapter, we examine the thermochemistry of polyvinylidene fluoride, $(CH_2CF_2)_n$, as a showcase of many of our various mimicry methods used in concert. We take as a given quantity the heat of formation of the solid polymer to be -113.5 kcal/mol, from Reference 2. Following from our earlier biases favoring gas phase energetics, we need its heat of sublimation. Paralleling our earlier published analysis[4] of the heat of formation of gaseous polytetrafluoroethylene and assuming the heat of sublimation of these two polymers are equal, we immediately derive a heat of formation of gaseous polyvinylidene fluoride of -109 kcal/mol. Neglecting small end group effects and other imperfections, this polymer is composed totally of $C(H)_2(CF_2)_2$ and $C(F)_2(CH_2)_2$ groups. Making the assumption that —CH_2— groups are "normal" regardless of what they are attached to, we hereby deduce that the heat of formation of the $C(F)_2(CH_2)_2$ groups is ca. -104 kcal/mol, nearly the same as that derived earlier from the $C(F)_2(C)_2$ group, -105 kcal/mol. This is encouraging. However, we may still ask if there are any destabilizing or stabilizing interactions associated with two CF_2 groups located geminally on a CH_2 group. After all, one can imagine dipolar repulsions from the two attached positive carbons and also attraction analogous to gem-difluorination. We know of no other species with two CF_2 groups attached to a central CH_2 group for which the requisite thermochemical data is available. We do, however, know of a species with two CO groups so attached, the keto form of "acetylacetone," pure pentane-1,3-dione. In the gas phase its heat of formation is -89.4 kcal/mol[17], nearly identical to the value of -88.7 kcal/mol predicted by using the macroincrementation reaction[11]

$$CH_3COCH_2COCH_3 = CH_3CH_2CH_2COCH_3$$
$$+ CH_3COCH_2CH_2CH_3 - CH_3CH_2CH_2CH_2CH_3 \quad (12\text{-}10)$$

in which this 1,3-interaction has been explicitly neglected. This suggests there is seemingly no significant stabilizing or destabilizing interaction in the polymer of any consequence. We may make use of this pentane-1,3-dione data in yet another way. From the following macroincrementation reaction we can "synthesize" a polymer of ketene:

$$1/n[(CH_2CO)_n] = CH_3COCH_2COCH_3 - CH_3COCH_3 \quad (12\text{-}11)$$

The predicted heat of formation of polyketene is -37.5 kcal/mol. Is this reasonable? By the earlier difluoride, keto equivalence, we would derive a heat of formation of gaseous $(CH_2CF_2)_n$ by summing this polyketene quantity and the heat of formation of $H_2O(g)$ and correcting the sum by ca. 6 kcal/

mol per fluorine. The resulting value is -107 kcal/mol, to be compared with the value of -109 kcal/mol given above. Given the number of assumptions that needed to be made in this section, the agreement and thus conceptual self-consistency is astonishing.

To derive the heat of formation of polyketene, per se, is not particularly important. It might not even be particularly interesting. However, the importance of this thermochemical number, and hence this section, is to demonstrate that mimicry methods and estimation techniques may be consistently and reliably used to predict properties of organic compounds regardless of the difficulty of the property and/or species to study experimentally.

7. CONCLUSION

We conclude this chapter by reviewing our successful mimics and acknowledging some of their interrelations and their limitations. Recall we demonstrated that cyano simulates trifluoromethyl as a substituent in modifying proton affinities. How well does cyano do for the other properties discussed in this chapter? The comparative paucity of thermochemical data of nitriles[23a] parallels that of fluorinated species[23b] and so systematic investigation is thwarted. However, to the extent[24] that single substitution by groups such as cyano introduce a nearly additive correction over that for the parent hydrocarbon, ie,

$$\Delta H_v(RX) = \Delta H_v(RH) + [\Delta H_v(X)] \qquad (12\text{-}12)$$

$$\Delta H_v(RX) = 1.1\ \bar{n}_C + 0.3 n_Q + 0.7 + [\Delta H_v(X)] \qquad (12\text{-}13)$$

then the —CN containing species will have a higher heat of vaporization than their —CF$_3$ analog by ca. $[\Delta H_v(\text{—CN})] - 1.1$, where the 1.1 arises from equating the common 1-carbon contribution of CF$_3$ and CH$_3$. Whether these relations are true for polysubstituted species remains virtually unexplored if for no other reasons than that even the simplest di and polycyanocompounds are solids with unmeasured heats of fusion and vaporization while many of the corresponding trifluoromethyl species are liquids that likewise lack thermochemical characterization.

Likewise, recall the finding that replacement of F by OH results in a nearly constant effect on heats of formation, even when extended to the formal replacement of >CF$_2$ and —CF$_3$ by >CO + H$_2$O and —COOH + H$_2$O, respectively. Does this equivalence carry over to the other properties of interest? Apparently not for proton affinities. CH$_3$F has a proton affinity numerically comparable to CH$_4$ suggestive of C—H bond protonation[25] while CH$_3$OH protonates on oxygen to form the simplest alkyloxonium ion, CH$_3$OH$_2^+$. Protonation of the few trifluoromethyl-containing compounds that lack more basic sites often form HF,[26] suggesting the intermediate is a carbonium ion rather weakly solvated by hydrogen fluoride. By contrast, carboxylic acids protonate to form resonance stabilized ions of the type $RC(OH)_2^+$. With regards to heats to vaporization and of sublimation, recall Woolf's earlier success was generally for condensed phase species. As such, the above formal replacements and resulting mimicry must apply to these quantities as well, eg,

$$\Delta H_f(l, RF) = \Delta H_f(l, ROH) + k_l^{FO} \qquad (12\text{-}14)$$

$$\Delta H_f(g, RF) = \Delta H_f(g, ROH) + k_g^{FO} \qquad (12\text{-}15)$$

$$\Delta H_v(RF) = \Delta H_v(ROH) + k_l^{FO} - k_g^{FO} \qquad (12\text{-}16)$$

Recall that Woolf found the constant we call here k_l^{FO} to be approximately 0 while we found k_g^{FO} to be approximately equal to 6 kcal/mol. The resulting 5 kcal/mol difference in Equation 12-16, characteristic of gas phase formal replacement processes, may be recognized as largely arising from the new intermolecular hydrogen bond found in the oxygen-containing alcohols when in their condensed phase. The analogous 6 kcal/mol per fluorine for the F_2 and F_3 cases arises from the analogous hydrogen bonds in their oxygen-containing products in their condensed phase.

Finally, we turn to the fluorine–hydrogen equivalence for heats of vaporization and sublimation. This clearly cannot be valid for proton affinities—fluorine generally decreases proton affinities with a magnitude that depends on the separation from and orientation relative to the cationic site. For example, from Table 12-1 we find that fluorine decreases the proton affinity of pyridine by 10.2, 6.5, or 4.2 kcal/mol depending on whether the site of substitution is on the 2-, 3-, or 4-position on the ring. Furthermore, we have earlier implicitly shown that this F vs H mimicry cannot be extended to heats of formation either. The earlier discussed gem-difluoro effect and the difference of the $C(F)_2(CH_2)_2$ and $C(F)_2(CF_2)_2$ group increment contributions amply document that Equation 12-17 cannot be valid:

$$\Delta H_f(g, RF_n) = \Delta H_f(g, RF_{n-m}H_m) + mk_g^{FH} \qquad (12\text{-}17)$$

This is further illustrated by the established thermochemistry of alcohols[27] since Equations 12-15 and 12-17 are numerically incompatible. For example, the heats of formation of gaseous n- and iso-propyl alcohol differ by 4.2 kcal/mol, a quantity expected to be precisely zero were OH and H groups equivalent.

It is thus clear that chemical regularities such as found in this chapter facilitate the study of fluorine compounds. Considerably more thermochemical data is needed to test and extend our understanding.

ACKNOWLEDGMENTS

The author wishes to thank Dr. James S. Chickos, Dr. William R. Dolbier, Jr., Dr Arthur Greenberg, Dr. Sharon G. Lias, and Dr. Deborah Van Vechten for numerous discussions as to the science, the style, the substance, and even the sociology of mimicry approaches and substituent effects for the understanding of chemical phenomena. He also wishes to thank Ms. Patricia Gagné for help in preparation of this manuscript.

REFERENCES

1. (a) Lias, S. G.; Liebman, J. F.; Levin, R. D. *J. Phys. Chem. Ref. Data* **1984**, *13*, 695.
1. (b) A comprehensive update to this reference is in progress and it is from this latter source that all proton affinity data in the current review is, in fact, taken.
2. Taft, R. W. *Prog. Phys. Org. Chem.* **1983**, *14*, 248.
3. Liebman, J. F. *J. Fluor. Chem.* **1973/4**, *3*, 27.
4. Liebman, J. F.; Dolbier, W. R., Jr.; Greenberg, A. *J. Phys. Chem.* **1986**, *90*, 394.
5. (a) Greenberg, A.; Liebman, J. F.; Dolbier, W. R., Jr.; Medinger, K. S.; Skancke, A. *Tetrahedron* **1983**, *39*, 1533.
5. (b) Greenberg, A.; Tomkins, R. P. T.; Dobrovolny, M.; Liebman, J. F. *J. Amer. Chem. Soc.* **1983**, *105*, 6855.
6. Benson, S. W. "Thermochemical Kinetics", 2nd Ed. Wiley: New York, 1976.
7. Benson, S. W. *Chem. Rev.* **1978**, *78*, 23.
8. See, for example, the following studies of A. A. Woolf: (a) *Adv. Inorg. Chem. Radiochem.* **1981**, *24*, 1; (b) *J. Fluor. Chem.* **1978**, *11*, 307; (c) ibid., **1982**, *20*, 627; (d) ibid., **1986**, *32*, 453.
9. Unless otherwise said, all thermochemical data in this chapter is from Pedley, J. B.; Naylor, R. D.; Kirby, S. P. "Thermochemical Data of Organic Compounds", 2nd Ed. Chapman & Hall: London, 1986.
10. Terlouw, J. K.; Lebrilla, C. B.; Schwarz, H. *Angew. Chem. Intl. Ed.* **1987**, *26*, 354, reported the synthesis of gas phase H_2CO_3 and presented both semiempirical quantum chemical calculations and Benson-type analysis (cf. Ref. 6) for its heat of formation.
11. For a discussion of macroincrementation reactions, their use and associated philosophy, see Liebman, J. F. In "Molecular Structure and Energetics: Studies of Organic Molecules", Vol. 3; Liebman, J. F.; and Greenberg, A., Eds.; VCH Publishers, Inc.: New York, 1986. For a comparison of them with Benson-type analyses and "exchange incrementation," see Liebman, J. F.; Van Vechten, D. In "Molecular Structure and Energetics: Physical Measurements", Vol. 2; Liebman, J. F.; and Greenberg, A., Eds.; VCH Publishers, Inc.: New York, 1987.
12. This quote is from Woolf, op. cit., Ref. 8(c), p. 628.
13. See the analysis and accompanying literature citations in Smart, B. E. In "Molecular Structure and Energetics: Studies of Organic Molecules", Vol. 3; Liebman, J. F.; and Greenberg, A., Eds.; VCH Publishers, Inc.: Deerfield Beach, New York, 1986.
14. This conclusion was deduced by an altogether different type of reasoning and data bank than used here by Liebman, J. F. In "Molecular Structure and Energetics: Biophysical Aspects", Vol. 4; Liebman, J. F.; and Greenberg, A., Eds.; VCH Publishers, Inc.: Deerfield Beach, New York, 1987.
15. Majer, V.; Svoboda, V. "Enthalpies of Vaporization of Organic Compounds: A Critical Review and Data Compilation". Blackwell Scientific Publishing Co.: Boston, 1985.

16. This is a textbook example of the consequences of substitution on atomic electronegativities and bond polarities, eg, Huheey, J. E. "Inorganic Chemistry: Principles of Structure and Reactivity", 3rd Ed. Harper and Row: New York, 1983, in particular p. 151.

17. Hacking, J. M.; Pilcher, G. *J. Chem. Thermo.* **1979,** *11,* 1015. These authors disentangled the heat of formation of the "keto" and "enol" forms of acetylacetone in both the gaseous and condensed phase.

18. Chickos, J. S. In "Molecular Structure and Energetics: Physical Measurements", Vol. 2; Liebman, J. F.; and Greenberg, A., Eds.; VCH Publishers, Inc.: Deerfield Beach, New York, 1987.

19. Chickos, J. S.; Annunziata, R.; Ladon, L. H.; Hyman, A. S.; Liebman, J. F. *J. Org. Chem.* **1986,** *51,* 4311.

20. Chickos, J. S.; Hyman, A. S.; Ladon, L. H.; Liebman, J. F. *J. Org. Chem.* **1981,** *46,* 4294.

21. This conclusion may be derived from the above assertion and "the difference enthalphy of melting between fluorinated compounds and their hydrogen analogies is negligible," Erastov, P. A.; Kolesov, P.; Igumenov, I. K., *Russ. J. Phys. Chem.,* **1984,** *58,* 1311, citing Kolesov, 1970.

22. Joshi, R. M. In "Encyclopedia of Polymer Science and Technology", Vol. 13; Mark, H. F.; Gaylord, N. G.; and Bikales, N. M., Eds.; Wiley: New York, 1979. We have corrected the value in this source by 2.5 kcal/mol in that we have accepted the heat of polymerization of vinylidene fluoride from Joshi and the heat of formation of the monomer from Pedley, Naylor, Kirby, op. cit., Ref. 9.

23. (a) Chu, J. Y.; Nguyen, T. T.; King, K. D. *J. Phys. Chem.* **1987,** *86,* 443.

23. (b) See Smart, B. E., op. cit., Ref. 13.

24. Chickos, J. S.; Annunziata, R.; Braton, M.; Hesse, D. G.; Ladon, L. H.; Hyman, A. S.; Panshin, S. Y.; Liebman, J. F. Unpublished results based on Chickos, J. S.; Hesse, D. G.; Liebman, J. F.; Panshin, S. Y., *J. Org. Chem.* in press.

25. (a) McMahon, T. B.; Kebarle, P. *Can. J. Chem. 63,* **1985,** 3160.

26. Vogt, J.; Beauchamp, J. L. *J. Am. Chem. Soc.* **1975,** *97,* 6602.

27. Wilhoit, R. C.; Zwolinski, B. J. *J. Phys. Chem. Ref. Data, 2* **1973,** Supplement 1.

Addendum

CHAPTER 11

New fluorinated nucleoside and nucleotide analogs and more efficient synthetic routes to known biologically active fluoronucleosides continue to appear in the literature.

There have been further developments in the synthesis and antiviral studies of pyrimidine nucleosides substituted at C-2' by fluorine. A new high yield synthesis of 2'-fluoro-2'-deoxy-β-D-arabinofuranosyl nucleosides in five steps from commercially available 1-O-acetyl-tri-O-benzoyl-β-D-ribofuranose has been published.[1]

1-Methyl-5-(2-deoxy-2-fluoro-b-D-arabinofuranosyl)uracil, a C-nucleoside isostere of FMAU (**33**, figure II-3) has been synthesized and was found to be a less active antiviral agent than FMAU.[2]

A convenient synthesis of 3-deoxy-3-fluoro-D-ribofuranose by reaction of n-Bu$_4$NF with a protected xylofuranoside C-3 triflate provides access to 3'-fluoronucleosides.[3]

C-4' derivatives of nucleoside are relatively rare, but Ajmera and coworkers have reported the synthesis of 5'-deoxy-4',5-difluorouridine as a superior prodrug of 5-fluorouracil.[4] Another new difluorinated analog that shows significant antitumor activity is 5-fluoro-6-fluoromethyluridine.[5] Fluorine is introduced by way of the reaction between CF$_3$OF and 6-methyluracil followed by sugar attachment. 5,6-difluorouracil derivatives have also been reported to have significant anticancer activity.[6]

Urata et al. have prepared some interesting C-6 linked palladium complexes of 1,3-dimethyl-5-fluorouracil with the ultimate aim of constructing molecules that would effect DNA and RNA synthesis both through metal binding and through release of 5-fluorouracil.[7]

Introduction of branched chain fluoroalkyl substituents at C-5 of 2'-deoxyuridine gave derivatives with relatively low antiviral activity.[8]

Reefschlaeger, Pein and Cech have now published more detailed information on synthesis and biological activity of 4-O-difluoromethyl analogs of pyrimidine nucleosides.[9]

En route to oligonucleotides in which the phosphodiester linkage is replaced by a phosphorofluoridate diester linkage, Dabkowski et al. have described the transformation of phosphoroazolides to phosphorofluoridates by reaction with benzoyl fluoride.[10]

Nucleoside triphosphate analogs in which one of the bridging phosphate oxygens is replaced by the CF$_2$ group are beginning to find use in mechanistic studies. The slow rate of macrocyclic polyamine catalyzed hydrolysis of the ATP analog App$_{CF2}$p in comparison to ATP and Ap$_{CH2}$pp, provides evidence for a S$_N$1P type mechanism.[11]

REFERENCES

1. Howell, H. G.; Brodfuehrer, P. R.; Brundidge, S. P.; Benigni, D. A.; and Sapino Jr., C. *J. Org. Chem.* **1988,** *53,* 85–8.
2. Pankiewicz, W. K.; Nawrot, B.; Gadler, H.; Price, P. W.; and Watanabe, K. A. *J. Med. Chem.* **1987,** *30,* 2314–16.
3. Noyori, R.; Hayakawa, Y.; Uchida, K.; Yasuda, A.; and Morisawa, Y. **Jpn. Kokai Tokkyo Koho JP 62 81,397,** 14 April 1987. Cf. *Chem. Abstr.* **1988,** 108: 75784r.
4. Ajmera, S.; Bapat, A. R.; Stephanian, E.; and Danenberg, P. V. *J. Med. Chem.* **1988,** *31,* 1094–8.
5. Felczak, K.; Kulikowski, T.; Vilpo, J. A.; Giziewicz, J.; and Shugar, D. *Nucleosides and Nucleotides* **1987,** *6,* 257–260.
6. Shimokawa, K.; Yamamoto, S. **Jpn. Kokai Tokkyo Koho JP 62,138,481,** 22 June, 1987. Cf. *Chem. Abstr.* **1987,** 107: 176420d.
7. Urata, H.; Tanaka, M.; and Fuchikama, T. *Chemistry Letters* **1987,** 751–54.
8. Mel'nik, S. Y.; Bakhmedova, A. A.; Yartseva, I. V.; Kochethova, M. V.; Preobrazhenskaya, M. N.; Sviridov, V. D.; Chkanikov, N. D.; Kolomiets, A. T.; and Fokin, A. V. *Bioorg. Khim.* **1987,** *13,* 934–9.
9. Reefschlaeger, J.; Pein, C. D.; and Ceck, D. *J. Med. Chem.* **1988,** *31,* 393–7.
10. Dabkowski, W.; Cramer, F.; and Michalski, J. *Tetrahedron Letters* **1987,** *28,* 3561–2.
11. Blackburn, G. M.; Thatcher, G. R. J.; Hosseini, M. W.; and Lehn, J.-M. *Tetrahedron Letters* **1987,** *38,* 2779–82.

General Index

A

Acidity, fluorine and keto-enol equilibria, 88–91
Adenosine 5'-fluorophosphate, biological activity, 283
AgF-HF-H$_2$O system, 184
Aldol reactions,
N,N-Dimethylfluoroacetamide and, 130–133
enol silyl ether formation, 133–136
fluorination effects on, 124–145
Alkenes,
1,1-difluoroalkene formation, 11–12
epoxides and, 14–17
Alkoxide-promoted dehydrohalogenations,
mechanisms of, 114–116
PKIE, 114–116
Alkylation, of deprotonated chiral fluoracetone imines, 138–141
Alpha-fluorinated enolates, 128–141
N,N-Dimethylfluoroacetamide and, 130–133
Ammonium fluoride-hydrogen fluoride (NH$_4$F-HF) system, 183
Ammonium fluoride-hydrogen fluoride-water (NH$_4$F-HF-H$_2$O) system, 183
Amphoteric behavior, 191
Antiviral activity, 292–294
Aquo complexes, 184
Arenes,
intermolecular reactions with polyfluoroarenes, 24, 26, 30, 33, 37, 38
nonbonded interactions between arenes and polyfluoroarenes, 20–38
polyfluoroarylation of, 36
polyfluoroarylazides reactions with, 36–37

Aromaticity, 45, 51
Aryl hydrocarbons, polyfluoroarylazides reactions, 36–37
Arylation, of polyfluoroarenes, 36
1-arylperfluoropropenes, preparation, 158–159
Atomic carbon,
1,1-difluoroalkanes formation, 11–12
molecular orbital calculations and, 3–5
olefinic trapping agents and, 13–17
oxygen and reactions with fluorocarbons, 12–13
reaction of arc generated with CF$_4$, 5–7
reactions with fluorocarbons, 1–17
Azaenolates, fluorinated, 137–142

B

Back-donation, 45, 52
Basis set superposition error, 55, 59
Benzene,
complex with hexafluorobenzene, 20–22
reactions with CF, 9–10
Benzene ring, benzenoid aromatics, 47–52
Benzene-dimer, 21
Benzenethiolate, 84–87, 110–112
Benzenoid aromatics, 45–52
bond order considerations, 49–51
geometrical changes and, 45–47
ortho-para effects, 45
r$_e$ and r$_o$ structures, 46
Benzyltrimethylsilanes, 106–107
Binding energy, of NF$_4$, 250–251
Biological/biochemical probes,
^{19}F Nuclear magnetic resonance studies, 302–304